Maldynia
Multidisciplinary Perspectives on the Illness of Chronic Pain

T0392487

Maldynia

Multidisciplinary Perspectives on the Illness of Chronic Pain

Edited by
James Giordano

CRC Press is an imprint of the
Taylor & Francis Group, an **informa** business

CRC Press
Taylor & Francis Group
6000 Broken Sound Parkway NW, Suite 300
Boca Raton, FL 33487-2742

First issued in paperback 2019

ISBN-13: 978-1-4398-3630-9 (hbk)
ISBN-13: 978-0-367-38325-1 (pbk)

Library of Congress Cataloging-in-Publication Data

Maldynia : multidisciplinary perspectives on the illness of chronic pain / edited by James Giordano.
p. ; cm.
Includes bibliographical references and index.
Summary: "An in-depth examination of maldynia, also known as chronic pain, this book explores pain as a biocultural phenomenon that necessitates reexamination of medical philosophy, ethics, education, and practice. It provides a historical account of pain and then frames it with contemporary neurobiological perspectives. This book supplies a foundation upon which to illustrate (1) how what we know about pain could and should influence medical education, and the scope, and value(s) of medical practice, and (2) how a knowledge of the history and present considerations of pain might help to construct meaningful, patient-centered medicine."--Provided by publisher.
ISBN 978-1-4398-3630-9 (hardcover : alk. paper)
1. Chronic pain. 2. Chronic pain--Social aspects. I. Giordano, James J. II. Title.
[DNLM: 1. Pain. 2. Chronic Disease. 3. Culture. 4. Education, Medical. 5. Sociology, Medical. WL 704]

RB127.M345 2011
616'.0472--dc22 2010043045

Visit the Taylor & Francis Web site at
http://www.taylorandfrancis.com

and the CRC Press Web site at
http://www.crcpress.com

Dedication

To Barbara,

Semper fi

and Sherry

...und für die Tiere

Contents

Acknowledgments

This volume is a direct result of the enthusiasm and support of my acquisitions editor Barbara Norwitz. She has been a fan, advocate, guide, and friend from this book's inception to completion; and her diligence, fortitude, and great humor have been as much a part of this book as is the text. Norwitz shared my interest in depicting the broad impact and effects of chronic pain, not simply as a clinical condition, but as a manifest illness, a condition I refer to as *maldynia*—with a very deep nod of gratitude to my colleague Phillip Lippe who originally coined the word albeit in a more purely clinical sense. To be sure, there are numerous books on pain that address "why it hurts, where it hurts, and what kinds of things can be done to fix the hurt," and many are quite good. In fact, I am honored to have contributed to several of them, and written or edited others (although I leave it to the readers to determine if mine have any merit). But this book was not designed, developed, or implemented to take that approach. Norwitz and I spent a good bit of time trying to decide if this book "fit" into the science or humanities category. It does, and by intent strives to take a "both/and" rather than an "either/or" approach to the topic, and in so doing, might thus be considered to assume a "medical humanities" orientation.

And right that it should. The original idea for such a book grew from long conversations with my colleague and friend, Tom Cole, about the importance of the humanities in and to pain medicine, while I was a Fellow at the John McGovern, M.D., Center for Health, Humanities, and Ethics at the University of Texas Health Science Center at Houston. As director of the center, Cole spent hours talking with me and debating the relative roles of neuroscience, philosophy, ethics, and the arts in pain research and care. Being involved with this group during the early part of the *Decade of Pain Control and Research* provided ample opportunity to work on the problem of pain with scholars and clinicians from a variety of medical and humanities disciplines; and I am grateful to all of my colleagues from my days at the University of Texas. Among those to whom I am especially indebted for their help, support, and contributions are Alessandro Chaoul, Joan Engebretson, Kay Garcia, Julia Pedroni, Stan Reiser, and Vic Sierpina.

This project continued while I was Samueli-Rockefeller Professor, and Director of the Center for Brain–Mind and Healing Research at Georgetown University Medical Center, Washington, D.C., and I appreciate the L.S. Rockefeller Trust and the Samueli Institute for their support. I went to Georgetown for a very simple reason—to work with Professor Edmund Pellegrino, an experience that has been, and continues to be intellectually rich, professionally fulfilling, and just plain fun. I count myself as being most fortunate to have enjoyed his interest in my work, contribution to it, mentorship, support, and friendship. While at Georgetown, I had the pleasure of working with wonderful colleagues, such as Hakima Amri, Jim Duffy, Kevin FitzGerald, Carlos Gomez, Adi Haramati, John Collins Harvey, Paul Hutchison, Guillermo Palchik, Andy Putnam, Hans Martin Sass, and Rachel Wurzman; and hosting and collaborating with some outstanding visiting scholars, including Roland

Benedikter, Lucia Galvagni, Niko Kohls, Peter Moskovitz, Alessandra Valadas, and Giusi Venuti, all of whom have contributed to this project in significant and very much appreciated ways.

Visiting professorships sponsored by the American Academy of Pain Medicine, and the Neuroscience, Neurophilosophy, and Neuroethics of Pain, Pain Research, and Pain Treatment (N3P3) Program enabled this work to continue and expand while at the Messer-Racz Pain Center of Texas Tech University Health Sciences Center in Lubbock; and the Rheinische Friedrich Wilhelms Universität, Bonn, Germany, respectively. Thanks to Mark Boswell, John Hall, Gerhard Höver, Heike Baranzke, and Hans Werner Ingensiep for their encouragement, friendship, and efforts in these projects. Special thanks also to Gerd Höver for inducting me into the *Grappa Bund.*

The book was completed as part of my ongoing work at the Center for Neurotechnology Studies of the Potomac Institute for Policy Studies, Arlington, Virginia; Krasnow Institute for Advanced Studies of George Mason University, Fairfax, Virginia; and the Wellcome Centre for Neuroethics and Uehiro Centre for Practical Ethics at the University of Oxford, United Kingdom; with funding from the Institute for Biotechnology Futures (IBTF), Nour Foundation, and the William H. and Ruth Crane Schaefer Endowment of Gallaudet University, Washington, D.C. I thank Mike Swetnam of the Potomac Institute, Richard Rass of the Nour Foundation, Carol Erting of Gallaudet University, and Julian Savulescu and Neil Levy of the Wellcome and Uehiro Centres at Oxford, for their friendship, support, collaboration, and collegiality.

As the *Decade of Pain Control and Research* was closing, I had the honor of becoming involved in the planning, articulation, and, ultimately, direction of the *Neuroethics, Legal and Social Issues (NELSI)* program of the proposed *Decade of the Mind*; in many ways, this book seeks to bridge these agendas by conjoining their foci on those often ineffable aspects of being—consciousness and pain—that define experience, and I am grateful to James Olds, Director of the Krasnow Institute of Advanced Studies of George Mason University, for both involving me in this initiative and bringing me to the Krasnow Institute.

Last, but certainly not least, is the fact that this volume involved considerable writing, rewriting, editing, and reediting. For all of the typing, collating, and revising, anticipation, excitement, humor, and patience, I am thankful beyond words to my wife, Sherry. Nobody is allowed to look that beautiful while working furiously at two in the morning…and I am, without doubt, a very lucky man.

James Giordano
Oxford, United Kingdom

Editor

James Giordano, Ph.D., is the Director of the Center for Neurotechnology Studies and Chair of Academic Programs at the Potomac Institute for Policy Studies, Arlington, Virginia; Research Associate at the Wellcome Centre for Neuroethics and Uehiro Centre for Practical Philosophy at the University of Oxford, United Kingdom; and University Affiliate Professor of Molecular Neuroscience at the Krasnow Institute for Advanced Studies, George Mason University, Fairfax, Virginia. He is also currently William H. and Ruth Crane Schaefer Distinguished Visiting Professor of Neuroscience and Ethics at Gallaudet University, Washington, D.C.

Professor Giordano is Editor-in-Chief of the journals *Philosophy, Ethics and Humanities in Medicine*, and *Synesis: A Journal of Science, Technology, Ethics and Policy*; Associate Editor for the international journal *Neuroethics*, Neuroscience and Ethics Editor (and former Deputy Editor-in-Chief) for the journal *Pain Physician*, and Executive Editor-in-Chief of the book series *Advances in Neurotechnology: Ethical, Legal, and Social Issues* (also published by Taylor & Francis/CRC Press).

Professor Giordano's ongoing research addresses the neuroscience and neurophilosophy of pain, and the neuroethics of pain research and treatment.

Professor Giordano is the author of more than 150 publications in neuroscience, pain, neurophilosophy, and neuroethics. His recent books include *Scientific and Philosophical Perspectives in Neuroethics* (with Bert Gordijn, Cambridge University Press, 2010); *Pain: Mind, Meaning, and Medicine* (PPM Press, 2009); and *Pain Medicine: Philosophy, Ethics, and Policy* (with Mark Boswell, Linton Atlantic Books, 2009).

Contributors

Mark V. Boswell, M.D., Ph.D., MBA
International Pain Center
Texas Tech University Health Sciences Center
School of Medicine
Lubbock, Texas

Nathan Carlin, Ph.D., M.Div.
John P. McGovern, M.D., Center for Health, Humanities, and Ethics
University of Texas Health Science Center at Houston
Houston, Texas

Thomas Cole, Ph.D.
John P. McGovern, M.D., Center for Health, Humanities, and Ethics
University of Texas Medical School at Houston
Houston, Texas

Rosemary Feit Covey
Studio 224
Torpedo Factory Art Center
Alexandria, Virginia

F. Daniel Davis, Ph.D.
Center for Clinical Bioethics
Georgetown University
Washington, D.C.

Lucia Galvagni, Ph.D.
Fondazione Bruno Kessler
Trento, Italy

Carlos Gomez, M.D., Ph.D.
District of Columbia Pediatric Palliative Care Collaboration
Washington, D.C.

Hans Werner Ingensiep, Prof. Habil. Phil. Dr. Rer. Nat. Dipl. Biol.
Universität Duisburg-Essen
Essen, Germany

Scott L. Karakas, Ph.D.
Office of Curriculum and Instruction
Florida Gulf Coast University
Fort Myers, Florida

James D. Katz, M.D.
Division of Rheumatology
George Washington University
Washington, D.C.

Nikola Boris Kohls, Ph.D.
Human Science Center
University of Munich
Munich, Germany
and
Peter-Schilffarth-Institute for Sociotechnology
Bad Tölz, Germany

Peter A. Moskovitz, M.D.
George Washington University Medical Center
Washington, D.C.

Edmund D. Pellegrino, M.D.
Georgetown University Medical Center
Washington, D.C.

Elaine Peterson, D.M.A.
Department of Music
Mississippi State University
Mississippi State, Mississippi

Michael E. Schatman, Ph.D.
Pain and Addiction Study Foundation
Bellevue, Washington

Victor S. Sierpina, M.D.
Department of Family Medicine
University of Texas Medical Branch
Galveston, Texas

M. Alexandra Valadas, Ph.D.
Department of Philosophy
Michigan State University
East Lansing, Michigan

Giusi Venuti, Ph.D.
Cognitive Science Department
University of Messina
Sicily, Italy

David B. Waters, Ph.D.
Department of Family Medicine
University of Virginia
Charlottesville, Virginia

1 Maldynia—The Illness of Chronic Pain

James Giordano

CONTENTS

INTRODUCTION

What is *maldynia*? As originally defined, it is intractable, chronic pain (1). This is a valid definition, but one that I believe leaves considerable room for speculation about what such pain involves, and why, how, and in whom such pain occurs. Such conjecture must acknowledge the biological, psychological, and social dimensions of both cause and effect. Maldynic pain is not merely chronic pain; there is a natural history and *purpose* to chronic pain. It engages a hierarchy of physiological mechanisms that promote rest, repair, recuperation, recovery (ideally), and, it might be argued, resilience (2). Perhaps, this might be regarded as *wild-type* chronic pain—a process through which the organism recuperates and recovers, or if not, succumbs to the ravages of progressive pathology (caused, at least in part, by alterations in the immunological and endocrine systems), loss of social conjoinment, or predation.

Instead, maldynia is pain without purpose. It occurs when the disease process that evokes pain—or, in some cases, that *is* pain—not only produces malaise but progresses to multidimensional illness. It is a durable event and experience of the lived body and life world. If we consider maldynia to literally represent *bad* pain, then its *badness* is not simply the noxiousness of pain itself, but rather its nonpurposiveness, its prolongation—not toward ends of recovery but toward escalating severity and increasingly manifest cognitive, emotional, and behavioral expression. The mechanisms may involve sensitization (and often spontaneous activity) of peripheral and central substrates to evoke a constellation of features—a syndrome of pain as both disease and illness (3). But these neither foster recovery nor are fatal. To be sure, modern medicine has done much to defer the mortality of many diseases (4). Yet, these very means have resulted in an increased prevalence of chronic morbidity, inclusive of pain (5). In this light, we may come to regard maldynia as an iatrogenic illness—evoked by the capacity of medical technology to allow an extended life span with chronic pain, while being only nominally prepared to address its broadly biocultural consequences.

In many ways, maldynia defies the medical model. It is not wholly objectifiable by third-person and technological means. Lacking the tools for objective assessment, medicine relies upon the patient for insight to the subjective nature of the maldynic experience. But pain opposes language (6). This is reflected in the difficulties inherent to clinically evaluating chronic pain. Due to these semantic problems, such pain often is beyond patients' communicative scope, places the patient in diametric opposition to many tools of medical assessment, and may lead to failure of disease-based, curative paradigms of intervention. This is because the patient is no longer simply suffering the painful manifestations of a disease process, but rather becomes a person for whom a new sense of self (and perhaps its representational consciousness) is defined by the illness experience of pain (3,6–8). Yet, pain compels its communication (9), and history and culture are rich with examples of such strivings.

Thus, this book acknowledges these "problems" of maldynia, and seeks to depict pain as an event and experience of the person in pain. Such an account must consider the role of brain–mind in the phenomenal experience, identification, and articulation of the first-person *self* in which pain is embodied. But it must be equally cognizant of the environment(s) in which the self is embedded. Thus, the challenge—if not the enigma—is to understand the experience of pain-as-illness in persons who are nested in time, place, and situation.

The problem of pain compels using various disciplines in an ongoing exploration of dynamic relationships, not only of body, brain, mind, and environment, but also of patients and clinicians, individuals and cultures, knowledge and values, and morals, ethics, guidelines, policies, and laws. Yet, the question remains as to how to approach these variables, not singularly, but in ways that recognize and uphold their relationality. Thus, if we are to consider the impact of pain, we must also contemplate the nature of the being who is in pain.

Pain is a complementarity—it is an interactively physiological event and psychological experience of living person that is expressed in, and often defined by, social contexts and constructs. Thus, in the strictest sense, pain is biopsychosocial, and such a biopsychosocial orientation dictates a complementary approach to the ways that pain is studied, and how the patient in pain is treated. In this light, it becomes important (if not necessary) to speculate on what such complementarity might entail, and how it could be enacted in pain care.

Any authentic approach in this regard would involve both the sciences and the humanities (9). The goal is not a mere homogenization of disciplines and specialties, but more effective discourse that provides differing lenses through which various perspectives may be focused, acuities sharpened through coparticipation, limitations overcome through collaboration, and outcomes enhanced through multidisciplinary integration. This complementarity enables the philosophy, ethics, and practices of pain management to be factually based, yet free to embrace new ideas and methods in an original, creative, and resourceful way (10,11).

The epistemic, anthropologic, and ethical questions of philosophy have given rise to scientific investigation, social thought, and moral consideration throughout history. These questions have exerted (and continue to exert) profound and perdurable influence. Discourse upon the phenomenal reality of pain has been explicit since Socrates—and as science further investigates mechanisms of the brain–mind, such inquiry reflects Socrates' inquiry of Phaedrus: "where have you come from, and where are you going" (12)? If we allude to the allegory of the cave as a metaphor for current neuroscientific and philosophical understanding, we, too, like those in the cave, must address what we believe to be real, what we discover to be the nature of reality, the implications of this reality for the future, and the challenges and obligations to make this reality apparent to others. But, as for the prisoners in the cave, the elucidation of reality and workings of the mind must ultimately be directed toward understanding and seeking *the good.*

This is the work of ethics, in general, and neuroethics more specifically, as relates to the capabilities, roles, and limitations of neuroscience in pain care (13). I believe that what it means to have a brain yet *be* a mind will have significant implications for individual and social values, contexts, and conduct (e.g., redefining what is "normal" and "abnormal," and what constitutes "treatment" or "enhancement"), and the progressive integration of neurotechnology into various aspects of medicine and culture (14).

These are the challenges that pain medicine must face as both profession and practice. But then we must ask, what is pain management? To be sure, it is situated within the larger fabric of medicine, as a whole, and as such may be considered to be the art and science of treating those made vulnerable by injury, disease, and illness (15). Clearly, the purpose of medicine is to render care to the person who is the patient. But it is less clear what defines medicine as art or science, and how these definitions are related to the premises and practicalities of treating pain as disease and illness.

The natural sciences establish a foundation upon which medicine can be based, for the practice of medicine involves an understanding of bodily functions and dysfunctions as events that are knowable according to laws of nature (16). Certainly, contemporary medicine's reliance upon scientific experimentation, the expansion of scientific knowledge, and evidence-based practice has fortified this relationship. Much of this understanding has been gained through the use of technology, and it is easy to see how and why the regnant medical model has become so deeply associated with applied biotechnology (17). But as the philosopher Hans Georg Gadamer noted, "the science of medicine is the one which can never be understood entirely as a technology, precisely because…its own abilities and skills belong(s) to nature" (18).

The practical, combinatory use of objective information and subjective knowledge to include the generalizability, precision, and explanatory value of science, and the wisdom, insight, and subtlety of art is what relates the profession of medicine to its practice—the treatment of persons made vulnerable by disease and illness. To be sure, disease and illness are not the same and thus cannot be approached and treated as such. Discerning pain-as-illness from pathophysiology is not simply a matter of differentiating effects from cause but necessitates an appreciation of the person who is the patient. This preserves the humanitarian dimensions of medicine that compel the acquisition and use of particular knowledge, inform decisions, and ultimately guide actions to sustain what Paul Ramsey calls "moral art" (19). This approach is dialectical and allows for seemingly different perspectives to be communicated, discussed, and reconciled in a synthesis that is reciprocally informative and precipitate learning and positive change.

Achieving this balance requires communication within the pain management community, at large; thus, responsibility to foster discourse rests not only upon the individual practitioner, but upon their professional organizations, as well. Many professional pain organizations exist, each with somewhat differing missions and agendas. However, a harmonizing task might be to reflect upon the intent and outcome of the past 10 years' *Decade of Pain Control and Research*, and look ahead to how progress in the sciences and humanities could create improved opportunities for understanding, ethics, and care.

It is often far too easy to look back and comment about the "darkness" of unfulfilled intentions and tasks left incomplete. The real challenge in facing the future, at least with any sense of purpose, is to proverbially "light a candle" so as to illuminate the potential paths, pitfalls, and possibilities that lie ahead. Toward these ends, this book will examine the neurobiology and phenomenology of maldynic pain, its expression as narrative, transcendent experience, and art, and the role that these dimensions play in the ethics and practice of pain medicine. Exploring how a contemporary science of pain impacts and is affected by the humanities may help to refine our current knowledge of the brain, mind, self, and society, and may help to chart a course forward in pain care. It is my hope that this volume will provide a useful tool with which to reflect upon and guide discussion that is meaningful to the constituent disciplines of the sciences and humanities, the profession and practice of pain medicine, and, ultimately, to those individuals who suffer pain and those who treat pain.

REFERENCES

1. Lippe, P. 1998. An apologia in defense of pain medicine. *Clin J Pain* 14(3): 189–190.
2. Panksepp, J. 1998. *Affective neuroscience: The foundations of human and animal emotion. Part III: The social emotions.* New York: Oxford University Press.
3. Giordano, J. 2009. *Pain: Mind, meaning, and medicine.* Glen Falls, PA: PPM Press.
4. Magner, L.N. 2005. *A history of medicine* (2nd ed.). Boca Raton, FL: Taylor & Francis.
5. Morris, D.B. 1998. *Illness and culture in the postmodern age.* Berkeley: University of California Press.
6. Scarry, E. 1985. *The body in pain: The making and unmaking of the world.* New York: Oxford University Press.
7. Good, B.J. 1996. *Medicine, rationality and experience: An anthropological perspective.* Cambridge: Cambridge University Press.
8. Leder, D. 1990. *The absent body.* Chicago: University of Chicago Press.
9. van Hooft, S. 2003. Pain and communication. *Med Health Care Phil* 6: 255–262.
10. Giordano, J. 2008. Complementarity, brain–mind, and pain. *Forsch Komplementärmed* 15: 71–73.
11. Giordano, J., and M.V. Boswell. 2009. Prolegomenon: Engaging philosophy, ethics, and policy in, and for pain medicine. In: *Pain medicine: Philosophy, ethics, and policy*, eds. J. Giordano and M.V. Boswell, 13–20. Oxon, UK: Linton Atlantic Books.
12. Levinson, R.B. (ed.) 1967. *A Plato reader.* Boston: Houghton-Mifflin, pp. 53–112.
13. Giordano, J. 2010. From a neurophilosophy of pain to a neuroethics of pain care. In: *Scientific and philosophical perspectives in neuroethics*, eds. J. Giordano and B. Gordijn, 172–189. Cambridge: Cambridge University Press.

14. Gini, A., and J. Giordano. 2010. The human condition and strivings to flourish: Treatments, enhancements, science and society. In: *Scientific and philosophical perspectives in neuroethics,* eds. J. Giordano and B. Gordijn, 355–369. Cambridge: Cambridge University Press.
15. Pellegrino, E.D., and D.C. Thomasma. 1988. *For the patient's good: The restoration of beneficence in health care.* New York: Oxford University Press.
16. Porter, R. 2006. What is disease. In: *The Cambridge history of medicine*, ed. R. Porter, 71–102. Cambridge: Cambridge University Press.
17. Reiser, S. 1978. *Medicine and the reign of technology*. Cambridge: Cambridge University Press.
18. Gadamer, H.G. 1996. *The enigma of health: The art of healing in a scientific age.* Stanford: Stanford University Press.
19. Ramsey, P. 2002. *The patient as person: Explorations in medical ethics.* New Haven: Yale University Press.

2 A Short History of Pain and Its Treatment

M. Alexandra Valadas

CONTENTS

INTRODUCTION

Pain, although subject to different interpretations and subjectifications, is one of the most basic human experiences. But to *feel pain* is not only to perceive a set of physical responses to an injury. The experience of pain entails a subject, who through pain is transformed, transfigured. Hence, pain, *as* a human experience, is differently sensed and understood by each individual, and within each culture. The language of pain is always a translation, because "pain involves a codified form of social behavior" (1). Therefore, the expression of pain is unique and inimitable in its form and articulation.

As Roselyne Rey stated, there are "idioms of pain" (1) that not only depend on culture and societal standards, but that are touched by the subjectivity of the being. An exclusive and exceptional relationship is therefore created, in which pain and the individual coexist, mutually modifying and transforming each other. Thus, a history of pain is an arduous, if not impossible, task. "Pain passes much of its time in inhuman silence" (2) confined in a universe characterized by the uniqueness of the subject, isolating the being. For someone who is in pain is alone in that pain.

But pain is a powerful part of humanity, and "no matter whether the culture is ancient or modern, pain and illness cannot be disentangled from complex social and personal systems" (1), and these are inextricable from the history of man. This brief historical account seeks to provide a description of the different perceptions and understandings of pain through the different periods of Western culture. It may well be that to glimpse even a little of the history of pain is to create a chronicle of mankind.

FROM ANTIQUITY TO THE MEDIEVAL AGE

Contemporary Western thought has been built from cultural, social, and historical precepts of Classical Antiquity. In the Greek epic texts, pain inflicted in war and battle wounds are described in attentive detail, and the injuries are anatomically depicted. Pain is generally associated with the action that caused it, and the "Homeric warriors normally expire all at once in a black mist or in a bone-crunching clatter of armour; they groan, and grasp" (2). But even though the instruments and weapons of the battlefield that cause wounds and injuries are surgically described, there is a distinct omission of the pain they inflicted. It seems as if the Homeric warrior may have been severely wounded or even killed in a painless combat with the notable exception of a few scenes in book five of the *Iliad.* Here, in two different moments, Aphrodite and Ares, who intervene in the Trojan War, are wounded, and the dimensions of their pain are fully depicted (2). This account is particularly relevant since the Olympic gods, who belong to a divine dimension in which there is no suffering and death, are humanized, allowed to descend to an inferior realm, and thus feel pain as mere mortals. In Homer, pain is usually related to its instantaneous alleviation, generally through some medical treatment (e.g., the removal of weapons) or through divine intervention, such as in prayer to Apollo (the god of healing) by a *glaukos* (2), a "man in pain."

In Ancient Greece, the natural world was perceived as divine, and pain, although recognized as a biological process, had the additional meaning of an expression (viz., generally a chastisement or punishment) of the divine world. Rey's (1) extensive study of the vocabulary used to describe pain in Homer's *Iliad* shows overlap between the moral and physical dimensions (2). The absence of pain leads an indirect path to the characters' interior dimension, and it serves a social purpose: the Homeric account, extremely realistic in the dynamics of war and battlefield, shows that warriors should accept their fate, and pain, without fear; their destiny is in the hands of the gods.

Pain has a different expression in the Greek Tragic genre whose messages revolved around the notion of suffering and pain. A paradigmatic play in the exploration of pain, and particularly chronic pain, is Sophocle's *Philoctetes*, in which the tragic destiny of the character is shaped by physical suffering. Here, pain assumes a more savage bodily aspect. It consumes the subject, and there is a crescendo that depicts the ever-growing domination of pain and the exhaustion it brings. Pain exiles Philoctetes from the rest of humankind, to a place where communication (especially of his suffering) is not possible, arresting him in animalistic captivity. The tragedy is that the observers, while compassionately recognizing the intensity of such pain, cannot feel it; they are not able to share such pain. Thus, pain and suffering are depicted as the most individual and private of human experiences.

In Greek Tragedy, the drama of the body in pain—of everyday pain—portrays the power of pain, how it unravels the self to a mere state of painfulness. Perhaps for the first time, pain is depicted as subjective and objectively inaccessible (2). In the lyric genre, Pindar (fifth century BCE) uses the expression *nodynia* to refer to the alleviation of pain (1). Here we see direct expression and relation of both the pain as an isolative destructive experience, and a striving to relieve it. Thus, while Egyptian and Babylonian medical cultures showed some evidence of observational practices, it was

Greek medicine that speculated about symptoms including pain, and studies of their nature and origin were institutionalized and more fully developed. The *Hippocratic Corpus* exemplifies this developing thought. This set of treatises was written by a variety of authors, between 420 and 370 BCE, and includes works addressing the origin of diseases, epidemics, and gynecological studies. Some of these were based upon accurate observation, and others were mere speculation.

So, while a patient has insight into his or her bodily experiences and symptoms, it is the doctor's duty to interpret, explain, and try to intervene in order to cure, or at least to alleviate such symptoms. In the Hippocratic collection, pain and pain vocabulary were related to clinical situations, especially as focal to description of those disorders that could cause pain. In the text *Of Art*, the duty of the physician is stated to be the alleviation of pain and suffering, and an obligation to prevent pain (whenever possible). In many of the Hippocratic texts, the main emphasis is upon prognosis, incorporating diagnostic and prognostic techniques of Babylonian and Egyptian practices.

Because illness was perceived as process (1), it was considered possible to predict a relationship between the pathological processes and the evolution of symptoms, inclusive of pain. The progress of an illness could be linked to the intensity of pain, and the localization and origin of pain could be directly indicative of specific diseases. The etiology of pain was amply discussed throughout the treatises, even though in some, there was disagreement about origin, course, and effect. Thus, the different texts provide varying methods of treating and curing pain (e.g., the use of bleedings; the cure by opposites or by similarities) (1).

Although Hippocratic medicine was based on differing precepts and principles that tended to make generalizations difficult, it is possible to infer that pain was commonly accepted as a natural phenomenon (in health and illness). The Hippocratic works supported observations that led to a valorization of both symptoms and the relationship between physician and patient. Thus, the need for interpersonal engagement between physician and patient became an important construct and tool by which to gain access into the more subjective dimensions of illness—and most certainly this was true for pain.

With Greek conquests, Hippocratic knowledge spread throughout Asia Minor, and most of the known world. Hellenistic cities were developed, and Alexandria, in Egypt, became the apotheosis of Greek culture. It was a social and political center, and given its location, gathered scholars and intellectuals from the entire Mediterranean. The fusion between Hellenistic and Egyptian cultures led to medical practices characterized by autonomy and liberty. Consequently, and due in no small way to increased use of vivisection and postmortem examinations (that were constrained in the Greek system), anatomical knowledge was greatly augmented.

Herophilus (335–280 BCE) was particularly interested in the nervous system, distinguishing nerves from blood vessels, and motor from sensory nerves. His work* is considered to be progressive as compared to the *neurological* studies of his predecessors. His contemporary, Erasistratus of Chios (330?–250? BCE), a cofounder of the school of anatomy in Alexandria, developed extensive dissective studies of the

* His works were lost but were much quoted by Galen (second century AD).

brain, distinguishing between motor (*kinetika*) and sensory (*aisthetika*) functions. Erasistratus' works were challenging in that they contradicted many of the previously established doctrines: he rejected the theory of humors as well as the teleological view, in favor of a corpuscular, mechanistic perspective.

From 250 BCE and onward, the Greek empire, in accentuated decline, fell under the power of Rome, which controlled Italy, southern France, Spain, Portugal, and by the beginning of the millennium, part of the Britannic Islands, the Rhine and Danube, extending the empire to the Middle East. In the Roman world, especially in the East, Greek medicine thrived, while in the West, medicine, generally practiced by a small and disregarded group, had as its paradigmatic figure Aulus Cornelius Celsus (25 BCE–AD 50). Although Celsus is known by his extensive work *De Re Medicina* (AD 40), considered one of the most important volumes on medicine in Antiquity, there is uncertainty whether Celsus was in fact a physician. Nevertheless, the eight books of *De Re Medicina* form an elegant and detailed oeuvre that is clearly based on a rationalist approach.

Reflective of the Greek perspective, Celsus perceived pain to be a symptom of a disorder; therefore, it was directly related to the causal disease. Pain could predict the evolution of an illness, and its intensity and localization were fundamental to prognosis. As in Hippocratic medicine, it was perceived as an important diagnostic and prognostic factor, and alleviation of pain was a critical element in any determinant to successful diagnosis and treatment. This reflected the social view that pain could not have a positive value, and thus health was construed as the lack of pain and suffering.

A century later, Galen of Pergamum (AD 129–200) established a *corpus* of medical knowledge that would lay the groundwork for medical theory and practice until well into the 1600s. Galen lived most of his life in Rome, mastering (Greek) literature and philosophy before deepening his knowledge of medicine—mostly as derived from the Hippocratic tradition. His *On the Elements According to Hippocrates* describes the system of four bodily humors: blood, yellow bile, black bile, and phlegm, which were identified with the four classical elements and, in turn, with the seasons.

Based upon Platonic theory, Galen posited that there were three bodily systems—heart, liver, and brain—and contested the Aristotelian idea that the mind was in the heart, claiming instead that it was situated in the brain. Adopting Erasistratus' distinction between motor and sensory nerves, he classified them, respectively, as hard and tender nerves. Galen developed important investigations on perception and sensation, establishing that the brain had a dual nature, being simultaneously *soft* (so as to harbor functions of imagination and intelligence) and *hard* (to serve motor and bodily functions). Galen theorized that bodily functions were teleological—that is, designed to serve a particular purpose.

Pain was considered to be an impression of the tactile sense that could be due to two different causes: an internal state of transition or an evoked response to external change. Irrespective of cause, pain was seen as a warning sign, a means of protection. In this sense, Galen classified different types of pain, attributing qualitative characteristics as "pulsating" or "lancinating," according to a complex and deductive analysis that would identify all sensory impressions and scrutinize the localization of the "pains." The Galenic dual nature of pain and its attendant classifications persisted, mainly because they proved to be effective, based on a rational system that could be linked to observed phenomena. The Galenic tradition, as established during his lifetime, appealed to a unitary and contextualized medicine and was widely disseminated through considerable translations and summaries. From AD 313 onward, Christianity became the official religion of the Roman Empire, which led to internal conflicts, and ultimately to a schism between the East and West. During the fifth century AD, doctrinal divergences and inconsistencies within the Christian church led to the political, economic, and military downfall of the Roman Empire.

In the expanding Arabic world, medicine was mainly practiced by Christian and Jewish physicians, who perpetuated Galenic medicine and teleology that was easily reconciled with the monotheistic nature of Islam. The Christian Arabic scholar Hunain Ibn Ishaq travelled extensively in the Greek Byzantine Empire searching for the Galenic treatises, and became one of the most important medical figures in the Arab world. Ibn Ishaq and his pupils undertook the monumental task of translating Galen's works, first to Syriac and subsequently to Arabic, communicating successfully the Galenic tradition to the Islamic medical community.

From the tenth to the thirteenth centuries, the Arabic world saw an extraordinary proliferation of medical writings, and the development of the Arabic encyclopedias, that included the authors ar-Razi (Rhazes), al-Majusi (Haly Abbas), and Ibn Sina (Avicenna). Their works were aligned with both Galen's systematization as well as the precepts of Aristotelian deductive logic. Ibn Sina's *Canon of Medicine*, translated into Latin, became the authoritative medical textbook in Europe for more than six centuries.

Although not viewed in the same diagnostic perspective as in the traditional Galenic system, pain still retained prognostic value in Arabic medicine. Ibn Sina discussed at least 15 types of pain, denominated as "boring, compressing, corrosive, dull, fatigue, heavy, incisive, irritant, itching, pricking, relaxing, stabbing, tearing, tension and throbbing" (3). His discourse about agents that would alleviate pain is subtle but indicative of the overall level of sophistication, efficacy, and knowledge that both characterized Arabic medicine of the time and served to greatly influence Western medicine in the centuries to follow (3).

In medieval Europe, Galen's work remained a major medical influence. Galenic constructs were passed on through recopied manuscripts, which led to progressive alterations due to various copyists' personal perspectives. According to Rey "dissections were abandoned…there was an increasing disregard for anatomy, theoretic discussions began to predominate over observations…deductive reasoning and *disputatio*—already prevalent in Galen—then took over" (1). Many medical physicians acquired their first contact with Hippocratic knowledge through Galen's work, and even if the Galenic tradition was somewhat defused by iterative interpretations, some features persisted—namely, the theory of the four elements and correspondent humors. Accordingly, a humoral notion of pain was generally prevalent in the medical theory and practices of medieval Europe.

By the mid-fifteenth and early sixteenth centuries, the church had become a main sociocultural force throughout much of Europe, and religious dogma and value systems increasingly influenced public and scientific thought. In light of this, pain was widely viewed as having religious connotations, and these were evidenced in meaning, expression, and treatment. Pain was viewed within a dogma of salvation, such that bodily pain and suffering were seen as a consequence of humanity's "sinful actions." Therefore, pain was considered to be a worldly punishment for religious disobedience. In Catholicism, the devotion of martyrs became widespread, with mortification of the flesh through goads, hair-shirts, and fasting, being perceived as a form of purifying the soul (4). However, at the same time, ideals of philanthropy and charity became popular and practiced. Thus, while pain and suffering were seen as retribution for sin and desired as means for salvation, it became more widespread to help those in pain.

FROM RENAISSANCE TO ENLIGHTENMENT

Until the middle of the sixteenth century, life in general was shadowed by poverty and disease. The effects of the plague were continuous, and the horror of disease became a strong social force. Quarantines were frequent, and this inherent fear amplified the sociocultural view of pain as a portent of human suffering that could only, in the most fortunate cases, be provided temporary relief against a backdrop of mortal finitude. Baroque poetry portrays a world of contrasts, where pain is amalgamated with joy, yet still accompanied by feelings of abandonment and inevitability.

The Renaissance saw a renewed focus upon physical aspects of the body. At the same time, while Protestantism captured the concept of the body from the Church, Catholic notions of the submitted body—that was punished by pain and redeemed by penance—persisted. Mystical experiences subjugated and spiritualized pain through suffering and "apprenticed" agony (1), with the purpose of attaining eternal life and the love of God.

In literature, self-awareness as analysis of the individual body emerged in the works of authors such as Montaigne. The body was perceived not as prison for the soul or as an object of mortification to reach salvation, but as a natural entity, understood in its magnificence and with all its misery. The knowledge of oneself and of the respective body led to a self-understanding, a sign of noninstitutionalized sensibility that implied not only moral but also a personal responsibility to attend to pain.

An additional consequence of this focus on the individual was the revival of anatomy, derived in part from a need for a more systematic medicine that was based in a mechanistic knowledge of the body. Prior to this, dissection was prohibited and disregarded, for the human body was seen as sacred and inviolable. After the Black Death, *postmortem* examinations were permitted by Papal ruling, but it was only with Pope Clement VII that teaching anatomy by dissection became formally adopted. Andreas Vesalius (1514–1564), greatly contributed to the eminence of anatomy in scholarly works with the publication of *De Humani Corporis Fabrica* (*On the Structure of the Human Body*) in 1543. Based upon systematic inquiry, Vesalius called for the establishment of observation as a scientific method. In the *Fabrica*, Vesalius presented accurate anatomophysiological descriptions and illustrations of the skeleton, the muscles, the nervous system, the viscera, and the blood vessels, so as to depict an "architectonic vision of the body" (1).

The French surgeon, Ambroise Paré (c. 1510–1590), integrated Vesalian anatomy into his conceptualization of the three components of surgical capability: "the revival of anatomy, the architectonic conception of the body, and the rehabilitation of *téchne* or surgical skill as an instrument of knowledge" (1). Paré translated parts of Vesalius' work into French in his *Anatomie Universelle du Corps Humain* (1561), making the anatomical teachings more accessible to the community of barber surgeons (who were not versed in Latin). This changed the *status quo* of the barber-surgeon profession and initiated first steps toward its fusion with medicine as practiced by physicians. By 1523, firearms were widespread, and extensively used in battle. This introduced new types of trauma and pain: firearms "bruised, lacerated and tore the flesh" (1). Concerning war wounds, Paré substituted the hot-oil cauterizing of open injuries with an ointment made of egg yolk, oil, and turpentine, which would, in the majority of cases, prevent pain and inflammation. This treatment was of particular importance in a time in which the majority of wounds were treated with incisions and burning, provoking excruciating pain, and in many cases, death.

Paré had a concept of pain similar to that held by Galen: a disruption of the continuity or a change in the humors or qualities that constituted the body. Due to a relatively short life expectancy, and the social, political, and economic aspects of the period, the most feared pain was acute pain (derived from trauma of normal daily life or from the battlefield), rather than chronic intractable pain. The prevalent intellectual view ascribed pain as attributable to and reflective of Galen's hierarchy of humors. Anodynes—drugs to reduce pain—were prepared according to this view; so, for example, "cold" anodynes would prevent pain by suppressing the "animal spirits" (1) and, therefore, induced insensibility in the affected area. While incipient, this view of pain-specific treatment gave rise to a more detailed chemistry from the constraints of alchemy and was ultimately important to the development of a viable pharmacopeia that could be used against pain.

Valerius Cordus (1515–1544), a German botanist and apothecary, was the first to describe the synthesis of ether from sulfuric acid and alcohol. Cordus traveled extensively, visited many universities, and was widely acclaimed by his colleagues for his systematic and rigorous study of botany. His work, *Historia Plantarum* published in 1544, was unique in the European history for its balanced analysis and thoughtful study of plants, which was a value not only to botanists, but to apothecaries and herbalists as well. *De Arteficiosis Extractionibus Liber*, published by Conrad Gessner after Cordus' death, provides the earliest known account of the synthesis of *oleum dulci vitrioli* (diethyl oxide) from the "sour oil of vitriol" (1). Cordus found that "sweet vitriol" was highly volatile and an excellent solvent for many substances, but did not discover its potential use as an anesthetic.

Paracelsus, however, refers to the numbing qualities of sweet vitriol in the *Paradoxes*, but the true nature of the vitriol that he mentions remains obscure (3). Paracelsus' work was disregarded for several years, and pain continued to be treated with a diverse number of potions and medical preparations based, for example, on lettuce sap, barley, chamomile, sweet-clover, periwinkle, and water lilies, in decoctions, fomentations, ointments, and liniments.

Pain was of central concern in Renaissance medicine, especially in surgery, in part because by this time, there was established knowledge that pain and inflammation were directly related. One of the first known references to an anesthetic agent is made by William Bullein, in *Bulwarke of Defence againste all Sicknes*: "The juice of this herbe pressed, and kepte in a close earthen vessel, accordyng to arte: this bryngeth slepe, it casteth men into a trauns on a depe terrible dreame, until he be cutte of the stone" (3). And there are several references to the effects of some potions with poppy and mandragora in the works of Shakespeare and Marlowe.*

But it was Montaigne's perspective that shaped how pain was viewed in Renaissance Europe: simply, pain was personal, and body and soul would therefore determine pleasure and pain in a distinctive and idiosyncratic way. Pain became part of the individual, both in its physical and emotional aspects. This was explored by writers in a quest for an expression of the individuality of their suffering, and by surgeons and doctors, who, transitioning from the mysterious ways of alchemy to a systematic scientific method, tried to control and alleviate patients' pain.

In the seventeenth century, William Harvey's work furthered the emergence of medicine as a science. Until then, prevailing theories of illness were based in the Hippocratic equilibrium of the humors. A new analytical attitude in the fields of astronomy and physics was coupled to the ever-expanding mechanistic perspective of the natural world, setting the ground for a new medicine. This revolutionary setting discarded obscurantism and occultism, and viewed the human body as a complex machine, guided by physical principles. Harvey (1578–1657) established a theory of circulation derived not from conjecture but upon a series of observations and experiments using cautious and methodic reasoning. Based upon Aristotelian natural philosophy and teleology, Harvey conceptualized the body's structure as designed to execute particular functions that would enable ultimate purpose. This causal understanding of the body was to play a significant role in conceptualizing the mind–body relationship, and a concomitant view of pain.

René Descartes (1596–1650), a French philosopher, mathematician, and scientist, developed a framework for the natural sciences that were steeped in skepticism. Discarding perception as unreliable, Descartes admitted only deduction as a method. In *Dioptric*, Descartes explored the senses and perception, especially pain, so as to establish the relationship between "body and soul." Descartes, correlated the pain "felt" in phantom limbs with "perception of the soul"—that is, by experiencing this supposed pain, it was possible to validate the existence of the duality of soul and body, and at the same time, the existence of external bodies (1). In the Cartesian perspective, pain was "one mode of action of the animal spirits involving the nerves of touch" (1) and was mediated within the brain by the pineal gland that Descartes referred to as the "seat of the soul" (5). For Descartes, the fact that "the soul being united with *all* other parts of the body was the single, indivisible nature of *the body*, whose parts were independent" (1).

Descartes' interpretation of pain in animals established a new philosophical concept: the *animal machine*. Simply, Descartes asked how could animals, having "no soul," feel pain? Descartes makes an important distinction between voluntary and involuntary activities (i.e., pain belonging to the category of involuntary acts) and claims that animal pain is possible as an involuntary reflex but is not, per se,

* W. Shakespeare, *Romeo and Juliet*, Act 4 Scene I—One of the most famous scenes in *Romeo and Juliet* is the description given by Friar Laurence, to what will happen to Juliet when she drinks the potion: "Take thou this vial ..."

identical to the pain of the human *soul*. This fundamental distinction has historically served as a basis for rationalizing against both the validity of pain in animals, and the bias against forms of animal consciousness which has only recently been contested and refuted.

The teachings of Harvey and Descartes were enthusiastically embraced and served as inspiration for further physiological inquiry, in both England and on the continent. Thomas Willis (1621–1675), who later became a founding member of London's Royal Society (1662) and Sedleian Professor of Natural Philosophy at Oxford, was an English physician who pioneered the study of the anatomy of the brain, the nervous system, and their diseases. Willis' account of the brain and nervous system established the localization and specialization of different parts of the cerebrum and cerebellum. For Willis, the problem of pain was addressed in the study of involuntary movements and their relationship to the brain. To explain reflexes, Willis evoked Descartes' notion of the *animal spirit* as essential to the transmission of sensations. However, Willis' animal spirits were like fuses that, through a radiance flux, transferred the sensations perceived by the senses to the brain. Willis tried to correlate conscious cerebral activity and the mechanical regulation of involuntary movements so as to depict pain as a result of this interaction.

Thomas Sydenham (1624–1689), a clinical physician, favored observational rather than theoretical knowledge in the practice of medicine. His focus was upon epidemic diseases, which he believed were due to atmospheric conditions. Sydenham became famous for the preparation of laudanum* that was used not only for the relief of pain, but also in the treatment of dysentery as well as nervous system disorders. Sydenham was an enthusiastic defender of the virtues of opium. According to his medical observations, opium alleviated chronic pain and should be taken moderately, with small increases over the years to overcome the effects of habituation.

Based upon his studies of nervous disorders, Sydenham explored the problematic relationship between body and soul, introducing the notion of an *internal man*. In Sydenham's view, the "external man" would be a sphere of different parts, which would be the receptacle of external stimuli, while the "internal man" could be seen as a set of "arrangements of the spirits" (1). In this context, pain reflected both a neuromuscular reflex, as well as an expression of an inner "life" that otherwise remained concealed.

Despite such pioneering work, adoption of the scientific worldview was not uniform, particularly in those regions of the continent that sustained some primacy of Papal authority. For example, Harvey's medical systemization was not highly regarded by French physicians who tended to be highly conservative and faithful to the Galenic teachings. Due to the strong Catholic orientation prevalent in France, pain was associated with the Catholic tradition of repentance. Religious attributes were ascribed to both physical pain and transcendental suffering. Blaise Pascal (1623–1662) in *Prière pour demander à Dieu le bon usage des maladies*, describes pain not only as bodily burden, but as something with mystical meaning, which

* Paracelsus had already experimented with the medical value of opium, recognizing its medical and analgesic value, calling it *laudanum* (*laudare*, to praise; or from *labdanum*, the term for a plant extract).

persons should endure and understand in order to comprehend the meaning of life and being (2).

However, by the eighteenth century, understandings of pain underwent profound conceptual shifts that accompanied the appearance of the *Enlightened Rationalization*. While the seventeenth century evidenced the rise of a *New Science*, the 1700s extended this new way of looking at the natural world to both aspects of society, and life in general. Research in general anatomy, following Vesalius' premises, extended the form/function debate to organs, bones, joints, and muscles, as well as to microscopic structures. This fascination with the mechanics of the body did not necessarily reflect a reductionist approach. In fact, there was still debate as to the nature and importance of the "human soul" as the intrinsic constituent of the being. Increasingly, however, scientific and, in particular, medical inquiry were being considered as depicting the physical reality of the body. The notion of the sentient being, rather than a concept of purely a metaphysical or psychological realm, was now more deeply explored by physiology; thus, it was held that sensations ought to be measured and quantified.

According to John Locke (1632–1704) and Etienne Bonnot Condillac (1715–1780), it was via sensation that knowledge could be acquired. In other words, there was no intrinsic information prior to that provided by sensations; therefore, the study of the senses enabled study of knowledge. This natural philosophical framework accompanied a skepticism derived from the (earlier) societal milieu of seventeenth-century England and Europe. Within this worldview, pain was seen as a more physiological event—there was a secularization of pain. This is not to infer that its religious element was disregarded, rather only that the debate diminished.

To some extent, pain had always been perceived as a warning sign of disequilibrium of the internal or external environment. Generally, a description of pain served as a diagnostic tool, having a useful purpose (1). Sometimes, pain was also interpreted as a positive manifestation, a sign that the body was "fighting" to regain its inner balance. However, by the 1700s, there was a clear distinction between pain that accompanied a crisis with a positive meaning, and pain that seemed to have no positive (or existential) value. Pain was still classified according to a somewhat antiquarian scheme that differentiated among tensive pain, gravative pain, pulsating pain, and pungative pain (the last two could be divided into lancinating and terebrant pain, according to their intensity). These categories did not include all types of pain; thus, minor modifications were introduced through the years.

Although some physicians followed a nosological orientation and tried to associate different classes of pain to various illnesses, the idea of the pain/illness relationship would not arise until much later. Pain and its signs (localization, rhythm, and temporal dimension) were becoming increasingly important to diagnosis and prognosis, but as Rey stated, "whether pain was the symptom or the effect of the illness, the essential question for the physician was to know whether a particular type of pain could allow one to determine that a given part of the body was affected and/or that a particular illness had befallen the patient" (1).

By the time of the Enlightenment, it became possible to identify at least three schools of medical thought regarding pain. First, was Mechanism, a legacy from the previous century that continued to look at the body as a machine, and consequently, pain and its etiology were seen as alterations in the stasis of the body. Pain was a standard occurrence in any clinical situation, and there was no further inquiry to its causes. Mechanical models would therefore even explain some mental pathologies. Pain remained associated to a psychological dimension and was linked to the concept of *internal man* inherited from the Sydenham model (later explored by Boerhaave and Gerard Van Swieten, 1700–1772). Man has a dual nature, and while the external facet could recollect the impressions of reality and suffer pain, the inner sphere—the *internal man*—governed by reason, would suffer painlessly.

Within the Mechanistic movement, Friedrich Hoffman, a German physician (1660–1742), developed a different explanatory model of pain. Following rigorous geometrical precepts, Hoffman held that "spasms and atonicity of the fibers were the principal causes of illness" (1). Hence, pain was as a signal constituent to pathology and would be manifested through spasms. It would be possible, then, to study and quantify the mechanisms and regulation of pain. Life, from Hoffman's perspective, was a "mutual correspondence between parts" (1), a model of action and reaction, which would explain pathology, dysfunction, and pain.

A second school of thought, Animism, primarily attributed to the teachings of Georg Stahl (1660–1734), a German chemist and physician, arose in opposition to the Mechanicist constructs of Boerhaave and Hoffmann. Animism posited a substance inside the body that is reliant upon interaction with the soul, to exert particular functions necessary for the enacting and sustenance of life. Even though Animism integrated something of the Mechanicist theory of pain, it upheld the need for a psychological dimension in its mechanisms. Thus, the *soul* would intervene directly in the functioning of tissues, organs, and systems. By drawing a direct association between body/soul, Animists could explain several illnesses (including mental disorders) as conflicts that had arisen between the "will of the soul" and the "will of the body" (1).

Not unlike Animism, a third school, Vitalism, was based in the belief that life forces were active in living organisms, and thus, life could not be explained solely by mechanisms. Within Vitalist doctrine, this element was referred to as *élan vital*—the vital spark, an energy that could be equated with the soul. This movement, however, was not without contention, and Albrecht von Haller (1708–1777), a Swiss anatomist and physiologist, was one of the premier figures in provoking and advancing the debate. In *Elementa Physiologiae Corporis Humani*, von Haller, following Boerhaave's work demonstrated Glisson's theory that contractility was an inherent property of muscular fibers, while sensibility was a characteristic of nervous fibers. Thus, von Heller established a fundamental division according to fibers' reactive properties: the sensibility of nervous fibers was due to the responsiveness to painful stimuli, and the irritability of muscle fibers was caused by the contractibility in reaction to stimuli. According to his theory, von Haller strongly defended a rigid dichotomy between sensibility (deeply connected with consciousness) and irritability (independent of consciousness), established through observations of the contraction of muscles (even *postmortem*).

Sensibility became the central question of Vitalism, and soon harsh attacks against von Haller arose within the Vitalists' core. In von Haller's view, sensibility was a property intrinsic to the nervous fibers, and nothing more, while to the Vitalists, sensibility was a property unique to organic beings, an evidence of life, capable of various characteristics and magnitudes. As relates to pain, the issue established distinct definitions and values for pain as a physical occurrence, symptom, and sign. These were reflective of larger, more encompassing constructs of body, mind, and soul, and thus by extension, inferred meaning to pain on scientific, medical, and social levels.

Pierre-Jean George Cabanis (1757–1808), a preeminent physiologist and Vitalist, drawing upon Sydenham's concept of the internal man, believed that sensations could arise spontaneously within the brain, causing pains that were real, that were not resultant from apparent external stimuli, but instead were the product of memories or imagination (6). Cabanis' efforts focused upon the demonstration that pain required cerebral activity (7). Thus, pain became central to the relationship of the mental and physical dimensions of organisms. Pain, in Cabanis' view, was a temporal and complex process, in which sensations "competed" to reach consciousness (6).

However, while Cabanis lacked physiological and anatomic data to support his theory, François Xavier Bichat (1771–1802), a French anatomist and physiologist, sought careful and reliable experimental data to validate his conclusions regarding the difference between animal functions and vegetative functions. Animal functions entailed conscious and voluntary activity, while vegetative functions involved only unconscious regulation. Bichat concluded that there was an interaction between the two functions, but this relation was indirect and complex in origin. Bichat established a set of differences between the nervous system of animal function (constituted by the brain, the spinal cord, and its emerging nerves) and of vegetative life (divided into the sympathetic and parasympathetic system). According to these divisions, Bichat differentiated sensibility into various forms, inclusive of pain. Sensibility had differing expressions, according to what functional part of the nervous system received the stimuli (i.e., animal or vegetative).

By the end of the eighteenth century, the concept of sympathy—that pain travels from an injured part to different parts of the body—was systematized. Sympathies were aligned with Vitalist theory by Paul Barthez (1734–1806), a French physician, physiologist, and encyclopedist. Barthez distinguished synergy—an interactive cooperation between the forces of different organs—from sympathy, a transmission of an *ailment* that was governed by the laws of mechanics (1). For Barthez, three situations defined sympathy: First, there was no clear relationship between the interacting organs. Second, there was a link between the organs. Third, the affected organs were similar both in structure and in function.

Depending upon the situation, pain could be differentially propagated. Through sympathy, it was possible to interpret sensibility as a dynamic force in a body composed of different pathways and communicative structures. Sympathy was somewhat of a *bridging theory*, linking knowledge of eighteenth-century Vitalism to the great discoveries of the nineteenth century, and paving a path for the physiological observations that would lead to frank scientific analysis of the neurological system.

Despite the various conceptualizations of pain, pain care evolved little from the previous century. Contrary to the 1700s, opium became widely used to treat pain but was also employed more generally in the treatment of several other conditions. Opium was the object of a well-organized and efficient trade, due not only to its medicinal uses, but also to its overall popularity. Yet, the use of opium as an analgesic was part of a growing enterprise of medicaments (and technologies) that appealed to physicians' established duty to relieve pain, and a nascent but important social expectation for same.

THE NINETEENTH CENTURY

The nineteenth century is known as a true age of science. The scientific heritage of the seventeenth and eighteenth centuries was advanced and systematized, promoted by the state and universities, and fostered by the second Industrial Revolution. The principles of anatomy and physiology, well established in the previous century, explained the laws of life, and gained increasing intellectual authority. Medical societies became prominent in society, constituting a generally conservative group, although bespeaking different positions and orientations to medicine. By the middle of the century, there was a defined increase in the importance and, subsequently, the socioprofessional recognition of surgery as a valid medical practice.

Dominique-Jean Larrey (1766–1842), a French military surgeon, known as a highly skillful battlefield amputator, was a paradigmatic figure. Larrey developed the first effective ambulance, believing in the direct correlation between prompt surgical intervention and the chances of survival. Due to the nature of the wounds he treated, Larrey used amputation widely, always aware of the pain caused. He believed that rapidity of amputation could diminish the pain of the procedure and was one of the first to support Gérardin's reports of the use and benefits of anesthesia.

Opium, long known to have analgesic properties, was now the focus of chemical analysis. In 1806, Friedrich Wilhelm Adam Sertürner (1783–1841) synthesized the alkaloid salt of opium, recognizing its soporific qualities, and called it *morphium* after Morpheus, the Greek god of dreams. In the following years, morphine was widely used in medicine. Twenty years later, Bally, a French physician, in his memoir, *Observations sur les Effets Thérapeutiques de la Morphine ou Narcéine*, established doses, therapeutic indications, and effects of morphine in different bodily systems, mentioning for the first time the problem of addiction.

There was considerable progress made in the chemical analysis of gas, following the work of Priestley. Sir Humphrey Davy (1778–1829), a disciple of Thomas Beddoes, and trained in Beddoes' treatment of disease by inhalation of gas, became interested in nitrous oxide after reading Mitchill's *Doctrine of the Septon* (3), in which nitrous oxide was described as capable of inducing amazing (and terrible) effects when inhaled by animals. After several experiments in animals (3), Davy tested nitrous oxide in man: some of the subjects did not react favorably, but many felt a sensation of "warmth and joy." Davy reported: "From the strong inclination of those who have been pleasantly affected by the gas to respire it again, it is evident that the pleasure produced is not lost, but that it mingles with the mass of feelings, and becomes…hope." (3). Davy experienced the effects of the gas himself, discovering that it caused loss of consciousness, and this implied its possible use in surgical operations. Nevertheless, Davy's discovery (and its surgical value) was to remain unappreciated for several years.

In 1828, Emile Gérard attempted to induce anesthesia using ether, in a presentation before the Académie de Médecine in Paris. His presentation provoked suspicion and even some disdain among French academics and was supported only by a few, including Larrey (*vide supra*). In March 1842, William Crawford Long (1815–1878), an American physician, extracted a tumor from the neck of a patient who was anesthetized with ether (3). The surgery was successful and without pain, but Long was unaware of the far-reaching implications and importance of his accomplishment and was denied the honor of his discovery.

In 1844, Horace Wells (1815–1848), an American dentist, after carefully planning the use of nitrous oxide, was himself the subject of a painless tooth extraction under anesthesia. Wells advocated the use of anesthesia as heralding a new era in medicine and dentistry, but skepticism related to his discovery and experiments, as well as the subsequent development and use of ether by William Morton, led Wells to commit suicide in 1848. Twelve days after his death, Wells' family received a letter from the Societé Médicale de Paris, proclaiming him to be the original "discoverer" of anesthesia (3).

In France, the debate about the use and value of anesthetics continued. In 1847, Alfred Velpeau, in a report to the Académie des Sciences, declared the use of ether as one of the most important discoveries ever made, with implications for all branches of medicine; this actually contradicted his previous declarations that pain and surgery were inseparable. Only François Magendie, a physiologist, and Ludger Lallemand, a military surgeon, remained renitent to the discovery. In the following months, ether was used more widely and successfully in a variety of surgical applications, reflecting the progressive abatement of professional dissension, and a growing enthusiasm for agents capable of reducing the scourge of pain. The success and growing acclaim for ether prompted investigation of other potentially anesthetic substances.

The structure and chemical action of chloroform were defined by Jean-Baptiste Dumas (1800–1884), in 1834, after analyzing the novel substance that had been described (independently) by the American physician Samuel Guthrie (1782–1848), French chemist Eugène Soubeiran (1797–1859), and Justus von Liebig (1803–1873) in Germany. Chloroform's action was quicker and more durable than ether, and this allowed broader use. In 1847, James Young Simpson (1811–1870), a professor of obstetrics in Edinburgh, used chloroform to induce anesthesia during childbirth. After some initial resistance, the use of chloroform in obstetrics expanded rapidly thereafter in Europe, and in a year's time, Simpson used chloroform to induce anesthesia in over 150 deliveries.

However, the use of chloroform was not without controversy: several cases of chloroform-induced deaths led to reports, inquiries, and protests. Yet, temperance prevailed, and it was noted that instances of harm or death reflected possible misuse or counterindications of the anesthetic agent. Clearly, anesthesia had completely changed both the medical and public communities' relationship with pain, and in such terms, the risk was seen as almost negligible compared with the advantages offered.

The increasing use and acceptance of anesthesia became the impetus for the development of new anesthetics. Alkaloids of cocaine were first isolated by the German chemist Friedrich Gaedicke in 1855, who described the principal alkaloid as a stimulant. In 1859, Friedrich Wöhler (1800–1882), a German chemist, received coca leaves from South America, and he and his doctoral student, Albert Niemann (1834–1861), developed an improved purification process to isolate and extract viable alkaloid. The process was the topic of Niemann's doctoral dissertation, *Über eine neue organische Base in den Cocablättern (Regarding a Novel Organic Base in Coca Leaves)*, published in 1860, which became something of a standard and established the process by which cocaine could be made readily available as a pharmaceutical agent.

Sigmund Freud (1856–1939) published a well-known essay, *De la Coca*, which provided one of the best historical accounts of the use of coca leaves in South American Indians. In Europe and the United States, cocaine was increasingly used as a substitute for morphine and alcohol for a multitude of therapeutic indications, yet it took almost 20 years from the time it was first synthesized for it to be used as an anesthetic. In 1884, Carl Koller (1857–1944), an Austrian ophthalmologist, employed cocaine as a local anesthetic in ophthalmic surgeries. In the same year, William Stewart Halsted (1852–1922), an American surgeon, anesthetized the cubital and orbital nerves using cocaine, and in this way demonstrated its general utility as a local anesthetic.

The ability of cocaine to block pain at the site of induction gave rise to considerable academic debate within the medical community about whether pain was an event of the peripheral or central nervous system. In the past, this debate was additionally fueled by contemporary elucidation of tactile and thermal sensory receptors in the skin. The *peripheralist-centralist* debate was to endure for over a century, influencing doctrinal orientations to treating pain, and was only resolved (through a reconciliation that pain involved both peripheral and central functions) with some authority in the latter part of the twentieth century.

Additional progress that was important to the understanding and treatment of pain included the discovery of the tactile corpuscles by Fillipo Pacini (1812–1883), an Italian anatomist, in 1840; the finding of other tactile receptors in the superficial cutaneous tissues by the German anatomists Rudolph Wagner (1805–1864) and Georg Meissner (1829–1905); the separation of the physiology of pain and touch by the French physician, Joseph H.S. Beau (1806–1865); the experiments of Augustus Waller (1816–1870) that isolated ascending and descending neural tracts; the French physiologist Claude Bernard's (1813–1878) work both with *curare,* and on the distinction of sensory nerves from motor nerves; the crossed transmission of sensations at the spinal cord level demonstrated by British physiologist and neurologist, Charles-Édouard Brown-Séquard (1817–1894); and the theory of sensory specificity of Max von Frey (1852–1932) that stated that to each sensation there was a specific receptor. Taken together, this body of work advanced scientific constructs relevant to depicting pain as a focus of medical intervention. Coupled to the experimental turn in psychology, such approaches established pain as involving activity of the nervous system, as well as being an experience of conscious meditation.

PAIN IN THE TWENTIETH CENTURY

Both experimental urge and medical progress were enabled by technology in the twentieth century. Two convergent influences were instrumental in the speed and extent of this advancement. The first is that technology in science and medicine was becoming evermore an apologia against the technological turn in warfare. Simply put, there was an increasing need to study, and repair, the ways that technology could inflict harm to the human body. Much of twentieth-century medicine can be seen as a response to the ravages of war. Second, technology was (and continues to be) used with increasing sophistication in scientific and medical research. Thus, the conflation of these trends created fertile ground for ardent strides in pain research and treatment.

At the beginning of the century, two British scientists, Edgar Douglas Adrian (1889–1977) and Keith Lucas (1879–1916), focused on the neurophysiological analyses of neural excitability in order to determine the neural conduction velocity of pain (1). In 1927, two American scientists, Herbert S. Gasser (1888–1963) and Joseph Erlanger (1874–1965), established a classification of nerve fibers according to their conduction velocity. Nerve fibers could be divided into three main groups (A, B, C fibers) of which the first, A-fibers, was further subdivided to demonstrate that the A-delta type (myelinated nerves that are rapidly conducting, with velocities ranging from 5 to 100 m/sec) was specific to subtending (acute, or "first") pain, while C-fibers (small, unmyelinated sensory nerves) have conduction velocities below 2 m/sec and transmit chronic or second pain.

Erlanger and Gasser also showed how this highly differentiated system with its three main types of fibers was distributed over the in- and outgoing fibers of the spinal cord, the sensory and motor roots, and established the basic physiological and anatomical properties of these systems, receiving the Nobel Prize in Physiology and Medicine in 1944 for their work. These studies paved the way for later investigations by a number of researchers that demonstrated the receptive fields, spatial and temporal summative properties, and differential activational thresholds of these fibers, and subsequently depicted the anatomic and physiologic basis of differing types of pain.

In 1937, Sir Thomas Lewis (1881–1945), a British cardiologist and scientist, discovered that certain chemical substances were released at the site of tissue injury, radiating outwardly from the affected point, describing the relationship between the inflammatory process and pain (as evidenced by the *Lewis triple response* of heat, redness, and pain). In the same year, René Leriche (1870–1955), a French surgeon, published *La Chirurgie de la Doleur (On the Surgical Treatment of Pain),* based upon his experiences with pain control using surgical ablation of the sympathetic nervous system as employed during World War I. Leriche presented a comprehensive study of pain in various diseases and its treatment, and included discussion of pain as an abstract concept. In his perspective, surgery was in evolution—from a technical and mechanical skill to a more comprehensive physiological science (1).

As a humanist and activist against the "dolorist" trend (1), Leriche was totally opposed to the idea that pain was a necessary evil. He maintained that medicine attempted to treat pain only by suppressing its effects, but without understanding its causes, and thus would remain unsuccessful. To be sure, this argument proved to influence the worldview of pain care both for the remainder of the century and for the burgeoning field that was to become pain medicine. Anesthesia enabled improvements in surgery, which facilitated survival of terrible wounds inflicted by new and devastating armaments of World War I, considered at the time to be "the war to end all wars."

Yet, this was not to be the case. World War II gave rise to horrors as yet unforeseen, and while medical innovation enabled patients to survive trauma and surgery, it also left many in chronic, and oftentimes intractable, pain. From the conundrum of chronic pain was born the field of pain medicine, championed during the 1940s and 1950s by John Bonica, an American physician (who continued to be an important figure in establishing the International Association for the Study of Pain (IASP) and the mission and directions for pain medicine as a specialty field until his death in 1994). As a developing field, pain medicine was initially subsumed within other medical specialities—namely, anesthesiology, neurology, physiatry/rehabilitative care, and psychiatry. In this latter regard, the psychopharmacological "revolution" of the late 1950s and early 1960s did much to fuel research into biochemical and anatomical bases of pain and analgesia.

Perhaps one of the most influential premises in pain research and treatment was the concept of the *gate control* theory, as devised by Ronald Melzack and Patrick Wall in 1965 (8). Melzack and Wall proposed that pain transmission involved hierarchical activation of neural circuits from the periphery to the brain. As well, it was posited that pain could be mitigated (i.e., *gated*) at various levels of the peripheral and central neurological axes, and that under normal conditions, these anatomical sites engaged various biochemical gates to suppress noxious impulses. The prescience of this work is nothing short of amazing; bear in mind that advanced pharmacologic and anatomic techniques (such as radioisotope labeling and neuroimaging, respectively) were not available to these researchers at that time. Thus, while aspects of the original "Gate Control Theory of Pain" may be inaccurate, or seem a bit anachronistic, that the general schema has withstood over 50 years of scientific scrutiny, and served as the template for almost all subsequent depictions of the pain and analgesic system, is testimony to these scholars' insight and wisdom.

The field of neuroscience gained prominence in the 1970s, and in bringing together disciplines of anatomy, physiology, biochemistry, and pharmacology, created a nexus for further mapping of the pain and analgesic system by Alan Basbaum and Howard Fields (9), William Willis (10), and Linda Watkins and David Mayer (11), among others. As well, neurochemical and pharmacological research in the latter 1970s and early 1980s defined opioid peptides and receptors, and further elucidated the heterogeneous and complex nature of the pain and pain-modulating systems (12–15). By this time, pain medicine was established as a discrete medical specialty, and many view the period from the 1980s to the 1990s as a golden time in pain care, reflective of far-reaching and well-financed services, and momentum gained by scientific research. The Human Genome Project of the 1980s, together with the Decade of the Brain (1990–2000) afforded considerable insight to the genetic bases of pain, and novel pharmacotherapeutics. Together, these endeavors sparked the Congressionally declared Decade of Pain Control and Research (2000–2010), a national agenda focused on advancing pain research and, perhaps more importantly, translating such research into clinical practices.

Yet, it remains to be seen if and how these efforts will be implemented in both medicine and sociocultural conceptualizations of pain. Pain medicine has not yet become a unified field, and economic factors have affected insurance provision to negatively impact the range, type, and duration of care available to chronic pain patients. Although it is inarguable that we have come far in our understanding and treatment of pain, it is evident that the road ahead is long, and perhaps difficult (16). In this way, the history of pain is still being enacted. Much remains to be investigated and discovered, as many questions remain unresolved. The increasing clarification of pain mechanisms has shown that an interdisciplinary approach is necessary, both to research and treatment, as pain is more than a physiological occurrence. Pain is an occurrence of a multifaceted kind, it is physical and psychological/phenomenological, and any description of pain must account for these aspects (16). In this way, we must acknowledge that we determine the present and future history of pain, and thus it remains a work in both evolution and, hopefully, progress.

ACKNOWLEDGMENTS

The chapter author acknowledges and is most grateful to Professor James Giordano and Dr. Pierre LeRoy for their assistance and insight in developing and writing this chapter.

REFERENCES

1. Rey, R. 1995. *The History of Pain*. Cambridge and London: Harvard University Press.
2. Morris, D.B. 1991. *The Culture of Pain*. Berkeley: University of California Press.
3. Robinson, V. 1936. *Victory over Pain*. New York: Henry Schuman.
4. St. Catherine of Sienna. 1980. II Dialogo della Divina Provvidenza ovvero Libro della Divina Dottrina [S. Nofke trans.]. *The Dialogue*. New York: Paulist Press.
5. Descartes, R. 1985. Treatise on man (1664). [J. Cottingham trans.]. *The philosophical writings of Descartes, Vol. 1*. Cambridge: Cambridge University Press.
6. Cabanis, P.J.G. 1980. Rapports du Physique et du Moral De l'Homme. In: M. Staum, ed. *Cabanis, enlightenment and medical philosophy in the French Revolution*. Princeton: Princeton University Press.
7. Porter, R. 2004. *Cambridge Illustrated History of Medicine*. Cambridge: Cambridge University Press.
8. Melzack, R., and P.D. Wall. 1965. Pain mechanisms: A new theory. *Science* (150) 699: 971–979.
9. Basbaum, A., and H. Fields. 1984. Endogenous pain control systems: Brainstem spinal pathways and endorphin circuitry. *Ann Rev Neurosci* 7: 309–338.
10. Willis, W. *Control of nociceptive transmission in the spinal cord*. Berlin: Springer-Verlag.
11. Watkins, L., and D. Mayer. 1982. Organization of endogenour opiate and nonopiate pain control systems. *Science* 216: 1180–1192.
12. Goldstein, A. 1977. *Solubilization of an opiate binding component from mouse brain. Neuroscience research program bulletin*. Cambridge: MIT Press.
13. Kosterlitz, H., and S. Patterson. 1981. Characterization of opioid receptors in nervous tissue. *Proc Roy Soc, London* 210: 113–122.
14. Hughes, J. 1975. Isolation of an endogenous compound from the brain with pharmacological properties similar to morphine. *Brain Res* 88: 295–308.
15. Snyder, S. 1975. Opiate receptor in normal and drug altered brain. *Nature* 257: 185–189.
16. Giordano, J. 2009. *Pain: Mind, meaning, and medicine*. Glen Falls, PA: PPM Press.

3 Pain Does Not Suffer Misprision

*The Presence and Absence That Is Pain**

James D. Katz

CONTENTS

INTRODUCTION

Suffering speaks to the core of human experience. Philosopher Hannah Arendt (1906–1975) wrote from personal experience, having escaped from Nazi internment in southern France during the German occupation of World War II. Her writing suggests that within the ether of human reality, there exists a communal understanding of suffering:

> Only pain is completely independent of any object, only one who is in pain really senses nothing but himself; pleasure does not enjoy itself but something besides itself. Pain is the only inner sense found by introspection which can rival in independence from experienced objects the self evident certainty of logical and arithmetical reasoning. (1)

To Arendt, the inescapable phenomenon of pain is the motivating impulse for the historical development of philosophical introspection as a mechanism of worldly escape. In short, pain does not allow for misreading: it does not suffer misprision. It has essence without existence.

As a prelinguistic phenomenon, we are at once afraid of and, at the same time, in awe of pain and suffering. Cassell tackles this concept and begins by defining it

* This chapter has been adapted, with permission, from Katz, J.D., 2004. Pain Does Not Suffer Misprision: The Presence and Absence That Is Pain, *Med Humanities* 30: 59–62.

in negative terms. He distinguishes suffering from physical distress (2). He correctly asserts that suffering is not confined to physical symptoms. In developing his thesis, he introduces the idea that suffering may be conceived as a response to a threat to the integrity of the individual. These are helpful tools for medically *operationalizing* difficult concepts. In this chapter, I also strive to call into play a philosophical perspective. To this end, I would further assert that suffering resides outside of the physical realm. It is neither emotion nor thought. It is not a state of lesser perfection, as Spinoza would have it. Rather, as we will see, suffering is the cynosure, the point of attraction, at the center of the whirlpool of linguistic instability.

Even pain without suffering inhabits a metaphysical zone. Arendt hints at this radicalization of the interpretation of pain. She goes so far as to suggest that the most intense human experiences (including specifically pain) challenge our *assurance* of reality. In *The Human Condition,* she ponders the intimate core around which public and private life tensely revolve:

> Indeed, the most intense feeling we know of, intense to the point of blotting out all other experiences, namely, the experience of great bodily pain, is at the same time the most private and least communicable of all. Not only is it perhaps the only experience which we are unable to transform into a shape fit for public appearance, it actually deprives us of our feeling for reality to such an extent that we can forget it more quickly and easily than anything else. There seems to be no bridge from the most radical subjectivity, in which I am no longer "recognizable," to the outer world of life. Pain, in other words, truly a borderline experience between life as "being among men" (inter homines esse) and death, is so subjective and removed from the world of things and men that it cannot assume an appearance at all. (1)

To Arendt, pain exemplifies a *radical subjectivity* that both relates and separates a social species. It relates humanity by virtue of being a common experience, and it separates humanity by virtue of being a private experience. Martin Buber recognizes this futility of the *pain narrative* and emphasizes, instead, the visceral comprehension of the nature of pain. Pain is not thought; rather, it is experienced. Buber feels that the pain narrative is a distancing phenomenon. "The nature of pain," he writes, is not recognized as if one was standing at a distance from it, sitting in a box, and watching the drama of pain as an unreal example. The man who does this may have all sorts of brilliant thoughts about pain, but he will not recognize the nature of pain. This is recognized by pain being discovered in very fact (3).

In essence, Buber attributes a *spiritual* dimension to pain. It is likely that Arendt would agree that although pain is not subject to skepticism, and hence cannot be revealed by interpretation, it retains the power to define humanity and human intercourse. In other words, pain has the capacity to generate emotion. It is an impetus for action. It is precisely this essence of motivation that makes illness and suffering a prime candidate to serve as a hermeneutical principle. Religion intuitively capitalizes on the power of pain. Western religion, for example, motivates with the imagery and raw violence of crucifixion.

Even as the religious requirements for redemption may be clear, however, the agony of eternal damnation remains vaguely conceptual. This is the irony of pain. Although pain consists of a generally appreciated experience, pain remains a poorly communicated experience. Nowhere is this view of the instability of language underscored more heavily than in the attempts at the expression of pain. The communication of pain in no way approximates the mundane. Specifically, the language of pain is necessarily that of metaphor and metonymy (substitution or association). We are left to describe our hurt in comparative terms. We displace the language of pain onto the language of dysfunction. More to the point, the "narrative of pain" can never satisfy the demand for convention imposed by the "ordinary language" of Wittgenstein. To assume that a communal "sense making" of the pain narrative ultimately reflects textual realism is to ignore the fact that pain can be both cause and effect. In particular, pain can be the effect for a political cause of the injured; for example, a call for redress, such as the cause for greater research funding into orphan diseases, can be bolstered by the testimonials of personally suffering advocates. Hence, Wittgenstein's "ordinary language" is none other than a claim of "dominant convention" vulnerable to deconstructive skepticism (4). This is, in essence, Arendt's motivation. She emphasizes pain, illness, and suffering as completely subjective experiences. To take this one step further, the corollary to the subjectivity of the pain experience is that it is not possible to *testify* to another being's pain. No one can presume the expertise to objectify that which is subjective through and through. Therefore, the universal understanding of pain remains inextricably coupled with the universal understanding that the pain of another is ultimately unknowable (5). The complete comprehension of the pain of another is no more an achievable end than the structuralists' search for total intelligibility of communication.

Not knowing the pain of another does not, however, necessarily dictate that our response marginalize the sufferer. In fact, we go to great lengths to alleviate pain. This is exemplified by the institutionalization of altruism. It is here that we comprehend the pain of another in a meta-experiential fashion; we know such pain as our own potential pain. Hospitals and organizations dedicated to the welfare of various afflicted individuals are obvious examples of the ethics of altruism. Admittedly, this communal ethic suffers a shortcoming, for it does little to illuminate the personal emotions attached to pain. This is a lack that is attached to pain. Specifically, it is a lack of language. This lack of language is the property of pain that renders it inaccessible to the deconstructive challenge. Essentially, pain has meaning that exists both within and outside of communication. In other words, although pain is both a linguistic and a societal cynosure, the essence of pain remains recondite (6).

POSITIVE AND NEGATIVE ATTRIBUTES OF PAIN

Conceptualizing pain in terms of positive and negative attributes (a presence and a lack) is entirely dependent upon a history of experience of pain. One must consider the duration, tempo, and magnitude of *experienced pain* as well as calculate whether historical pain is immediate-past, past, or remote. This escalates the complexity of a rigorous understanding of pain at both personal and collective levels.

In science, such a challenge is normally met by the fundamental construction of a matrix of *understanding*, which derives from a defined language creating both positive and negative distinctions, polarities which of course include that between centralization and marginalization of thought. Polarities force hierarchical thought.

Hierarchical thought exposes presupposition (7). Uncovering such bias is embodied in the reversal questioning technique of deconstructionists. Thus, is it necessarily true that the absence of pain opposes pain? And, is the state of not-pain a state of insensibility? If so, does this not result in isolation and in turn the pain of loneliness? The very moment that we employ speech or writing to communicate pain or capture hurt is at once the moment that all meaning is lost. The extraconceptual nature of chronic pain stymies the expressive process as we become constrained by the comparative and figurative nature of language. Pain at once embodies and exceeds the nonlinguistic medium of social psychology. Such complexity is augmented by the fact that, even though pain does not suffer misprision, the personal passion of hurt varies from individual to individual.

PHENOMENOLOGY OF PAIN

A brief foray into the fragile latticework of rheumatology quickly exposes how our limited language for pain does not adequately communicate the private experience of arthritis. Though we, ourselves, may be unafflicted by the gnarled grip of arthritis, it still reaches into our souls to tear away a swatch of empathy. Consider the unconscious berth that we carve out for an individual with a limp. Consider, too, the seat we yield to the stranger obviously bent with osteoporosis. Recall the visceral reaction we suppress at the sight of deformity and disfigurement when our basic nature touches upon a sense of dread. These corporeal feelings are often encapsulated by the secret prayer, "there but for the grace of God go I." This internal comprehension of pain possesses a fundamental meaning for us, both personally and collectively.

It is precisely because pain does not favor words over deeds (the concept termed *logocentricism*) that it can be analyzed from a phenomenological perspective. The pain of a broken heart is as real as a third-degree burn and may not be any different from the same awareness experienced 2500 years ago. Pain is the true "fullness of presence." Yet, because pain cannot be objectified, it cannot be subject to skepticism. Therefore, the tautology of the Western presupposition of the existence of hermeneutics as a mechanism for approximating truth does not invalidate the use of pain as an organizing principle for interpretation of human thought, feeling, and interaction. Otherwise stated, even though pain cannot be objectified, it can be generalized (8).

That pain is a universal motivator is an integral component of the Western medical model: almost by definition, patients hurt. How patients individually respond to pain determines how they relate to suffering. Whether that pain is physical or psychological is moot. Dysphoria is not easily ignored. It presents in many guises; some are more recognizable than others. A paradigm for the study of chronic pain is arthritis. It is a vivid portrayal of private hurt made public. At such times, the narcissistic struggles of everyday life fall away. Merely to witness rheumatic vulnerability and relentless pain touches upon our subconscious sense of the collective. Arthritis illuminates radical subjectivity. The philosophical irony of arthritis is that it emulates the senescence that is our individual fate.

As observers of the natural experiment of suffering, rheumatologists are witnesses to the recurrent human drama of struggle with pain and dependency. Rheumatologists fully comprehend that the onset of arthritis is like watching the ontogeny of existential awareness in slow motion. Chronic disease is a window into the process of transforming what Søren Kierkegaard termed "unconscious" despair into "conscious" despair (9). Chronic disease allows us to observe the spectrum of failures of repression. We see regression, and we confront denial.

A HUMANISTIC INTERPRETATION OF PAIN

A more modern perspective on the interpretation of pain can be found in the writings of the humanist, Erich Fromm. He echoes the existential premonitions of Kierkegaard.

Specifically, chronic pain parallels the psychological process of developing individuation, whereby the being "becomes aware of being alone, of being an entity separate from all others. This separation from a world, which in comparison with one's own individual existence is overwhelmingly strong and powerful, and often threatening and dangerous, creates a feeling of powerlessness and anxiety" (10). The humanist must first, therefore, embrace existentialism. For Fromm, success in negotiating this tension requires reconciliation with psychoanalytic theory. Kierkegaard, during the pre-Freudian era, viewed this as the evolution into conscious despair. Ultimately, as originally exposited by Kierkegaard, conscious despair unlocks "passion." Consequently, by drawing a parallel between dread and suffering, this suggests that embracing pain may invigorate the passion for life.

What, then, are the stages of pain maturation? First and foremost, the pained individual seeks validation. Whether that validation derives from an internal or external source, the same question must be answered: "Is this really happening to me?" Close upon the heels of validation crystallizes an internal authentication of personal distress. In turn, this is what ultimately allows anger to germinate. A sparring spirit is called into play processing the threat to personal integrity. Eventually, a sense of futility takes hold, and then profound anguish is experienced. This, in turn, is a kind of grieving process that is often described in surrealistic terms and associated with a sense of hopelessness. Eventually, for a more grounded outlook to be achieved, acceptance and humility must then lead down a path to at least a partial closure. And ultimately, for true redemption to occur, simple joy in any remaining autonomy must be realized.

In this manner, chronic pain takes us on a journey. It is a reproducible travail characterized by predictable stages. Any misstep down the path of maturing pain renders happiness derailed. This is especially the case when hidden fears about mortality contaminate the fuel feeding the maturing of pain. Denial at any one of these necessary stages results in a stalled individual, one who is sputtering and not progressing. Such a person who is unable to carry on is in search either of validation or of hope. That individual is arrested in a stage characterized by endless digging. This is a process of ever reinventing denial and never uncovering peace. This is maldynia.

I, as a physician, am charged with the responsibility of defining and validating distress. This is fundamental to any helping profession. Acknowledging suffering affirms humanity. It says, "your pain is/was/will be my pain." Martin Buber feels that this is a preeminently spiritual act (3). "Only then does his own pain in its ultimate depth light a way into the suffering of the world. Only participation in the existence of living beings discloses the meaning in the ground of one's own being"

(3). But acknowledging suffering also entails a process of searching. The patient is searching for relief; I am searching for an answer.

Sometimes when I divine the correct diagnosis, the relief of the patient is palpable. The devastation of not knowing is suddenly reversed. Newfound knowing becomes an opportunity to regroup and insert some ordered direction to the chaos of directionless disorder. In an instant, a label is put forth, and one aspect of suffering is over. But self-fulfilling meaning is no different from self-deluding explanation. Although the label is taken as an explanation, it does not, of course, explain anything. Explanations merely serve to allow us to classify the future—to better brace ourselves for the insurmountable and the irresolvable. We do not need to know why something has gone wrong or why we get sick. We need to know that we are as prepared as possible for the future. Knowing, itself, functions as a further constraint on the sense of randomness encompassing the essence of the future. Knowing functions as a reminder of the state of security rendered by mantic (that is, prophetic) comprehension. It is a modifier of hope. It is knowing what we are up against that makes pain more bearable. Therefore, self-prescribed meaning, when substituted for nescience, enables us to face uncertainty and move through dread.

Searching is, however, also painful. It confuses hope with expectation. Searching lends itself to the displacement of the need for certainty onto an uncertain existence. It inflates the bubble of false hopes and ultimately amplifies pain. At times it seems that physical pain must be amplified in order to enhance the acuity of purpose. The *telos* (ultimate end) of physical pain may be simply this: "enhance the acuity of purpose and thereby commandeer searching activity."

Thus, searching begets searching, and teleology gives rise to tautology. Ultimately, given that teleology is the language of medical science, it is no wonder that we displace goal-oriented language onto the need to understand what is unfathomable. This medical model is not unlike the Western belief in hermeneutics as a path to certainty. Here medicine presupposes that an underlying meaning can be discerned that, in turn, explains the multitudinous dysphoric states. Western medicine adheres to the empirical model of epistemology. Indeed, it appears that the success enjoyed by humanity over the previous 100 years justifies the belief that science is homing in on the ultimate *meaning* behind pain and suffering. Unfortunately (or fortunately), the flaw with this belief is that successive iterations of a medical model may just as likely adhere to Zeno's paradox. It may halve the distance without ever bridging the gap. Alternatively, this rationalist's model of epistemology may not be achievable simply because of tautology. For it may be that what we define as scientific advancement may predetermine our medical success. This is so if cheating death is, in truth, a value unto itself and not an end. In other words, if we find that we are motivated by an underlying sense of meaninglessness, then an internal tension is necessarily created. All too often, we delude ourselves into thinking that avoiding death is an end in itself, whereas, in truth, it is nothing more than deferred hope to die. To wit, we may be afraid to die, but we are equally afraid to live forever. This would suggest that the study of science for the purpose of revealing metaphysics is less productive than a study of interpretation. To take this one step further, we find, for example, that owing to the homology between the structure of language and the structure of the unconscious, an analysis of language is revealing when applied to human motivation and human dilemma. One need only consider a Freudian slip of the tongue to gain an appreciation of this relationship.

Here, interpreting language may help us interpret human motivation and the unconscious. A deeper analysis of this concept is accessible through the writings of the French thinker, Jacques Laçan. In his discussion concerning psychoanalysis, he states: "the symptom resolves itself entirely in the analysis of language, because the symptom is itself structured like a language, because it is from language that speech must be delivered" (11). Laçan is perhaps radical in that he believes that not only is the unconscious structured like a language, but in truth, the unconscious resides in language. This becomes philosophically problematic because if the deconstructionist premise is true that language is unstable, then the unconscious, let alone the self, may be fundamentally unstable as well.

Alvin Reines grapples with this instability by tackling existentialism head on. He defines the neologism, *asoteria*, as the state of meaningless existence engendered by the tension between the fact of personal death and the intense longing for infinite existence (and invulnerability) that in turn results in a "conflict of finitude." As he writes in *What Happens After I Die?* "failure to deal with one's finity—in particular one's own death—leaves unresolved a human's fundamental existential problem, the conflict of finitude and invites the continuing haunting problem of asoteria" (12, p. 133). In other words, we must each uncover our own mechanism for diffusing existential dread. We must, in essence, make peace with the denial of death. Alvin Reines feels this is the task of religion.

Erich Fromm takes this concept further. He asserts that the essence of humanity is the very fact of such questioning, let alone the need (demand) for an answer. Therefore, where Reines addresses the tension born of existence, Fromm addresses the tension born of essence. But Fromm does not refine his analysis, and Reines does not generalize his. At different times and various stages of life, we may invoke any number of responses to the challenge of asoteria. Our responses may range from displacement of dread onto dysfunction, to replacement of dread with pain and symbolism. At other times, our responses may be faith based or trust affirming. In the final analysis, our responses are never cavils and, hence, should not readily be dismissed. In order to achieve peace with the asoteric challenge, we invoke these responses to achieve the courage to consent daily to die (13).

It is not simply a platitude to assert that human pain and suffering encompass much more than disease. Pain and suffering are inextricably tied to life and forever pitted in tense struggle with hope. Although I would not go so far as to label the response to pain as an *answer* to the human existential dilemma, I agree with Erich Fromm that "there is one condition which every answer must fulfill: it must help man to overcome the sense of separateness and to gain a sense of union, of oneness, of belonging" (14). Succinctly put, the hierarchy of inquiry, as set forth here, has been organized around this type of radical humanism. In particular, through the analysis of the responses to the asoteric challenge, it becomes clear that happiness requires that we also find the courage to consent daily to live. In this manner, we are forced to make peace with the denial of life.

REFERENCES

1. Arendt, H. 1998. *The human condition* (2nd ed.). Chicago: University of Chicago Press.
2. Cassell, E.J. 1982. The nature of suffering and the goals of medicine. *N Engl J Med* 306: 639–645.
3. Buber, M. 1968. *Between man and man* (4th printing). New York: Macmillan.
4. Norris, C. 1993. *Deconstruction. Theory and practice* (rev. ed.). New York: Routledge.
5. Kaspar, J., M.V. Boswell, and J. Giordano. 2009. Assessing chronic pain: Facilitating objective access to the subjectivity of pain. *Prac Pain Management* 9(3): 55–59.
6. Giordano, J. 2009. *Pain: Mind, meaning, and medicine.* Glen Falls, PA: PPM Press.
7. Derrida, J. 1993. *Aporias* [T. Dutoit trans.]. Stanford, CA: Stanford University Press.
8. Giordano, J. 2010. The neuroscience of pain, and the neuroethics of pain care. *Neuroethics* 3(1): 89–94.
9. Kierkegaard, S. 1983. *Kierkegaard's writings, XIX: Sickness unto death: A Christian psychological exposition for upbuilding and awakening* [H.V. Hong and E.H. Hong trans.]. Princeton, NJ: Princeton University Press.
10. Funk, R. (ed.). 1999. *The Erich Fromm reader/Erich Fromm* Amherst, NY: Humanity Books.
11. Laçan, J. 1977. *Écrit* [A. Sheridan trans.]. New York: W.W. Norton.
12. Reines, A. 1990. Death and after existence: A polydox view. In: *What happens after I die? Jewish views of life after death.* eds. R. Sonsino and D.B. Syme, 126–141. New York: UAHC Press.
13. Becker, E. 1975. *The denial of death.* New York: Macmillan.
14. Fromm, E. 1964. *The heart of man.* New York: Harper & Row.

4 Understanding Suffering
*The Phenomenology and Neurobiology of the Experience of Illness and Pain**

Peter A. Moskovitz

CONTENTS

* This chapter is based, in part, on "A Theory of Suffering" (1) and "Understanding Suffering: A Context for Understanding Chronic Pain," presented to the American Academy of Pain Management, Nashville, TN (2008). I am grateful to Rosemary Bowes and Russell Stevenson who helped me organize this chapter and justify assumptions that cannot be grounded in the intuitively obvious.

INTRODUCTION

This book is about maldynia. Maldynia happens when pain—either directly or indirectly—evokes the experience of suffering. This chapter is about suffering. When we speak of suffering, we talk about an experience that *feels* bad. Suffering feels awful and terrible. Suffering is *painful*. Synonyms for suffering do not constitute a definition; but synonyms tell us something of its nature. Some synonyms for suffering are distress, misery, agony, anguish, torment, wretchedness, despair, excruciation, woe, helplessness, and hopelessness. I propose that an understanding of the experience of pain (and illness) requires an understanding of suffering.

The theory, presented here, proposes that there is something fundamentally the same about the experience of suffering regardless of its cause or context. There is a common, intuitive taxonomy of suffering that distinguishes among several types or sorts of suffering—medical suffering, spiritual suffering, existential suffering, emotional suffering, physical suffering, and psychological suffering—and many scientists, ethicists, and theologians (too numerous to cite) go to great lengths to justify an ontological separation between them. The same can be said for distinctions among categories of pain—physical pain, psychic pain, and spiritual pain—that are variously associated with suffering of different sorts. The unifying event that makes the infinitely variable experiences of suffering fundamentally alike is a unitary neurobiological event. I propose that the underlying neurobiology of suffering is the same, regardless of the cause or context of suffering.

In this chapter on the phenomenology and neurobiology of suffering, I distinguish between objects and events that are "objective," and the experience of them, in or out of awareness, which is "subjective." I used quotation marks here for three reasons: first, I use the terms in the sense that what is "objective" is measurable and verifiable, and what is "subjective" experience is immeasurable and unverifiable. Second, I ask the reader to accept nontraditional definitions of these terms. Third, the distinction between objective objects and events and the subjective experience of them is counterintuitive. The distinction is counterintuitive because objects of which we become aware and events, like the occurrence of suffering, *feel* substantive. They are *real* to us.

The characterization of emotions as being objective is also counterintuitive, probably more so. We think of our emotions as the "subjectivity of our lives." We use the words *emotions* and *feelings* interchangeably. The apparent paradox is resolved if we consider that we name an emotion by the *subjective* feeling or experience of its *objective* bodily effects, in much the same way that we name suffering by the emotions that are reactions to it—for example, the suffering of grief or the suffering of shame.* The emotion that we call anger is the experience of a collection of unique

* The words that we commonly use to represent objects and events are symbols for those objects and events. Recall Magrite's painting of a pipe under which he painted the declaration, *Ceci n'est pas une pipe*. "This is not a pipe." The painting is a representation of a pipe, not the pipe itself, and so it is for words. One need not understand Wittgenstein to seek precise definitions of a word so that it might be used consistently in any context, the only place where "meaning" counts.

bodily effects, such as a tightening of muscles, clenching or baring of teeth, and increased pulse and blood pressure. We intuitively recoil from the notion that the *experience* or *feeling* of an emotion is separate and distinguishable from the physiological events of the emotion in our bodies. In this chapter, where I endeavor to justify that distinction, the words *feeling* and *experience* are interchangeable.

Suffering is the feeling or experience of an interior state; suffering is, therefore, a state of consciousness. An understanding of how suffering happens includes a tentative understanding of how consciousness happens. Understanding *why* suffering happens is the province of either evolutionary theory or religion, or both.* This chapter presents an understanding of *how* suffering happens, not why. In the effort to propose a coherent theory of the experience of suffering, I assembled ideas for which there is reasonable evidence, although, at times, there must be speculation where the science is incomplete. The whole has not been validated. What follows is counterintuitive for most readers and may even be disturbing. I beseech the reader's patience.

The theory of suffering is based on nine interrelated premises that I endeavor to justify in the course of this chapter:

1. Perceptions† of objects and events are reducible to finite elements or *contents* and are measurable, and, therefore, *objective* (2). Emotions are biophysiological events that are perceived (2) and are as objective as other perceptions. Here, *perception* denotes the transduction, transmission, encoding, and storage of the neural representation of a stimulus, either internal or external. The feeling or experience of a perception, and of emotion, is a subjective state of consciousness.
2. Suffering happens when there is a perceived loss or threat of loss of the integrity of personhood (3,4,5), including, of course, the threat of death (6). The facets of personhood include any and all domains of the "self." The threat to the integrity of personhood can be real or imagined. The initial experience or feeling of grief or fear need not be in awareness (7,8).
3. Suffering is *subjective*. That is, suffering is irreducible and immeasurable. Suffering is a state of consciousness, an experience of perceptual contents, and the subjective, phenomenal awareness of an internal state.
4. Consciousness is the coherent, reverberating, resonant, synchronous oscillation of the bioelectrical pulsations of neuronal or local field activity (9,10).

* One explanation of why suffering happens proposes that suffering is a test or punishment visited by a supernatural power or intelligence. An alternative explanation proposes that the capacity for the experience of suffering is the basis for the emotional communication that connects a parent and a parent's neurologically immature offspring in the process of nurturing. Neither explains how suffering happens, which is the subject of this chapter.

† Here I ask the reader to suspend the commonly held definition of perception that includes both objective and subjective experience. These terms require more precise, narrower definitions that follow.

at about 40 Hertz (11–14),* in multidimensional, massively interconnected, reentrant and recursive circuits (15,16), networks, and systems, among highly ordered cell assemblies (17,18) of the perceptual and motivational apparatus, in disparate parts of the global neuronal workspace (19–21), within the bounded and situated body–brain (22), binding them around the perception of an object or event (14).

5. Suffering happens in states of perceptual or behavioral (homeostatic) conflict or struggle for the integrity of personhood in which the anterior cingulate cortex (23–25) is the epicenter of the synchronous oscillation of the body–brain.
6. Illness, including pain-as-illness, classified as maldynia (26), is a condition (a disease, injury, deformity, or pathophysiological condition) that evokes the experience of suffering.
7. Four types of events cause suffering: loss, threat, the frustration of a biologically based drive or appetite, and pain as nociception.
8. Resilience in the face of suffering is composed of coping skills (27–29) and coping reactions (30,31) that modify and moderate the onset, course, and outcomes of suffering.
9. The experience of suffering and, therefore, of illness (including, particularly, pain-as-illness) is understandable only through the first-person narrative of its evocative, perceptual contents (32). The story of suffering is the symbolic representation of its onset, course, and outcomes. That narrative, structured in the biopsychosocial model of Engel (33), is analyzed as *literary content* as implied or described by Eric Cassell (3), Arthur Kleinmann (34), Arthur Frank (35), Joseph Natterson and Raymond Friedman (36), Howard Brody (37), James Katz (38), particularly Rita Charon (32), and Robert Gatchel and Dennis Turk (39).

PERCEPTIONS, EMOTIONS, AND THE EXPERIENCE OF SUFFERING

1. Perceptions of objects and events are reducible to finite elements or "contents" and measurable, and are, therefore, *objective* (2). Emotions are biophysiological events that are perceived (2), and are as objective as other perceptions. Here, *perception* denotes the transduction, transmission, encoding, and storage of the neural representation of a stimulus, either internal or external. The feeling or experience of a perception, and of emotion, is a subjective state of consciousness.

When observations of an object or event are controlled and measured with independently validated instruments, we write about them in the third person, the voice of science—for example, "subcutaneous, fractionated heparin reduces postoperative

* A growing body of knowledge suggests that different frequencies of synchronous oscillation are associated with the degree of awareness a perception obtains and how a perception that remains out of awareness affects other perceptions, one's behavior, or the transfer of the perception into long-term memory. Phase-specific oscillation occurs over small areas of the perceptual apparatus. Theoretical models and experimental observations are expanding the range and scope of synchronicity in neuronal networks. The science is incomplete.

thromboembolism compared to placebo." We know this to a reasonable degree of scientific certainty. Similarly, one can, within definable limits, objectively observe the events of perceptions, of emotions, and of the perception of an emotion's bodily effects.

Perception is defined in this chapter as the transduction, transmission, encoding, and storage of stimuli—both external and internal stimuli that include neurobiological events. In common usage, however, we include a much broader range of phenomena in the definition of —for example,

> [1] the act or the effect of perceiving. [2] insight or intuition gained by perceiving. [3] the ability or capacity to perceive. [4] way of perceiving: awareness or consciousness; *view: advertising affects a customer's perception of a product.* [5] the process by which an organism detects and interprets information from the external world by means of the sensory receptors.*

Naturally, the definition of perception for this chapter excludes concepts of *insight, interpretation, intuition, awareness,* and *consciousness.* The intuitive usage of *perception* places it between *sensation* and *consciousness*, the latter denoting only awareness, such as the awareness of the visual sensation of a red, ripe tomato. I ask the reader to limit *perception* to the transduction, transmission, encoding, and storage of a stimulus, which includes the process of sensation with all of its peripheral modalities and their filters.

Neural representation is commonly used where I use *perception*, but *perception* includes the process by which the neural representation was acquired—the process of transduction, transmission, encoding, and storage, not just the product of the process—the neural representation of a stimulus. *Percept* is commonly used to denote the product of perception, much as a *concept* is the product of conceptualization (least we confuse it with *the product of conception* in obstetrical usage). The limited definition of perception† has four purposes: first, to include the *process* of acquiring the neural representation of an object or event that is an essential component of the *experience* of an object or event; second, to better serve an understanding of how consciousness happens as proposed in Premise 4; third, to separate the terms used to denote pain—nociception and the experience of nociception that includes eudynia and maldynia; and fourth, to remove the redundant meaning of the common expression *perceptual experience*.‡

---

* See *Collins Dictionary of the English Language*, 2nd ed., 1986, Hanks, P. (ed.), Glasgow: William Collins Sons.

† The definition used in this chapter is, in fact, broader than the dictionary definition in the sense that it includes the representation of internal stimuli from memory and other neurobiological events that are not dependent on peripheral sense organs.

‡ The expression perceptual experience implies that there is another kind or sort of experience; one presumes it to be conceptual experience or a sort of epiphenomenal experience. This distinction assumes either different processes of experience or different contents of experience. I decline the notion that experience has contents. As I propose a theory of how consciousness happens, later in this chapter, I also decline the idea that the experience of a represented or perceived object or event, in the neurobiological process of the experience of it in consciousness, depends on whether it is acquired from an internal or from an external source, whether it is from an abstract stimulus or from a concrete one, whether it is veridical or nonveridical or whether it is the representation of a concept or of a percept. A discussion of the validity or lack of validity of distinctions between such categories of experience is beyond the scope of this chapter.

For example, if I bump my head on a doorframe, the injury puts mechanical force on various sense organs in my scalp. The sense organs excite certain classes of neurons (transduction) that carry a signal to my spinal cord, where some reflexive motivation and filtering take place, and then to my brain (transmission). The signal is encoded in highly ordered cell assemblies and stored, bioelectrically, chemically, or structurally, in a cortical assembly or in a deeper brain structure. Very few of the perceptions, as neural representations, that are transduced, transmitted, encoded, and stored ever enter awareness, though they may affect other perceptions or one's behavior. The bump on my head, nonetheless, is likely to be noticed in awareness—briefly but instructively, one hopes.

The bump on my head is transduced in *pain* sense organs. A-delta or C-fiber neurons of pain pathways transmit the signal, and the encoded representation is, therefore, identifiable as a pain perception, or nociception. Nociception is the sensory and neural substrate of pain, following the present definition of *perception.* Nociception is the precursor for the *experience* of pain. Nociception happens when the stimulus and the perception of it mimics or results from tissue damage. I have no objection to calling the process *sensation,* but, again and arguably, any distinction between sensation and perception, as here defined, is unnecessary and confusing. Sensation is a component process of perception. Sensation is worthy of study, per se, but not in the effort to understand the experience of suffering and of pain as illness.

The experience of an emotion happens by the same process as the experience of pain. An emotion is a biophysiological event that includes neural, humeral, immune, and hormonal events that affect the body proper. If our instruments were sensitive enough, we could record the inciting event and the perception of it. We could measure each of the physiological events that occur, from the neural, humeral, immune, or hormonal changes to the bodily disturbance they cause, to the perception of those bodily effects.

The distinction between the biomarkers of an emotion and the experience or feeling that names the emotion is important. The experience or feeling of an emotion is the experience (not commonly in awareness) of the *effect* of the emotion on the body (2). For example, one experiences an event as threatening when the perception of facial movement, piloerection (hair standing on end), vasoconstriction or vasodilatation, increased intestinal motility, somewhat edgy sphincter control, and much more enters consciousness, though, again, not necessarily into awareness. The awareness is not of the parts, but of the composite, of the *experience or feeling* of the component parts. One is seldom aware of the biomarkers of the emotion; one is aware, in this example, of the *fear.*

2. Suffering happens when there is a perceived loss or threat of loss of the integrity of personhood (3–5), including, of course, the threat of death (6). The facets of personhood include any and all domains of the *self.* The threat to the integrity of personhood can be real or imagined. The experience or feeling of grief or fear need not be in awareness.

Eric Cassell defines *suffering* as the experience of a threat to the integrity of any of the facets of personhood (3). I choose to characterize personhood as the composite of the existential domains of the autobiographical self, that permanent residue of dispositions and memories that we protect so assiduously (2,40). Personhood and the autobiographical self are roughly equivalent to Freud's term, *ego structure* (5). Discussions of psychoneurosis focus on the threats to the integrity of ego structure in much the same way that Cassell describes the loss or the threat of loss of various facets of personhood.

Suffering turns a condition into illness that the sufferer describes using the 11 words listed in the introduction (distress, misery, etc.). Suffering can turn an experience that was an annoyance into crisis. The domains of the self (personhood) that Cassell describes include cultural background, habitual and accustomed behaviors and roles, life experiences, expectations, attachments to family and friends, real or imagined secret relationships, thoughts, aspirations, bodily function, and more.

The complexity of Cassell's facets of personhood tempts one to separate types of suffering based on the existential domain that is subject to threat or loss, such as the loss of earning capacity or the loss of one's dreams of retirement, or according to any number of other criteria. I propose that the suffering of unemployment and penury is no different, at the phenomenological and neurobiological level, from the suffering of divorce, a broken leg, or the death of a loved one. Such distinctions, as between medical suffering and spiritual suffering, are deceptive and obscure the fundamental, unitary nature of suffering, even though an effort to understand any one person's experience of suffering requires an understanding of the infinitely variable causes and contexts of suffering.

3. Suffering is *subjective*. That is, suffering is irreducible and immeasurable.* Suffering is a state of consciousness, an experience of perceptual contents, and the subjective, phenomenal awareness of an internal state.

Suffering, unlike emotion, is immune to scientific validation. Suffering is interior and personal. Emotions may *feel* entirely internal; but, counterintuitive though it may be, emotions are external in that they are biomarkers and behaviors that are

* Earlier, I presented 11 words that sufferers use to describe their experience. One might take any of those words and create a scale, for example, "the agony scale" or "the misery scale," but it would not be an objective measure of suffering. Although one could rank-order experiences as associated with more or less "agony," the resulting visual analogue scales would not be subject to external validation. Clinicians use the 11 words, and others, to interpret or deconstruct a patient's narrative of suffering (38). That is far from an objective analysis. Jamie Mayerfield, in spite of earnest scholarship, could not devise an objective measure of suffering (41).

Others have made the effort to measure suffering. After the authoritarian exploitation, war, cruelty, and suffering that dominated the first half of the twentieth century, a group of concerned scientists and humanists searched for ways to prevent cruelty—the infliction of suffering. Ralph Siu, working with David Langmuir, Kenneth Boulding, and others, developed a unit measure of suffering, the dukkha, after the Sanskrit word, central to Buddhist culture, for which "suffering" is only a rough translation (42). Their work was termed Panetics, a contraction of Pan Ethics, or a coalescence of Eastern and Western philosophy and values. The dukkha is useful in following trends in cruel behavior and the stigmata of suffering, such as malnutrition, ignorance, epidemic disease, confinement, and death, within and across cultures. Even though there is near universal agreement (among civilized societies) about what constitutes cruelty and when suffering is present, Siu's efforts to validate the dukkha were self-referential. The dukkha does not measure the suffering of an individual.

measurable and observable and, therefore, objective. No one believes that understanding suffering requires a theory of how it happens. We intuitively think that we can recognize suffering in others even though its interior and personal nature makes it impossible to experience another's suffering or to *feel another's pain*.* The behaviors that express and reveal suffering are observable and measurable and may be described in the third person. An observer, for example, may say, "Mary looks sad." Mary's expression of sadness is recognizable to the observer, but Mary's experience of sadness is unknowable to the observer.

The one who suffers can only describe the experience in the first person by relating the chain of motivation for the expressions of suffering, insofar as his or her perceptions of motivation enter awareness. For example, Mary, who appears sad, might say, "My grandmother died this week. She took care of me when I was little. I miss her terribly." Although facial expression, body posture, and a voice of grief make sharing another's suffering more accessible, the observer of one who suffers can only speculate what the experience *feels* like.

UNDERSTANDING CONSCIOUSNESS

Suffering, like hope, is the feeling or experience of an interior, and therefore, personal state. In this chapter, the terms *feeling*,† *experience*, and *state of consciousness* are virtually synonymous. This, as much in the nature of consciousness, is, once again, counterintuitive. The understanding of suffering requires an understanding of consciousness. Consciousness is the experience of perceptual contents.

The experience or feeling of a perception may be in or out of awareness. Perceptions and the experiences of them that do not enter awareness are variously called subconsciousness, preconsciousness, subliminal consciousness (50), and access consciousness (51).‡ In this chapter, I will simply use the expression *out of awareness* to denote perceptions and the experience of them that do not enter awareness. The experiences of perceptions that enter awareness or *phenomenal consciousness* are, simply, *in awareness*.

---

* There can be little doubt that humankind would exploit and destroy each other at a far greater rate than we do presently were we incapable of experiencing suffering and were we incapable of the skill that we think of as sharing the suffering of others. We refer to this intersubjective phenomenon variously as sympathy, compassion, identification, fellow feeling, and empathy. It appears to require the neurobiological substrate of motor mirroring (43–45), the repository of predispositions or values in various parts of the frontal and prefrontal cortex (2,46,47), and the perceptual and motivational error detection and conflict-resolving function of the anterior cingulate cortex (48,49). It is not the subject of this chapter.

† Like others, cited here, I exclude the tactile, sensory meaning of feeling, as in, "I feel her hand on my shoulder" or "It's so cold I've lost the feeling in my fingertips." In the discussion of consciousness and awareness, the feeling of the sensory perception is the subjective experience of that perception, not the objective events of the sensory apparatus (40).

‡ Freud called perceptions that remain out of awareness "the unconscious," although that expression now has different meanings depending on the sense of consciousness under discussion. Neurophilosophers refer to such terms, like the unconscious and the self, as folk psychology—molar expressions that can be reduced to more precisely defined elements. I, for one, find the term folk psychology a bit demeaning and choose to avoid it; but it has its uses.

SENSES OF CONSCIOUSNESS*

Consciousness as Wakefulness

Wakefulness is the state of arousal—being conscious as opposed to being asleep, in a coma, or unconscious. Wakefulness is a prerequisite for awareness, but it is not necessary for the transduction, transmission, encoding, and storage of perceptions. We are often awakened from sleep by sensory perceptions of bodily disturbance, such as the tingling fingers of nerve compression, by a bright light, or by the sound of the alarm clock. Sleep is also disturbed by the cognitive events of dream work. The perceptions enter consciousness even during sleep, but we must be awake before they enter awareness.

Consciousness as the Experience of Perceptual Contents

The principal sense of consciousness in the present discussion is the experience of perceptual contents. In awareness, the experience is the "stream of consciousness," the awareness of the "remembered present" (53), or "the feeling of what happens" (40).†

The boundary of the "realms" of consciousness that are in and out of awareness is under intense study (51,56). It appears that there are different neurobiological functions‡ associated with the functions of perceptions either in or out of awareness (45,50,56–58), such as how the perception affects other perceptions, how it affects behavior, and how it enters awareness, immediately or later, when the stimulus is remote. The science is incomplete.

---

* After Edelman (22), Zeman (52), and others.

† The definition of consciousness as "the experience of perceptual contents" is often misstated as "the contents of perceptual experience." The difference is important. The title, "The Center for the Study of Perceptual Experience," is not a misleading title, because its members study the experience of perceptual phenomena. I regret, however, that the expression "contents of...experience" entertains the possibility that "perceptual experiences" are the "contents" of consciousness. This error implies that experience and consciousness have contents, leading to the neologism of "qualia" (54). The intuitive notion that the imagined contents of consciousness can be reduced to something like "qualia" is the source of much wasteful polemic. It is intuitively unpalatable to think that consciousness is "about" something but has no content.

The idea that consciousness has contents suggests, to some, the possibility of the existence of a class of beings that are humanoid in every respect, with all the perceptual, associative, and behavioral capacities of humans, but without consciousness—without qualia (55). As the reliably sensible Adam Zeman (a Scot) put it: "if your concept of consciousness tempts you to consider the possibility of zombies, your concept 'needs fixed'" (52). Consciousness is wholly subjective and without "content." What is counterintuitive is not necessarily impossible. I, for one, think it unfortunate that it is almost always appropriate or necessary in philosophical discourse to ground one's assumptions in the intuitively obvious.

‡ Hypothetically, objects enter awareness when they oscillate at about 40 Hz, as described in Premise 4, with other gamma ranges associated with various realms of "the unconscious." The science is incomplete.

Consciousness as Reportable Experience

"Reportability" of experience is thought to be a separate sense of consciousness, if one imagines that consciousness has contents. Subjective as it is, consciousness can only be reported in symbolic representation, of which language is only one variety, an extensive and flexible one, to be sure, but not always the most efficient. Reportable consciousness need not be in awareness, as in parapraxis, commonly called a *Freudian slip. Lucid dreaming* may be another example. In general, however, reportability is a function of awareness. A recent example of *blind sight* reveals how perceptions may affect behavior while remaining out of awareness. Beatrice de Gelder studied a man who suffered a stroke to his occipital, visual cortex and could not see (59).* He could, however, navigate a maze with perfect accuracy without using tactile cues. He could not explain how he was able to avoid the objects in his path. His eyes "see" them (transduction), the signals are transmitted to his brain, his brain encodes and stores the perceptions as neural representations and then uses them to control behavior (a function of consciousness), yet the images of the objects do not enter his awareness and are, therefore, not reportable.†

Consciousness as Self-Awareness

Finally, there is self-awareness, the first-person ownership of consciousness. One feels, "This is *my* world; this is happening to *me*." Awareness of the perception of any stimulus, internal or external, has the subjective property of being unique in the first person. An experience or feeling, furthermore, is a perceptually competent stimulus, just as it is an emotionally competent stimulus (2,40), in the sense that the composite experience or feeling of the emotional effects, or biomarkers of an experience, such as suffering or the sweetness of a red, ripe tomato, can be encoded, stored (perceived), and experienced. "Self-consciousness" is, in part, being aware *of being aware*.‡

A THEORY OF CONSCIOUSNESS: HOW CONSCIOUSNESS AND AWARENESS HAPPEN

A proper theory of consciousness must answer at least three questions:

1. How does a perception enter awareness?
2. How is it that only a few perceptions enter awareness while the majority of perceptions that are transduced, transmitted, encoded, and stored remain out of awareness but are capable of influencing behavior and other

* The summary here is abstracted from reports in the popular press. A report in a biomedical journal had not occurred as of this writing.

† At the risk of repetition, an indulgence of this chapter, one intuitively thinks that if Dr. de Gelder's patient is not aware of objects in his path, he is not conscious of them—they do not enter his consciousness. In this chapter, no such distinction exists. A perception that affects behavior satisfies a defining characteristic of consciousness.

‡ Thinking is derived from self-consciousness. There is, therefore, no fundamental difference or ontological separation between thinking and feeling (14,22).

perceptions of objects and events, and of entering awareness later when the stimulus that evoked them is remote?
3. How is it that perceptions, including the perception of the biomarkers of behavior and of emotions, are *objective*, and that feelings and experience, and particularly the experience of suffering, are *subjective*?

THE MULTIPLE DIMENSIONS OF PERCEPTUAL CONTENTS

Consider the example of Mary, walking into a room in which there is a red, ripe tomato in a bowl on the table.* Even if she is not aware of the red, ripe tomato, she might be able to say later, if asked, "Yes, there's a red, ripe tomato in the other room." For the red, ripe tomato to enter Mary's consciousness, even out of awareness, multiple dimensions of consciousness coalesce in less than a third of a second. Some of the perceptual contents that are bound into the consciousness and awareness of the red, ripe tomato include the following:

1. The encoded representation of the retinal image of the red, ripe tomato
2. Categorization of the perception as "red," "ripe," and "tomato"
3. Associations of size, shape, and color of the red, ripe tomato
4. The perception of the somatic and autonomic motor activity that was needed to acquire the perception (2), such as body, head, and eye movement
5. Remembered representations of objects and events associated with red, ripe tomatoes, such as grandma's tomato sauce
6. The bodily disturbance as if the remembered representations were occurring presently (2,40), such as salivation or flaring of nostrils
7. Drive states and appetites (Mary will process the perception differently if she is more hungry than she is tired or sexually aroused.)
8. Mood (background emotional tone, such as depressed, anxious, or cheerful)
9. The relationship of the red, ripe tomato to body boundaries, and, therefore, to the perception of body boundaries (including, of course, Mary's skin and also boundaries of sight and reach, among others)
10. The state of arousal and wakefulness

All of the information listed above is reducible and measurable. It is objective information, assuming, again, that the instruments have sufficient sensitivity and resolution. Frequency (color) and intensity of the light falling on the retina and its cortical representation are measurable. Yet, when the perception of the red, ripe tomato enters awareness, now or later, it does so as subjective experience. Mary's experience of its redness is wholly subjective. One can measure the pH, sugar content, chemical composition, and physical properties of the tomato, along with the neural substrate of mastication, smell, and taste, but the experience of biting into the red, ripe tomato and tasting it are subjective. Only Mary knows what it is like for her to taste the red,

* Frank Jackson originally coined the example of a fictional Fred and his red, ripe tomato, and of Mary, the all-knowing neuroscientist (54). No doubt, he would see red at my use of them.

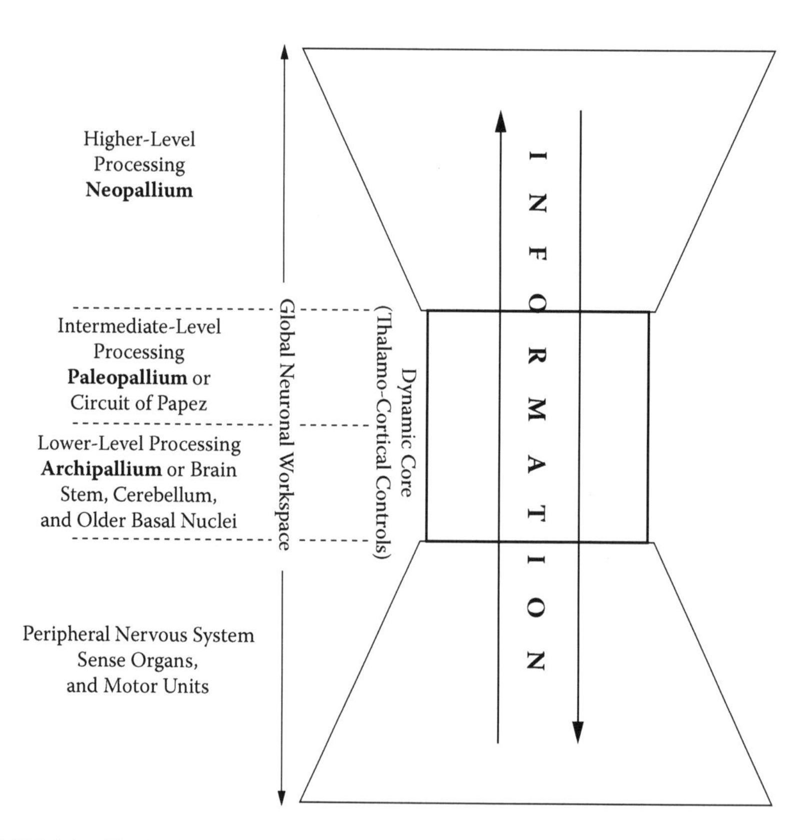

FIGURE 4.1 The body–brain in the image of an hourglass.

ripe tomato. An observer can speculate, but everyone's experience of her tomato is uniquely her own. The perception has contents, and the experience of it has none to be shared.

MODELING CONSCIOUSNESS—CREATING A DIAGRAM OF THE BODY–BRAIN

To help answer the three questions I posed earlier, it would be nice to create a diagram that illustrates the workings of the mind. It is difficult, however, to represent multidimensional phenomena in two dimensions. Any diagram, nonetheless, begins by considering consciousness as the processing of information to do purposeful work.

Figure 4.1 presents a model of consciousness in the image of an hourglass whose upper reservoir is the brain, and the lower reservoir represents the body. The model presents the idea that the body and brain are in continuous communication (2). The body and brain are inseparable; they are the *body–brain*, supporting the notion of mind–body unity. The flow of information is controlled by the arousal centers of the brain stem and by the reentrant dynamic core of the thalamus, of adjacent, interconnected nuclei and of the limbic forebrain (53). The four-part, hierarchical structure

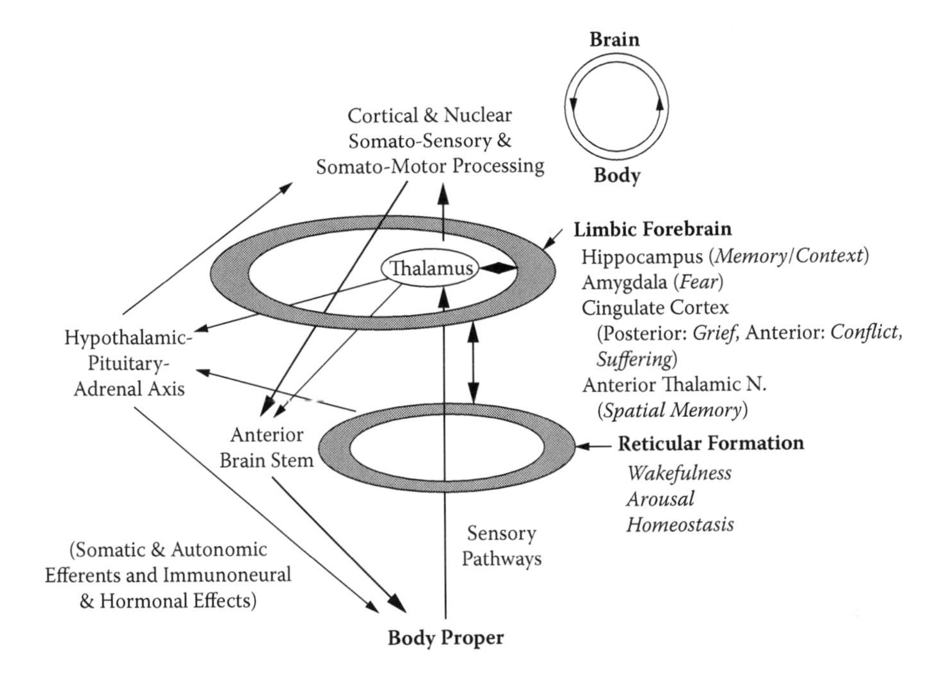

FIGURE 4.2 The body loop of consciousness.

of neural processing is taken from MacLean (60), who used it when considering the phylogeny of neurobehavioral evolution, and from Engebretson and Giordano (61), who used it to model spiritual experience.*

In a somewhat more nuanced diagram of the body–brain (Figure 4.2), the hourglass becomes a continuous *body loop* of afferent and efferent information following the somatic marker hypothesis of Antonio Damasio (2). In this model, the brain stem and the midbrain control and aim the flow of information like a pair of "gimbaled lenses," while adding considerable contextual and associative information to that flow. The body-loop model is both more flexible and more specific than the hourglass, but the model of consciousness must further represent the multiple dimensions of the information of the perceptual apparatus and of the motivational apparatus—the 10 types and sorts of perceptions associated with Mary's consciousness of the red, ripe tomato.

Figure 4.3 presents a model of consciousness that includes the perceptual apparatus and the motivational apparatus with its homeostatic functions that form a basic, though incomplete, sample of the components of the neural substrate of consciousness.† The model only hints at *how* consciousness happens. The hint is in the preponderance of *two-way* arrows in the diagram that give it interconnectivity

* This hierarchical structure is fast becoming outmoded, perhaps archaic. It remains intuitively useful but should not imply that a higher-order function like consciousness happens in the neopallium.

† There are many models with similar features. All are useful in a given context, but none can be complete or universally applicable (52,62).

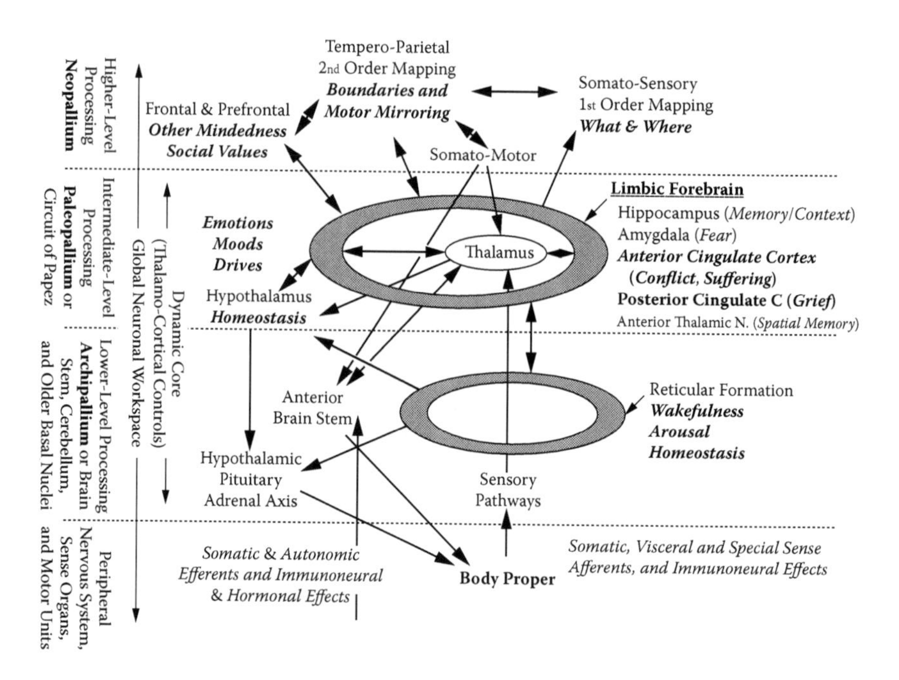

FIGURE 4.3 A neurobiological model of consciousness.

and reentrant or recursive properties. The unifying concept of how consciousness happens is *resonance.*

Most everyone has, at one time or another, used the expression, "that resonates with me." It is the feeling of recognition or comprehension of an object or event. The connection is very personal and subjective. Resonance of perceptual contents, I propose, is the subjectivity of experience. The theory of how consciousness happens in Premise 4 does not address the myriad complexities of how neuronal and local field excitation happens and how they are entrained in oscillatory resonance. The theory has not been validated. Parts of it are supported by good evidence; the whole is not, yet.

4. Consciousness is the coherent, reverberating, resonant, synchronous oscillation of the bioelectrical pulsations of neuronal or local field activity (9,10), at about 40 Hertz (11–14), in multidimensional, massively interconnected, reentrant and recursive circuits (15,16), networks, and systems, among highly ordered cell assemblies (17,18) of the perceptual and motivational apparatus, in disparate parts of the global neuronal workspace (19–21), within the bounded and situated body–brain (22), binding them around the perception of an object or event. (14)

Each of the 10 examples of perceptual contents that must resonate together to make Mary conscious of the red, ripe tomato is, by itself, a coherent resonance within a highly ordered cell assembly of the perceptual architecture. How many components of perceptual contents must resonate synchronously for the red, ripe tomato to affect

Mary's behavior while remaining out of her awareness, and what content is required for it to enter her awareness, is not clear. The science is incomplete.*

I am persuaded to think that *what* enters consciousness, the *object* of consciousness (what consciousness is *about*) from among the myriad perceptions from Mary's sensory inputs, is the epicenter or node of the resonant, synchronous oscillation, somewhat like the nodal point on a vibrating cello string that determines the pitch of its tone. The richness of the sound depends on secondary nodes of overtones and their relative strengths and phase as they reverberate in the sound box of the cello. Extending the metaphor, arguably too far, the neural, humeral, hormonal, and immune background of the perceptual apparatus is like the sound box of the cello, with all its variable qualities, on which the string resonates.

Consciousness does not happen in any one place within the body–brain, it happens throughout the body–brain (63). Functional magnetic resonance images (fMRIs) are used to study the neural substrates of consciousness; but the fMRI gives us the false impression that consciousness consists of focal centers of neuronal activity lighting up against a quiet background of the brain—that fear, for example, happens in the amygdala or that memory happens in the hippocampus. The parts of the brain that predominate in the synchronous oscillation determine *what* enters awareness *and* its subjective qualities. Subjectivity does not merely *depend* on resonant synchrony; subjectivity *is* the resonant, synchronous oscillation of bioelectrical activity.

There is no compelling evidence to suggest that widespread synchrony is consciousness. There is nothing yet that refutes the theory. There is good evidence that synchronous oscillation occurs in cortical columns of the perceptual and motivational apparatus, as well as in thalamocortical circuits (15,64). Many neuroscientists find the notion that synchronous oscillation is consciousness to be counterintuitive and unsustainable (65).

A LINEAR VERSUS AN OSCILLATORY MODEL OF CONSCIOUSNESS

Neurons are linear. Neuronal activation is the sum of excitatory and inhibitory stimuli. In a computational model of consciousness, circuits, networks, and branching loops interconnect in linear algorithms. Determined relationships are digitizable, and, one might conclude, consciousness could be reproduced by a computer of sufficient size. In an oscillatory model of consciousness, cell assemblies oscillate together in iterative, dynamical, multidimensional relationships that represent the uniquely subjective, irreducible, and immeasurable first-person qualities of consciousness. With current technology, reproducing consciousness is not possible. A

* The oscillating model of consciousness replaces an algebraic, linear, algorithmic, or computational model. A causal chain of neuronal activity that is the summation of excitatory and inhibitory impulses exists and appears to end in the transduction, transmission, and, possibly, encoding of stimuli in the perceptual apparatus. The difference between an oscillating model and a linear, algebraic model has profound implications for linguistics and artificial intelligence, but it is not the subject of this chapter.

Also, discussed earlier, there is suggestive evidence that perceptions that enter awareness do so when they are bound by synchronous oscillation at about 40 Hz, and that perceptions in other realms of consciousness that are out of awareness oscillate at other frequencies of the gamma range (~25 to 100+ Hz).

computer can mimic consciousness and appear to have emotions, but such a trick of linear programming relies on the erroneous assumption that emotions are states of consciousness, which they are not.* The experience of an emotion is a state of consciousness, and the computer that appears to *emote* only behaves as if its internal processes contain the biomarkers of fear or grief; but it does not experience fear or grief, envy or jealousy, or any other emotion; it only acts that way.

If the foregoing summary is only a rough approximation of how consciousness happens, it suggests how the immeasurable, subjective experience of suffering happens.

5. Suffering happens in states of perceptual or behavioral (homeostatic) conflict or struggle for the integrity of personhood in which the anterior cingulate cortex (23–25) is the epicenter of the synchronous oscillation of the body–brain.

The cingulate cortex is a large band of cerebral cortex on the inner surface of each brain hemisphere just above the swathe of white matter (nerve fibers) that interconnects the two halves of the brain. It appears grossly homogeneous but is not. The anterior cingulate cortex (ACC) is the forward-most portion of the cingulate cortex. A number of observations persuade me to think that the ACC is essential for the experience of suffering:

1. Christopher deCharms, in collaboration with Sean Mackey and others (23), trained subjects to control chronic pain by learning to decrease the activation of their ACC using real-time fMRI of their own brains. I propose that the subjects were not decreasing nociception† in this study; they were moderating the experience of suffering.
2. An extensive review of the centrality of the ACC in "the neurocognitive overlap between physical and social pain" by Naomi Eisenberger and Matthew Lieberman further supports the association between ACC function and the experience of suffering (66).
3. Anterior cingulotomy or prefrontal leucotomy—surgical injury to the ACC or separation of the connectivity to the ACC—is an effective treatment for chronic pain (67). After such treatment, patients report that the pain (nociception) is unchanged but it does not bother them (2). Loss of the function of the ACC, I propose, turns maldynia to eudynia by eliminating the capacity for the experience of suffering (a discussion of the three-part nature of pain follows).

* Face recognition is instructive. Mary, entering a room full of, say, 150 people, without expecting to know any of them, can, within a fraction of a second of her vision falling on the face of one of the occupants, recognize her uncle George. A computer, that can in milliseconds perform calculations that Mary cannot accurately perform in minutes, is incapable of recognizing George's face in many multiples of the time it takes Mary to do so. The difference exists between the capacity of a linear face recognition program and an oscillatory one.

† As noted earlier, nociception is the transduction, transmission, encoding, and storage of a stimulus. The experience of the perception may be painful, or not.

4. fMRI studies demonstrate increased activity of the ACC associated with the experience of illness due to complex regional pain syndrome and due to fibromyalgia, two conditions associated with considerable suffering.
5. Diseases, such as strokes or tumors, and injuries, either traumatic or surgical (as described in (3), above), that destroy the ACC cause an insensitivity to pain that, I propose, is not an insensitivity to nociception (see (1) and (3), above) but an inability to experience suffering in the face of nociception.
6. There is further reason to propose that the ACC is essential to the experience of suffering. It derives from a complex, theoretical postulate that prosocial behavior (schooling and flocking in premammalian species not withstanding) requires a capacity for altruism, which requires a capacity for empathy, which requires a capacity for the experience of suffering. Sociopathic and antisocial behaviors emerge after disease or injury that eliminates the function of the ACC (2).

Suffering does not happen *in* the ACC. Suffering happens when the ACC predominates in the coherent, synchronous oscillation of the body–brain. Nonetheless, understanding the neural correlates of suffering is useful, in the main, when trying to understand abnormal behavior in the face of neurological injury, disease, or deformity. When trying to understand the experience of pain when the patient is neurologically intact, one seeks to understand *how* suffering happens, not *where* it happens.

6. Illness, including pain-as-illness, classified as maldynia (26), is a condition (a disease, injury, deformity, or pathophysiological condition) that evokes the experience of suffering.

In this chapter, the term *illness* denotes the state of a person, not a medical condition.* A medical condition that does not evoke the experience of suffering is an annoyance, but it does not merit the name *illness*. Hyperglycemia is an example. A diabetic who takes her medication, tests her blood glucose, stays enjoyably fit and trim, sees her physician regularly, and does not otherwise concern herself with the course of her disease does not *feel ill*. She *suffers* the disease of diabetes in the sense that she puts up with it, but she does not *suffer* for it. It does not affect her quality of life. If, on the other hand, she worries that she will have a heart attack or stroke, go blind, or be on dialysis, that she will lose feeling in her feet, has painful peripheral neuropathy, has a leg amputated, despises having to take medication, or fears she will die before her grandchild marries, then her distress turns the disease of diabetes into a state of illness. Illness derives from the experience of suffering

* It would be incorrect, therefore, to say, for example, "an illness," as in "he suffers from an illness." At the risk of introducing another counterintuitive usage, I prefer, "he suffers illness from/of (any condition)." Alternatively, "(Any condition) makes him ill/unwell."

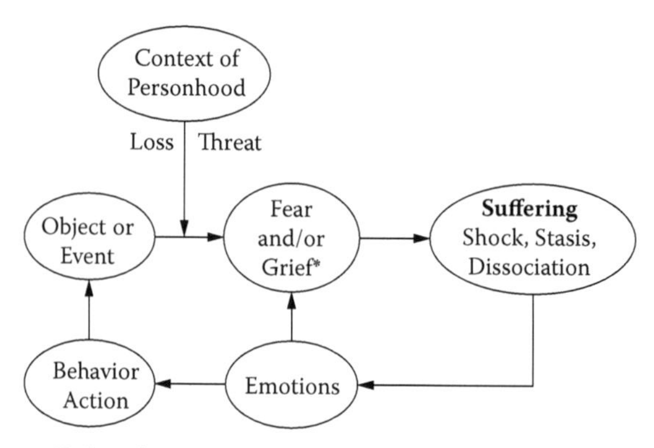

FIGURE 4.4 A partial phenomenology of suffering.

evoked by the medical condition. Suffering, therefore, defines illness. Illness does not define suffering.*

It would also be wrong to define "health" as the absence of suffering, or as the lack of anything else (e.g., the lack of disease or the absence of illness). The negation of an antonym does not constitute a definition. Health is the capacity to cope, to adapt, and to flourish in the face of suffering or a state that obtains the capacity to evoke suffering.

The proposed understanding of how suffering happens is a context in which to develop a coherent theory of how illness happens, particularly maldynia—pain-as-illness. Here it is useful to begin diagramming the phenomenon of suffering.

Figure 4.4 diagrams the process by which an object or event, when perceived in the context of the unique autobiographical self, is interpreted as loss or the threat of loss of the integrity of personhood. Loss and threat evoke the emotions of grief and fear. The perception of the bodily disturbance or biomarkers of grief and fear, before they enter awareness, is experienced as suffering, characterized, at first instance, by states of shock, stasis, and dissociation. The experience of suffering is not an emotion, but it is an emotionally competent stimulus (2). When the emotions evoked by suffering are predominantly grief and fear, then the grief and fear that initially evoked the experience of suffering are reinforced in a dysphoric feedback loop. Clinicians know these patients. They include the cancer patient who gives up and the chronic pain patient who becomes suicidal or hopelessly and helplessly abuses medication.

7. Four types of events cause suffering: loss, threat, the frustration of a biologically based drive or appetite, and pain as nociception.

* It would be incorrect, therefore, to say, for example, that suffering is "a particular patient's predicament in the face of illness." The expression uses illness to define suffering, a usage I decline. It would be equally incorrect to say that health is the absence of disease. Health is the capacity of the individual to cope, to adapt, and to flourish.

LOSS THAT EVOKES GRIEF AND THREAT THAT EVOKES FEAR

Suffering is caused by loss, as, for example, after the death of a loved one. Suffering is caused by threat and danger, as when one receives a diagnosis of cancer. Often, it is the threat of loss.* Grief and fear, though clearly different, are common causes of suffering, and they have two common properties. They are *primary emotions*† that have specific effects in the body proper along with unique facial expressions and vocal tone that make them recognizable across cultures and, to some extent, across species (7). Second, they are commonly thought to be mortal but are not: for example, "he died of a broken heart," or "I was frightened to death." One does not die of a broken heart, although a severe grief reaction is capable of evoking involution, anorexia, and starvation. Also, indirectly, fear can induce a catacholamine surge that, in the face of prior impairment, can cause a fatal heart rhythm or myocardial infarction. Such events are more common in fiction‡ than in experience.

The reader will want to add any number of events that can cause suffering. I propose that events such as nightmares, hopelessness, or empathic grief or fear evoke suffering only when they threaten the loss of integrity of personhood. The terror of a nightmare derives from the dream's ability to threaten one's self-image. Hopelessness is not a cause of suffering; hopelessness is synonymous with suffering and indistinguishable from it.§ Here, again, I ask the reader to reject the intuitive taxonomy that includes many different types and sorts of suffering. There is something fundamentally the same about the experience of suffering, regardless of its cause or context.

UNRELIEVED APPETITES AND DRIVE STATES

Unrelieved appetites and drive states, at least some of them, can, unlike grief and fear, threaten death. Drives and appetites are chemically and neurohormonally regulated functions of homeostasis, survival, and reproduction. A useful, though incomplete list includes thirst, hunger, oxygenation, elimination, thermal regulation, rest, lust, mastery/dominance (which probably includes curiosity, once thought to be an independent drive), and attachment. Intense, prolonged, and unrelieved states of activated drives and appetites cause suffering. For example, causes of suffering include

* Arguably, danger evokes fear only when there is anticipation of loss.

† Concerning the other primary emotions (joy, disgust, contempt, anger, and surprise—seven primary emotions in all), I propose that they are not causes of suffering. Joy or pleasure is not a cause of suffering, although a joyful event can have such poignancy that its context recalls events that were the cause of suffering. Disgust, contempt, and anger do not evoke suffering. To the contrary, as the phenomenological model of suffering develops, I will propose that disgust, contempt, and anger have a compensatory, ameliorating effect on the experience of suffering. Disgust and anger may become indirect causes of suffering by evoking the anticipation of threat or loss, as, for example, when vengeance threatens one's self-image as a beneficent, forgiving person. Surprise, by itself, does not evoke suffering. Surprise is an attentional stimulus focusing on an object or event in awareness (68).

‡ Two evergreen genres of popular culture are the tearjerker and the horror film. They permit the experience of suffering caused by grief and fear, respectively, in the safe environment of a theater or one's home and in a context that is impersonal and nonthreatening in a literal sense.

§ Hopelessness, though not a separate sort of suffering, is a particularly troublesome loss of resilience and failure of coping. The anticipation of successful coping is hopefulness. I will discuss resilience and coping later in this chapter.

the dehydration of thirst, the hunger of starvation, the frustration of sexual arousal (at least during adolescence, if not thereafter), the boredom of confinement, or the separation from a loved one.

PAIN

Most people, when asked, "What is the most common cause of suffering?" will say, "Pain." Some will say, "psychic pain," as though the suffering caused by physical pain and that caused by a psychological event were different. I consider it unfortunate that we use the word *pain* to denote both the perception of pain and the experience of pain. It is better to use three distinct terms for pain: nociception, eudynia, and maldynia (69).

Nociception is the transduction, transmission, encoding, and storage of a stimulus. As noted previously, one identifies the stimulus as painful when it mimics or results from tissue damage and is, therefore, transduced by sense organs that respond to tissue damage. The injection of capsaicin under the skin is painful, though it leaves no residue of damage; it only mimics tissue damage. The electrical stimulation of pain pathways in peripheral nerves evokes the perception and experience of pain without peripheral tissue damage.

Pain is also used to denote two types of experience: eudynia and maldynia. Eudynia, or "good" pain, is nociception that is corrective or instructive. The experience of eudynia is brief, and it does not leave a residue of distress. Maldynia, or "bad" pain, on the other hand, is described by Giordano as "pain-as-illness." In other words, maldynia is nociception that engenders illness to evoke the experience of suffering. When nociception, eudynia, and maldynia are all called *pain*, confusion is sure to follow. The International Association for the Study of Pain began the process of separating nociception from pain (70), and Giordano champions the terms *maldynia* and *eudynia* (26), coined by Lippe a decade ago (69).

Earlier in this chapter, I declared, "Suffering…is immune to scientific validation. It is interior and personal." The same must be said of the experiences of eudynia and maldynia. Nociception, on the other hand, like any perception that is transduced, transmitted, encoded, and stored, is reducible to finite elements and is measurable (assuming sufficient resolution of the instrument). The measurement can be recorded, shared among observers, and analyzed. This is a proper definition of objectivity—the philosophy and craft of science. Neither eudynia nor maldynia are measurable; they are, like all experience and states of consciousness, wholly subjective. It is because pain as nociception evokes suffering and because pain and suffering (when they enter awareness) share the properties of subjectivity and negativity that they are mistakenly linked in common usage.*

* Neurophilosopher Daniel Dennett put it well: "While the distinction between pain and suffering is, like most everyday, nonscientific distinctions, somewhat blurred at the edges, it is nevertheless a valuable and intuitively satisfying mark or measure of moral importance" (71, p. 162).

THE THRESHOLD OF SUFFERING

Pain, like other causes of suffering, is a threshold phenomenon. Nociception that is mild or has a limited and predictable natural history (eudynia) does not evoke suffering. Such pain is an annoyance, but it is not a disability. Eudynia does not degrade one's quality of life. When it is instructive and corrective, eudynia enhances the quality of life. Very much the same is true of drives and appetites: when we are hungry, we eat; when we are tired, we sleep; when we are aroused, we initiate the rituals of sexuality. We do not suffer for such states. When thirst turns to dehydration, it becomes a threat. When winded after climbing stairs, one resolves to get in shape, but at the cusp of asphyxiation one suffers cruelly. The threshold at which loss, threat, an unrelieved appetite, or pain evokes suffering varies greatly among individuals and for any individual according to the context in which loss or threat is perceived.

SUFFERING: SHOCK, STASIS, AND DISSOCIATION

The behavioral characteristic of the initial experience of suffering, before it evokes emotions (Figure 4.4), is collapse (shock, stasis, and dissociation)—that is, no behavior at all. Emotions—visceral and somatic motor activity—are evoked by suffering as an emotionally competent stimulus, while the experience of suffering, *by itself* motivates no action. The behaviors that express and reveal suffering are motivated by emotions that are reactions to suffering. We recognize and name emotions by the feeling of the bodily disturbance produced by that activity according to Damasio's somatic marker hypothesis (2). Though suffering is evoked by the experience of grief and fear, suffering is not an emotion or collection of emotions. Suffering is interior and involutional.

The phenomenon of suffering, before it evokes emotion and behavior, is observed in the facial and bodily expression of a person who, for example, has just been told terrible news, such as the presence of cancer or the death of a loved one—in fact, at first instance, there is a lack of expression. The first reaction is typically one of shock: the body is still, the face is blank, and there is no discernable reaction. There may be loss of motor tone with physical and postural collapse, from which experience comes the expression, "I was floored." If there is any vocalization, it is a short, soft grunt or moan that I, for one, interpret as signifying a closing down of body boundaries and with it a loss of comprehension of the event and its meaning in awareness. Only *after* the initial stasis is there outward expression, such as the keening of grief, the exclamation of startled fear, or a resistant contradiction, as in, "That can't be true!" Persisting stasis can result in states of dissociation, withdrawal, isolation, and denial.*

* I propose that the phenomenon of denial and the corollary phenomenon of dissociation derive directly from the shock and stasis of suffering. Denial is well known to oncologists and to palliative and hospice care specialists and counselors (72). Early attempts to understand denial portrayed it as the product of an immature, repressive personality (73). Recently, denial has been viewed as a natural, if not a necessary, stage along the path to acceptance and reconciliation in the face of dire threat and suffering in the presence of mortal disease or at the approach of death (30).

Coping Skills	Coping Reactions
Raise the threshold at which an object or event evokes suffering Coping Skills include: Connectedness Symbolic representation Belief system	Motivate behaviors that constitute "coping strategies" Coping Reactions include: Primary Emotions: Fear and grief, anger, disgust, contempt Secondary Emotions Indignation, guilt, shame, envy, and other social and value-dependent emotions Beliefs, explanatory concepts, spiritual practice
Coping Skills and Reactions modify and moderate the course of suffering. Coping Strategies are not always successful, or they moderate suffering at the cost of collateral damage.	

FIGURE 4.5 Resilience: coping skills and reactions.

The period of shock and stasis may be very brief, until suffering, as an emotionally competent stimulus, evokes emotions that break through to visible expression. The duration of the expression of grief or fear may be equally brief or may be preempted by the coping reactions of disgust, contempt, and anger, among many possible reactive emotions.

8. Resilience in the face of suffering is composed of coping skills (27–29) and coping reactions (30,31) that modify and moderate the onset, course, and outcomes of suffering.

RESILIENCE AND COPING

Discussion of the predispositions to suffering and the threshold at which any cause evokes the experience of suffering requires a brief discussion of resilience and coping. Resilience is "the ability to recover readily from, or resist being affected by, a setback, illness, etc." (74, vol. xIII, p. 714). The engineering definition of resilience is also illustrative in this discussion: "The amount of energy per unit volume that a material absorbs when subjected to strain" (74, vol. xIII, p. 714). The analogy to human experience is helpful in understanding the process of coping. Resilience in the face of grievous or threatening events requires coping skills and coping reactions that enable a person to contend with an object or event and to undo the loss or threat to the integrity of personhood.

Separation of resilience into coping *skills* and coping *reactions* may be artificial, but I propose it is useful. *Skills* are predispositions, behaviors, and capacities, both inborn and acquired, that raise the threshold at which a perception of any loss, threat, or so forth, evokes suffering, permitting an individual to experience a greater intensity of fear, loneliness, hunger, nociception, without the experience of suffering. *Reactions*, both innate and learned, permit the person to recover from the negative effects of suffering and, not always in an adaptive or corrective way, take action to the cause. The boundary between skills and reactions is permeable. Both coping skills and coping reactions enable an individual to find consolation, corrective action, or transcendence in the face of suffering.

Coping does not always have a positive outcome. Common clinical usage portrays *coping strategies* as a collection of adaptive or corrective behaviors or attitudes. Even the dictionary defines *to cope* as "to contend *successfully* with (an opponent, difficulty, situation, etc.), to deal competently with one's life or situation" (74, vol. xIII, pp. 903–904). Decreasing the intensity of suffering is often achieved at the cost of collateral impairments. For example, abused children commonly adopt coping strategies that interfere with mature personal and social relations.

Coping reactions are subject to operant conditioning, and such conditioning may produce the apparent paradox of reinforcing, rather than extinguishing, behaviors that cause suffering. Psychologist Robert Gatchel's discussion of "learned helplessness" offers an instructive example (75). Helplessness is a synonym of suffering; and Gatchel discusses how subjects learn to be "helpless" as a coping strategy. It is unfortunate that such coping strategies are commonly interpreted as improper or bad. Health-care practitioners are not immune to such moralizing. Patients who lack coping skills or who develop maladaptive coping strategies are seen as weak or blameworthy. Gatchel's work spanning three decades promotes Engel's biopsychosocial model as a context in which to understand, without moral judgment, how patients develop ineffective or maladaptive coping strategies in the face of chronic pain and to help them develop more effective ones.*

Most previous efforts to understand coping addressed reactions and skills with respect to an individual cause of suffering, for example, coping with pain, coping with death and dying, or coping with fear. In the developing model that began in Figure 4.5, coping skills and reactions have a unitary and unifying object: to contend with the experience of suffering, or the anticipation of suffering.

COPING SKILLS: CONNECTEDNESS

Connectedness is the attachment of an individual to another individual, to a family, to a community, to a vocation or avocation, or, even, to an identity, such as an ethnic, religious, or national one. The attachment can be to an imaginary individual, as occurs in childhood or delusional states. The value of connectedness in the prevention or amelioration of suffering is observed in ritual ceremonies surrounding death and grief (the gathering of the funeral and wake or sitting *Shiva*) and in the attachments that form in states of danger and fear (when people band together in otherwise improbable ways). Representations of such events abound in art and popular culture.

COPING SKILLS: SYMBOLIC REPRESENTATION

Symbolic representation can be as simple as giving the history of a twisted ankle or as complex as grand opera or the construction of a cathedral. These extreme

* One or even several citations here would give insufficient credit to Gatchel's body of work. Boothby et al. (31), Gatchel and Turk (39), and Gatchel and Proctor (75) will serve.

examples represent poles on a spectrum of abstraction by which symbols (language, music/drama, and architecture) represent the experience (not necessarily in awareness) of perceptual content. I do not propose that symbolic representation in all its forms are products of humankind's efforts to cope with the experience of suffering. I do propose, however, that only symbols (in art and architecture) can represent the subjectivity of the human condition, be it suffering or ecstasy and everything in between.*

COPING REACTIONS: DISGUST, CONTEMPT, AND ANGER

I propose that after the shock, stasis, and dissociation of suffering (Figure 4.4), and after a sometimes brief expression of grief or fear, disgust, contempt, and anger serve a compensatory function in decreasing the intensity of the experience of suffering. The context in which suffering is experienced determines how disgust, contempt, and anger are evoked, elaborated, and expressed as a variety of self-directed (inward, as in shame and guilt) and other-directed (outward, as in aggression and vengeance) emotions and predispositions to action.

The object of disgust, contempt, and anger can be real or imagined. There may be an identifiable cause of suffering. It may be an event, such as the flooding of hurricane Katrina, an object, such as the shoe that caused a painful blister, or a person, such as the faulty driver who precipitated an accident. Only brief reflection will convince the reader how often disgust, contempt, anger, and the myriad secondary, social emotions that follow grief and fear are inappropriate or obtain intensity out of proportion to the inciting event.

There is a rich literature discussing the inward, self-conscious social emotions, such as shame, embarrassment, and guilt, and the outward, other-critical emotions of disgust, anger, contempt, indignation (76), envy, jealousy, and more. A discussion of social emotions and predispositions to moral judgment and behavior is beyond the scope of this chapter.† I am persuaded by many sources to think that moral predispositions and judgments are based more on emotion than on reason. Hume is in; Kant is out. The structure of the present phenomenological model of suffering does not imply that a capacity for suffering is required for the development of moral intuitions, but that is what I propose.

---

* I propose that ecstasy and suffering are complementary experiences—a phase shift of a resonant, experiential event; two sides of the same neurobiological coin (61). One example (each reader will have a particular favorite) is the alternating grief and exultation of the third movement of Beethoven's Quartet in A minor, Opus 132, that bears this epigraph: *Heiliger Dankgesang eines Genesenen an die Gottheit, in der lydischen Tonart* (Holy Song of Thanksgiving by a Convalescent to the Divinity, in the Lydian Mode). He wrote it after he recovered from severe, painful illness. He was deaf from cranial Paget's disease.

† I recommend to the reader and refer often to Changeux, J.P. et al., 2005, *Neurobiology of Human Values*, Berlin: Springer-Verlag (14,25,43,46,47).

BELIEF SYSTEMS AND SPIRITUALITY

In the phenomenological model of suffering, I include belief system and spirituality as examples of coping skills and of coping reactions. Spirituality and belief systems are commonly linked, but they are not the same, and they are not necessarily related. Beliefs are representations of objects and events that obtain certainty without knowledge or external validation. There is a special irony in the differential use of the terms *to know* and *to believe*. We say that we know a *fact* about which we have reasonably certain evidence and that we believe something even when the evidence is shaky; but beliefs brook no contradiction, in spite of the evidence. Believers get a free pass on validation (77). It is, arguably, for this reason that belief systems provide better explanations for events that raise the intensity of grief and fear to the level of suffering. Beliefs help to explain *why* suffering happens. When the experience of suffering is "out of control," as in a state of helplessness and hopelessness, then one is not interested in *how* suffering happens. A rational explanation does not hold.

Belief systems are both a bulwark that raises the barrier across which fear and grief evoke suffering, and a balm that eases the painful, negative outcomes of suffering. Belief systems are commonly associated with religion but need not be. Because beliefs arise at the limits of knowledge, they are not dependent on observations of objects and events—that is, upon natural phenomena. Concepts of supernatural beings, objects, and events are the stuff of religious mythology and revealed truth. Skepticism is not my purpose here. I propose only that belief systems are universal, however varied they may be among cultures. The universality of some beliefs may be evidence of the validity of a substrate of unifying beliefs across cultures. It may also be evidence that belief systems are a natural residue of human learning (78) or, because suffering is the natural manure of the human condition, belief systems serve to ameliorate suffering regardless of the variations of cultural context; therefore, there is a substrate of beliefs that crosses cultures.

Spirituality and spiritual practice have an ameliorative effect on the experience of suffering. Downey offers the following definition of spirituality: "an awareness that there are levels of reality not immediately apparent and that there is a quest for personal integration in the face of forces of fragmentation and depersonalization" (79, p. 14). I propose that the "forces of fragmentation and depersonalization" of Downey's formulation are equivalent to the isolation and dissociation of suffering, as fear and grief threaten the integrity of any of the existential domains of personhood.

Discussions of spirituality often include three interrelated concepts: meaning, hope, and transcendence (29,80,81). Christina Puchalski outlines the elements of spiritual coping in the face of mortal disease that may be applied to the suffering of maldynia: "(1) Hope: (A) for a cure, (B) for healing, (C) for finishing important goals, (D) for a peaceful death; (2) Sense of control, (3) Acceptance of the situation; (4) Strength to deal with the situation; (5) Meaning and purpose: (A) in life, (B) in the midst of suffering" (numbering added) (29, Table 58.2, p. 636). Jerome Groopman proposes that hope is attainable without an organized belief system but admits that hopefulness appears to be easier when religious practices are strong (81).

Sam Harris, on the other hand, more concerned about geopolitical threat and global mortality than individual, medical mortality, argues that spiritual experience

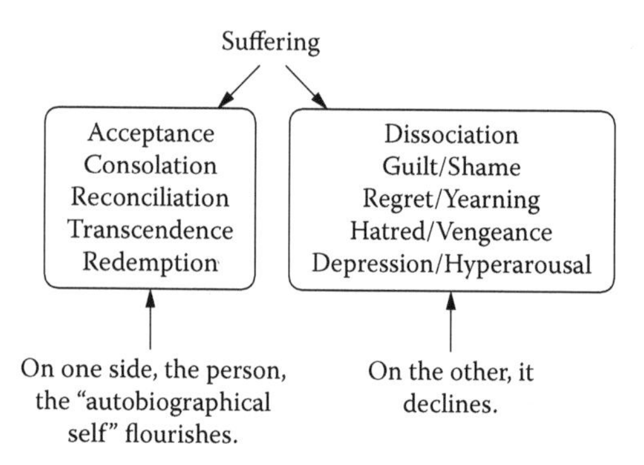

FIGURE 4.6 The outcomes and products of suffering.

is independent of religious belief systems and must be so if humankind is to survive (77). He argues that belief systems are used to justify exploitation and death far more often than to relieve suffering. This has always been so, he points out; only now the destructive methods are more efficient. I am persuaded to think, unfortunately, that Harris's prescription for belief-free spirituality requires rituals of contemplative practice that most people would find arduous and, therefore, undesirable. Furthermore, his path to transcendence requires a rejection of the intuitive concepts of supernatural or unnatural agency that all humans acquire in infancy if not in our genetic predispositions (78). I propose that few reasonably intelligent people can master Harris's counterintuitive reasoning, childhood religious education notwithstanding.

OUTCOMES AND PRODUCTS OF SUFFERING

Figure 4.6 divides the list of phenomena that constitute the outcomes of suffering into positive and negative *products*. The division is not complete. Not all negative outcomes are experienced as *painful* in the way that one would describe the experience of suffering. Dissociative states, much like the experience of denial, can be neutral, feeling neither good nor bad. Hatred, vengeance, and hyperarousal, particularly manic states, often obtain a positive emotional value even though the probability of a positive outcome, in a practical sense, is low. Revenge is more satisfying in fantasy than in reality, as its portrayal in popular culture* gives testimony. Real revenge stories are exceedingly rare, which explains the popularity of films that portray violent revenge in verisimilitude. Few people have the facility for put-downs, satire, parody, or withering, yet physically harmless, retorts often seem so satisfying. Shakespeare knew that veritable revenge is an unattainable fantasy, which is why Hamlet never understands the folly of seeking revenge.

* Action films are so common that even the most fastidious reader has probably encountered at least one. They are, with a few exceptions, more accurately denominated revenge films when they are not about domination and exploitation from the beginning. Revenge in such films is commonly justified by suffering inflicted on a noble protagonist, on his or her loved one, or on an innocent. Consider almost any example of the classic western.

Positive outcomes of suffering are not always joyful. Mediation, a method of dispute management, commonly requires compromise, conciliation, and sacrifice. Although a win-win outcome is the goal of mediation, sometimes a proper resolution to a conflict is when both parties are equally unhappy.* They are reconciled to their circumstances and to each other, but each feels that an element of grievance remains. The example of conflict resolution in small claims, civil disputes, and, increasingly, criminal matters† helps to illuminate the phenomena of reconciliation, transcendence, and redemption. I propose that the parties to a conflict first resolve the technical dispute (e.g., how much party A owes party B for what quantity of work B did for A). Then each party, after effecting the interpersonal reconciliation, reconciles himself or herself to the agreement, accepting it, ready to "move on" and transcending the stress of the conflict. A third stage is complete when the reconciled party feels properly and fairly paid off, which is a proper definition of redemption (74, vol. xIII, p. 412).‡

Figure 4.7 presents the assembled phenomenological model of suffering. It includes the causes of suffering, the internal milieu of the perceptual and associative apparatus by which suffering is experienced, along with the coping skills and reactions that modify and moderate the onset and course of suffering. The model includes the products of suffering and a useful, though incomplete, list of ameliorative and therapeutic modalities that address various neurobiological, cognitive, and experiential elements of the model. Each of these modalities has demonstrated some utility in reducing the intensity of suffering in various experiences of illness.§ The value of some therapeutic modalities may be grounded or validated only in their effect on the subjective experience of suffering and not by any objective measure or instrument. Understanding a patient's quality of life¶ in the face of illness requires an understanding of the experience of suffering, and so it is with the goal of understanding the experience of maldynia.

19. The experience of suffering, and, therefore, of illness (including, particularly, pain-as-illness) is understandable only through the first-person narrative of its evocative, perceptual contents (32). The story of suffering is the symbolic representation of its onset, course, and outcomes. That narrative, structured in the biopsychosocial model of Engel (33), is analyzed as "literary content" as

---

* Golten, M.M., CDR Associates, Boulder, Colorado, personal communication, after W. Ury and R. Fisher (1981).

† Offender-Victim Mediation is a tool of an international movement to ease the suffering inflicted by crime (82).

‡ I choose the example of conflict resolution to discuss transcendence and redemption in an effort to avoid confusion with the experiences of religious and spiritual practice. In religion, transcendence and redemption derive from identification with a transcendent being or emulation of a supernatural being or quality. Therefore, I discuss transcendence and redemption in terms that are found in common dictionaries of the English language. They do not permit the reader to know what it is like to transcend suffering or to feel redeemed, but they are a first approximation.

§ I direct the reader to the literature on multimodal pain management for a discussion of traditional therapies and what have come to be known as complementary and alternative modalities (CAM). The latter term is a bit pejorative, I propose, but it has stuck. Like the expression, "folk psychology," it has its uses.

¶ It is fortunate that, by and large, current research on the outcomes of treatment modalities for painful conditions includes measures for subjects' quality of life, such as the SF-36 and its components.

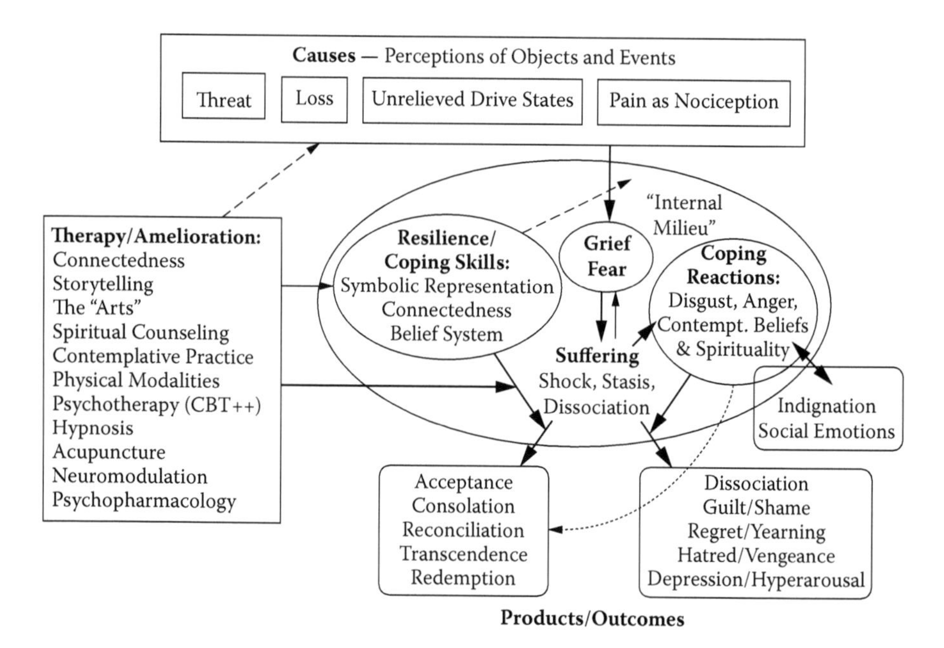

FIGURE 4.7 A phenomenological model of suffering.

implied or described by Eric Cassell (3), Rita Charon (32), Arthur Kleinmann (34), Arthur Frank (35), Joseph Natterson and Raymond Friedman (36), Howard Brody (37), James Katz (38), particularly Robert Gatchel and Dennis Turk (39).

Each *experience* of a red, ripe tomato, of nociception, or of suffering is unique. Even if the perceptual apparatus could be the same from one individual to another, the experience of the sight or taste of a red, ripe tomato can never be the same for any two people. So it is with suffering, even though there is a unitary neurobiological substrate of how suffering happens. The cause of suffering, the context in which it is evoked, and the coping skills and reactions that affect the onset and course of suffering are unique to each individual. The authors, cited here, cannot instruct the reader how to experience the suffering of another, neither in personal relationships nor in the patient–practitioner relationship. They offer, however, the intersubjective tools by which a clinician begins to understand experiences of suffering, of illness, and of maldynia.

Expressions of suffering associated with maldynia, such as *pain behavior*, may, at times, appear greater than expected for the observable physical impairment. The symptoms may appear exaggerated or amplified. The pain generator may be obscure; or, even when the condition is appropriately treated, sometimes the suffering persists. When the story of suffering appears inconsistent, the clinician is tempted to ask, "Is this an 'unreliable narrator'?" The moral role of the practitioner is to persevere in seeking an understanding of the subjective experience and of the objective condition—to trust the patient's narrative of suffering, even when trust is fragile. From the story of illness emerge therapeutic opportunities and realistic opportunities to hope.

REFERENCES

1. Moskovitz, P. 2006. A theory of suffering. *Pain Pract* 16(1): 74–81.
2. Damasio, A.R. 1994. *Descartes' error: Emotion, reason and the human brain*. New York: Grosset/Putnam.
3. Cassell, E.J. 1991, 2004. *The nature of suffering and the goals of medicine. II.* Oxford: Oxford University Press.
4. Engel, G. 1977. The need for a new medical model: A challenge for biomedicine. *Science* 196: 129–136.
5. Freud, S. 1923–1925. The ego and the id. In: *The standard edition of the complete psychological works of Sigmund Freud, Volume XIX,* ed. J. Strachey (1959). London: Hogarth Press.
6. Becker, E. 1973. *The denial of death*. New York: Simon and Schuster.
7. Ekman, P. 2003. *Emotions revealed: Recognizing faces and feelings to improve com munication and emotional life.* New York: Times Books.
8. Ekman, P. 2003. Emotions inside out. 130 Years after Darwin's "The Expression of the Emotions in Man and Animal." *Ann NY Acad Sci*. (Dec.) 1000:1–6.
9. Srinivasan, R., Russell, D.P., Edelman, G.M., and G. Tononi. 1999. Increased synchronization of neuromagnetic responses during conscious perception. *J Neurosci* 19: 5435–5448.
10. Wilke, M., Logothetis, N.K., and D.A. Leopold. 2006. Local field potential reflects perceptual suppression in monkey visual cortex. *PNAS* 103(46): 17507–17512.
11. Llinás, R., Ribary, U., Contreras, D. et al. 1998. The neuronal basis for consciousness. *Philos Trans R Soc Lond B Biol Sci* 353(1377): 1841–1849.
12. Llinas, R., and U. Ribary. 2001. Consciousness and the brain. The thalamocortical dialogue in health and disease. *Ann NY Acad Sci* 929: 126.
13. Singer, W. 2001. Consciousness and the binding problem. *Ann NY Acad Sci* 929(1): 123–146.
14. Singer, W. 2005. How does the brain know when it is right? In: *Neurobiology of human values*, ed. J.P. Changeux et al., 125–136. Berlin: Springer-Verlag.
15. Tononi, G., and G. Edelman. 1998. Consciousness and complexity. *Science* 282: 1846–1851.
16. Edelman, G., and G. Tononi. 1998. *At the frontiers of neuroscience*. Philadelphia: Lippincott-Raven.
17. Churchland, P. 2002. Catching consciousness in a recurrent net. In: *Daniel Dennett: Contemporary Philosophy in Focus,* eds. A. Brook and D. Ross, 64–81. Cambridge: Cambridge University Press.
18. Crick, F., and C. Koch. 1990. Towards a neurobiological theory of consciousness. *Semin Neurosci* 2: 263–275.
19. Baars, B.J., Newman, J., and J. Taylor. 1998. Neuronal mechanisms of consciousness: A relational global workspace framework. In: *Toward a Science of Consciousness II: The second Tucson discussions and debates,* eds. S. Hameroff, A. Kaszniak, and J. Laukes, 269–278. Cambridge, MA: MIT Press.
20. Dretske, F. 1995. *Naturalizing the mind*. Cambridge, MA: MIT Press.
21. Wallace, R. 2005. *Consciousness: A mathematical treatment of the global neuronal workspace model*. Berlin: Springer-Verlag.
22. Edelman, G. 2007. *Second nature: Brain science and human knowledge*. New Haven: Yale University Press.
23. deCharms, R.C. et al. 2005. Control over brain activation and pain learned by using real-time functional MRI. *PNAS* 102(51): 18626–18631.
24. Wilkinson, H.A., Davidson, K.M., Davidson, R.I. et al. 1999. Bilateral anterior cingulotomy for chronic non-cancer pain. *Neurosurgery* 45(5): 1129–1136.

25. Greene, J. 2005. Emotion and cognition in moral judgment: Evidence from Neuroimaging In: *Neurobiology of human values*, ed. J.P. Changeux et al., 57–66. Berlin: Springer-Verlag.
26. Giordano, J. 2008. Maldynia: Chronic pain as illness, and the need for complementarity in pain care. *Forsch Complementmed* 15: 277–281.
27. Bowlby, J. 1999. Attachment. *Attachment and loss, Vol. 1* (2nd ed.). New York: Basic Books.
28. Mercer, J. 2006. *Understanding attachment: Parenting, child care, and emotional development.* Westport, CT: Praeger.
29. Puchalski, C. 2006. Spirituality. In: *Principles and practice of palliative care and supportive oncology*, eds. A.M. Berger, J.L. Shuster, and J.H. von Roenn, 633–645. Hagerstown, MD: Lippincott Williams & Wilkins.
30. Kubler-Ross, E. 1969. *On death and dying*. New York: Simon-Schuster/Touchstone.
31. Boothby, J.L., Thorn, B.E., Stroud, M.W. et al. 1999. Coping with pain. In: *Psychosocial factors in pain—Critical perspectives*, eds. R. Gatchel and D. Turk, 343–359. New York: Guilford Press.
32. Charon, R. 2006. *Narrative medicine: Honoring the stories of illness*. New York: Oxford University Press.
33. Engel, G. 1997. From biomedical to biopsychosocial: Being scientific in the human domain. *Psychosomatics* 38(6): 521–528.
34. Kleinmann, A. 1988. *The illness narratives: Suffering, healing, and the human condition*. New York: Basic Books.
35. Frank, A.W. 1995. *The wounded storyteller*. Chicago: University of Chicago Press.
36. Natterson, J.M., and R.J Friedman. 1995. *A primer of clinical intersubjectivity*. Northvale: Jason Aronson.
37. Brody, H. 2002. *Stories of sickness* (2nd ed.). New York: Oxford University Press.
38. Katz, J.D. 2004. Pain does not suffer misprision: An inquiry into the presence and absence that is pain. *J Med Ethics; Med Humanities* 30: 59–62.
39. Gatchel, R., and D.C. Turk. 2008. Criticisms of the biopsychosocial model in spine care: Creating and then attacking a straw person. *Spine* 33(25): 2831–2836.
40. Damasio, A.R. 1999. *The feeling of what happens: Body and emotion in the making of consciousness*. New York: Harcourt Brace.
41. Mayerfield, J. 1999. *Suffering and moral responsibility*. New York: Oxford University Press.
42. Siu, R.H.G. 1993. Panetics and the Dukkha: An integrated study of the infliction of suffering, *Panetics Trilogy, Volume II*, Washington, DC, International Society for Panetics.
43. Rizzolatti, G., and L. Craighero. 2005. Mirror neuron: A neurological approach to empathy. In: *Neurobiology of human values,* ed. J.P. Changeux et al., 107–124. Berlin: Springer-Verlag.
44. Oberman, L.M., and V.S. Ramachandran. 2007. The simulating social mind: The role of the mirror neuron system and simulation in the social and communicative deficits of autism spectrum disorders. *Psychol Bull* Mar; 133(2): 310–327.
45. Sober, E., and D. Wilson. 1998. *Unto others: The evolution and psychology of unselfish behavior.* Cambridge, MA: Harvard University Press.
46. Damasio, A.R. 2005. The neurobiological grounding of human values. In: *Neurobiology of human values,* ed. J.P. Changeux et al., 47–56. Berlin: Springer-Verlag.
47. Davidson, R.J. 2005. Neural substrates of affective style and value. In: *Neurobiology of human values,* ed. J.P. Changeux et al., 67–90. Berlin: Springer-Verlag.
48. Greene, J.D., Nystrom, L.E., Engell, A.D. et al. 2004. The neural bases of cognitive conflict and control in moral judgment. *Neuron* 44: 389–400.
49. Bitvinick, M.M., Braver, T.T., Barch, D.M. et al. 2001. Conflict monitoring and cognitive control. *Psychol Rev* 108: 624–652.

50. Kouider, S., and S. Dehaene. 2007. Levels of processing during non-conscious perception: A critical review. *Philos Trans Royal Soc London B* 362: 857–875.
51. Block, N. 2004. Consciousness. In: *Oxford companion to the mind*, ed. R. Gregory. Oxford: Oxford University Press.
52. Zeman, A. 2002. *Consciousness: A user's guide*. New Haven: Yale University Press.
53. Edelman, G. 1989. *The remembered present: A biological theory of consciousness*. New York: Basic Books.
54. Jackson, F. 1982. Epiphenomenal qualia. *Philos Q* 32: 127–136.
55. Chalmers, D.J. 2006. Perception and the fall from Eden. In: *Perceptual experience*, eds. T.S. Gendler and J. Hawthorne, 59–125. Oxford: Oxford University Press.
56. Dretske, F. 2006. Perception without awareness. In: *Perceptual experience*, eds. T.S. Gendler and J. Hawthorne, 147–180. Oxford: Oxford University Press.
57. Chalmers, D. 1996. *The conscious mind*. Oxford: Oxford University Press.
58. Leopold, D.A., Wilke, M. et al. 2002. Stable perception of visually ambiguous patterns. *Nature Neurosci* 5(6): 605–609.
59. Sample, I. 2008. Blind man amazes scientists with his ability to detect objects he cannot see. *The Guardian* (December 23).
60. MacLean, P. 1990. *The triune brain in evolution: Role in paleocerebral functions*. New York: Plenum.
61. Giordano, J., and J. Engebretson. 2006. Neural and cognitive basis of spiritual experience: Biopsychosocial and ethical implications for clinical medicine. *Explore* 2(3): 216–225.
62. Decety, J., and C. Lamm. 2009. Empathy versus personal distress: Recent evidence from social neuroscience. In: *The social neuroscience of empathy*, eds. J. Decety and W. Ickes, 202. Cambridge, MA: MIT Press.
63. Llinas, R. 2001. *I of the Vortex*. Cambridge, MA: MIT Press.
64. Uhlhaas, P.J., Roux, F., Rotarska-Jagiela, A. et al. 2010. Neural synchrony and the development of cortical networks. *Trends Cogn Sci* 14: 72–80.
65. Lehrer, J. 2007. *Proust was a neuroscientist*. New York: Houghton Mifflin.
66. Eisenberger, N., and M. Lieberman. 2004. Why rejection hurts: A common neural alarm system for physical and social pain. *Trends Cogn Sci* 8(7): 294–300.
67. White, J.C., and W.H. Sweet. 1998. *Textbook of stereotactic and functional neurosurgery*. New York: McGraw-Hill.
68. Holland, P.C., and M. Gallagher. 2006. Different roles for amygdala central nucleus and substantia innominata in the surprise-induced enhancement of learning. *J Neurosci* 26(14): 3791–3797.
69. Lippe, P. 1998. An apologia in defense of pain medicine. *Clin J Pain* 14(3): 189–190.
70. Loeser, J.D., and R.D. Treede. 2008. The Kyoto protocol of IASP basic pain terminology. *Pain* 137: 473–477.
71. Dennett, D.C. 1996. *Kinds of minds*. New York: Basic Books.
72. Burgess, D. 1994. Denial and terminal illness. *Am J Hospice and Palliative Med* 11(2): 46–48.
73. Freud, A. 1948 (1936). *The ego and the mechanisms of defense*. London: Hogarth Press.
74. Brown, L. 1993. *The new, shorter Oxford English Dictionary on historical principles*. Oxford: Clarendon Press.
75. Gatchel, R.J., and J.D. Proctor. 1976. Physiological correlates of learned helplessness in man. *J Abnorm Psychol* 85(1): 27–34.
76. Kahnemann, D., and C.R. Sunstein. 2005. Cognitive psychology of moral intuitions. In: *Neurobiology of human values*, ed. J.P. Changeux et al., 91–106. Berlin: Springer-Verlag.
77. Harris, S. 2004. *The end of faith: Religion, terror, and the future of reason*. New York: W.W. Norton.

78. Bloom, P. 2004. *Descartes' baby: How the science of child development explains what makes us human*. New York: Basic Books.
79. Downey, M. 1997. *Understanding Christian spirituality*. New York: Paulist Press.
80. Frankl, V.E. 1959. *Man's search for meaning*. Boston: Beacon Press.
81. Groopman, J. 2005. *Anatomy of hope: How people prevail in the face of illness*. New York: Random House.
82. Johnstone, G., and D.W. VanNess. 2006. *Handbook of restorative justice*. Cullompton, Devon, UK: Willan.

5 How (Can) I Feel Your Pain

The Problem of Empathy and Hermeneutics in Pain Care

Giusi Venuti

CONTENTS

INTRODUCTION

Anyone who prepares to judge* problems within a given discipline falls victim to a feeling of unease, subtle and deceitful at the same time. This unease is bound to increase when the judge is not a specialist, but an academic: someone who, by definition, loves to be involved in the problems of his time. If, to echo Hegel, philosophy is qualified as "one's own time learnt with thought," it should not seem at all strange for a nonphysician to wonder about the crisis that for decades has beset medicine's knowledge and actions, despite the ability to enable a doubled life expectancy and improved health of the general public.

Although the biomechanical model of modern Western medicine (especially within the United States) has been heralded for enhancing the length and quality of human life, it has left many patients dissatisfied with the health care industry: "In spite of remarkable advances in medical therapy and in the development of fantastic diagnostic devices, American society appears increasingly disenchanted with the physician" (1, p. 311). So, why does the nagging unease remain; why is there the impression that biomedical thinking is always out of place? On one hand, we could reply that it is precisely the unease of never feeling at home that is the soul of real research; on the other, we may not hide the awareness of being unable to say anything really *new* on the overworked issue in medical progress. In the totally human desire to produce new arguments, this unease sows in the mind of the scholar fleeting doubt that because it is impossible to say something *new*, it is better not to say anything at

* Here intended as discern according to Greek etymology.

all. But then, reflecting further, why should we have to say something new? Perhaps, we must go back to old problems, to their roots in an attempt to reinterpret and decode the causes of a crisis. If we accept this approach, we may understand what David Clouser means when he says that ethicists are not reformers because their task is to shed light on the conflicts underlying ethical principles.

The ethicist helps in this search by "structuring the issues," or specifying conflicting ethical principles and isolating concepts that need clarification. This structuring reveals "where various arguments and actions lead, what facts would be relevant, what concepts are crucial, and what moral principles are at issue and probably in conflict [...] Ethicists are not reformers, nor do they provide solutions" (2, p. 328).

We thus need to be profaners, aware that beyond any legitimization that derives from knowledge, from a profession, or from the interdisciplinary nature of bioethics,* we each—philosophers, ethicists, physicians, and patients—need to assume responsibility for such questions. Being a patient is not only an identity and a role that we wear transitorily but is a possibility that belongs to the constitutive dimension of our being, insofar as we are mortal, and thus are, for this very reason, vulnerable.

Why, in this precise moment of history, does medical science seem more than ever unable to relate to its own object of study: the patient? And why is it that one of the most fundamental aspects of being a patient—pain—remains so inaccessible to medicine? Is it the fault of physicians that they should learn to see that object as something more than a body to heal? Or is it the fault of patients who, being excessively—and often badly—informed, presume that the physician's work can be reduced to mere performance, functional to their needs? Many scholars tip the balance on one side, convinced that the malaise is entirely due to medical science: the excessive technicalization, the increasing pace of life, and asphyxiating bureaucracy have surely contributed to undermining the very foundations of medicine, so much so that they often think of new paradigms for the clinical relationship. But in addition to these external causes, I believe a reason for this crisis is to be found in the intimate nature of relating to the patient in pain.

How can we overcome the impasse of technologization on one hand, and subjectivity on the other? Similarly, how are we to instantiate a reasonable model for pain medicine—a practice dedicated to a subjective condition—considering that Hippocratic paternalism seemed to treat the patient as "incapacitated (*incapacitado*) not only from a physical, but also from a moral point of view" (4, p. 68), that contractualism regards individuals as *moral foreigners*† whose only possibility of interacting is found in the procedural nature of the agreement, and that the *asymmetry* between the physician and the patient appears to be unavoidable? How can we bridge the gap between these philosophical, ethical, social, and clinical worlds? There is an awareness that it is necessary to abandon contractualistic logic in favor of dialogue articulated by compassion, and moreover, that compassion—and not mere technical skill—is an essential virtue in medical practice. A good physician is thus someone

* According to the definition, *bioethics* is "the systematic study of the moral dimensions—including moral vision, decision, conduct and policies—of the sciences and the health care, employing a variety of ethical methodologies in an interdisciplinary setting" (3).

† The expression is from H.T. Engelhardt.

who "does not just apply cognitive data from the medical literature to the particular patient.... Rather, the good physician co-suffers with the patient" (5, p. 78). A physician is, before being a specialist, a person who manages to make the patient's needs, problems, and suffering his or her own. Within this perspective of medical ethics, there are those who go even further, maintaining that to really understand the patient, particularly a patient in pain, it is necessary to identify with the patient to the point of becoming his or her *fellow of experience,** the patient's moral *friend.* From this, it can be deduced that when a doctor aspires to be, from a medical point of view, the friend of his or her patients, the practice of medicine could be defined by the following formula: "reliable donation of technical help to another ill me."† These precepts do not seem to cause any problem on a strictly theoretical level—seeing that it would simply be a matter of moving from one paradigm to another—but in terms of actual medical practice, there are doubts that test the application of such virtue as "habitual disposition to act in a certain way" (5, p. 78). The question thus becomes: how is it possible to identify with patients' subjectivity without losing objectivity—which is also an essential component of the art of medicine? What does being a friend mean in the sense of pain care?

Going even further to the root of the question, is it really possible to identify to the point of empathetically understanding‡ another person's pain? Is the other, the patient, simply another me in pain? Is the physician really given the possibility of understanding what pain provokes not only in the body (*Körper*) but in the embodied self (*Leib*) of the patient? And if identification were possible, would we not run the risk of reducing the other to a mirror in which to reflect our own image? Are we not falling into the trap of an overly human imagination that, failing to find a way to establish an authentic relation, thinks that this is the only way to bridge the distance that separates him from the world of the other? What value should be given to the expression: "I *realize* what pain you are feeling"? What "realization" or reality is possible in this sense?

Like any other human being, a physician finds himself or herself at the crossroads between authentic caring and unauthentic "taking care of." Unlike any other human being, however, the physician has in this situation an undeniable power, and finds himself or herself dealing with a person who is powerless. Implicit to vulnerability, the patient runs the serious risk of being reduced to an object. How can we consider the other in his otherness; that is, how can we recognize the patient as a subject? How can we respect the patient?

---

* The expression "fellowship in the experience" is from E.D. Pellegrino and D.C. Thomasma (5).

† The Spanish is *afable donacion de ayuda tecnica al semejante enfermo*, Lain-Entralgo (6, p. 239).

‡ I am referring to the most widespread definition of the term empathy whose roots date back to Romanticism. According to the supporters of this school of thought (we should mention T. Lipps for his attempt to systemize it in his work Einfühlung und ästhetischer Genuß (7), whereby empathy consists in feeling oneself in an object that is not us. Appropriation, insertion, transferral, and identification are all synonyms of what the Romantics meant by the term *empathy*.

CARE AND RELATION: EMPATHY FOR ANOTHER'S PAIN

I do not believe that the critical situation in which physician and patient have always found themselves can really be resolved simply by training clinicians on the specific nature of the *object* of their study, and hoping for a shift from the biomechanical to the biopsychosocial model. I do not believe that the inclusion in medical courses in medical humanities, as such, can really provide any valuable enhancement to doctors' knowledge and actions. The most recent studies, however, try to do just that. They tell us that medical science must finally learn to talk not of the body, but of corporeity, and that doctors must abandon paternalism, because it is disrespectful and self-centered; abandon the concept of patient autonomy, because it is indifferent to real needs; and assume an empathic attitude,* as the only attitude that bridges the world of the doctor and that of the patient, because it focuses attention on the vulnerable party. Moreover, it seems that empathy must be considered to be the foundation of the biopsychosocial model of medicine.

I argue that an empathic understanding of a patient's illness is certainly a necessary condition of the biopsychosocial model, because only through empathic understanding can the physician provide an effective diagnosis and treatment (8). In this sense, empathy is defined as the act of understanding, of being aware of the other's feelings, experiences, and thoughts. Empathy is a mode of perception that includes carefully listening, observing gestures and body language, and trying to understand the patient's unique medical situation. The goal of empathy is to perceive the subjective experience of another person without imposing external values on the data apprehended (8). Obviously, these are important requirements of pain medicine, given the nature of pain.

But, what leaves me perplexed is that the majority of scholars indistinctly use the terms *empathy, sympathy,* and *compassion* as if they all had the same meaning—namely, that of attention, identification, and sharing with the suffering of the other. I believe that compassion or empathy dictates that the patient's life is granted some sort of unique status, both quantitatively and qualitatively, if only for pragmatic reasons (9, p. 90). A compassionate-empathic physician is defined as one whose pattern of behavior reflects strong devotion to the *welfare* of the patient on the scientific technical *and* socioemotional levels, or, as it is often put, curing and caring (10).

This empathy is presented as a new paradigm, in which science and humanitarianism interact, to follow and replace inadequate and unsatisfactory models of the past, as if empathic comprehension, or compassion, was really the remedy for all ills. In this sense, it is not rare to find models, such as the one I am about to illustrate, aimed at exemplifying, to an exaggerated extent in my opinion, the situation.

Physicians firmly convinced of the validity and value of the biopsychosocial model, train students in empathy through role-playing. During the lessons, they are asked to *put themselves in the patient's shoes*, so that they can see what it means to be on the *other side*. Some medical schools developed programs that use trained laypeople to play the role of patients, to improve students' interviewing skills (11, p. 77). Empathy is thus presented as the essential characteristic of the physician who

---

* Commonly intended as emotional involvement, as penetration into the needs of the other.

wants to be seen as virtuous, who wants to act honestly according to Hippocratic precepts, and who would never dream of using power inappropriately. The virtues of medicine are those traits of character that predispose the physician habitually to act in a way that effects healing. But these virtues cannot be defined without some genuine grasp of the life-world of the patient by the physician, and of the physician by the patient (5).

To be a good physician, it is thus necessary to grasp and comprehend* the experience of the other as if it were the physician's own. When we experience empathy, we feel as if we were experiencing someone else's feelings as our own. We see, we feel, we respond, and we understand as if we were, in fact, the other person (12, p. 26). In these terms, empathy is defined as a subjective involvement that has the aim of entering the world of the other and sharing his or her pain: the good physician co-suffers with the patient (5).

Entering the world of the other and understanding the effect that pain has on the body and world of the patient means, according to a number of scholars in medical humanities, becoming skillful at using all the data of narration for the purpose of not only a correct diagnosis, but also of liberating medicine from methodological reductionism. Empathy and effective skillful communication increase patients' satisfaction, improve patients' compliance, and enhance physicians' ability to diagnose and treat their patients. But what is the best method of training physicians to become competent and fluent in the use of these techniques (13, p. 606)? Empathy thus becomes a technique, and those who use it obey Hippocratic precepts and earn the respect of the patient.

Cassel suggests that the physician uses personal experience of his or her lived body to gain an accurate understanding of the patient's experience. He also describes how, in attending the complaint of the patients, the physician can expand his or her knowledge when he or she is not previously acquainted with a particular body experience. Thus, it is not necessary for the doctor to suffer from all the symptoms to establish satisfactory communication with the patient (14, p. 311). The objective and objectivizing method, whereby the patients should not be listened to with one's ears, but with a stethoscope, is now abandoned in favor of the method based on the medical interview. Empathy, as the canon of the interview, assumes a guiding role. Training in empathy thus means becoming competent at obtaining human data in a more systematic way. By intending dialogue as technique, the interview transforms the relation into an instrument and means for interacting and exchanging information with the patient. The physician adopts empathy to obtain a given result.

But what happens to the clinical encounter? What happens to the request for help? If we are merely dealing with new reduction plus empathy, what happens to the phenomenological approach? In fact, despite the interactive element, the interview is the least dialectic thing imaginable, in the sense that its main aim is to collect information and provide data for diagnosis, treatment, and prevention (i.e., to know the other as a unilateral being). The interactive element that is sometimes referred to (i.e., the exchange of information) refers to the interview as the "principal vehicle of the information of various types transmitted to the patient" (14, p. 311). It is often said that the

* In the etymological sense of *compre-hendere*, take within oneself, include.

interview determines how the physician–patient relationship will evolve, in the sense that the quality of the relationship depends on the quality of information exchanged. The idea of interview, in the sense in which this concept is currently used, does not belong to the story of medicine.

Anamnesis is a collection of data relative to a patient, but in the sense of *memory*, understanding the patient as a *being-state*. The anamnesis is conceived as recovery of the past. Unlike the interview, it expresses an ontology of the illness, in this case, pain, as a recovery of its past. The interview is something else; it is generally a series of questions posed to a person and aimed at finding out about that person's opinions or tastes. Otherwise, it is a journalistic interview to obtain information. Even when the interview is conceived as part of an integrated approach, even when the doctor adopts an empathic, understanding, attentive attitude, its aim is to generate relevant data on the person and his or her symptoms. The protagonist of the meeting is always and only the physician, because it is he or she who asks the questions and asks them in a manner that orients and indicates the patient's responses.

To reiterate, many propose the biopsychosocial model to describe the person as an integrated set of biological, psychological, and social components. The integrated interview is valid as aimed at comprehension of both the disease and the patient. It represents a model that is both scientific and humanistic and is based on the premise that body and mind are inseparable and that the health care professional may understand the patient only by considering the patient in his or her entirety (15).

A person considered to be *integrated* is thus made to correspond to an equally *integrated* interview. But what meaning is attributed to this term? One has the impression that knowledge of the patient is intended as a *sum of the parts*, assumed in a certain order. This resembles the mathematical concept of integration (i.e., a process used to determine a value as sum of its infinitesimal parts). This implies that we consider the patient as a collection characterized by the composing variables. But what happens if the variables do not add up? The interview fails, the patient is classified as *strange,* and the possibility of a real clinical meeting is denied at the outset. Maintaining a distinction between an approach centered on the doctor from an approach centered on the patient is not as easy or decisive as one might think. From an epistemological point of view, the approach remains the same; what changes is the content of the information that may be more or less detailed, more or less persuasive. Not only is the distinction between the two approaches not decisive, but if the physician incautiously chooses one rather than another, continuing to reason on the basis of dichotomies and contrapositions, the physician encounters a series of *critical situations*, or, as we have already seen, identifies the symptoms but does not *meet* the patient. Alternatively, the physician will not identify the symptoms, because the answers mislead him or her, or can identify the symptoms but risks, due to excessive identification, compromising objectivity (5). Given the subjectivity of pain, what then is the (or at least "a") best approach?

The perplexities to which I refer do not belong only to my reflection, but arise within a meaningful discourse on the neurophilosophy, ethics, and clinical dimensions of pain and pain care. In sum, the discourse asks: how can we know pain in another—and can we instantiate ethically sound treatment upon such knowing?

Empathy has been the subject of two anthologies and multiple articles, literature proposing that empathic physicians are an ideal. Most critiques of empathy, however, argue that empathy is not an appropriate stance for physicians to assume with patients. There are many scholars who, after having described the positive characteristics of the empathic attitude, conclude by advancing the same doubt: what if this type of empathy was not really the solution, if the failure to distinguish between empathy, compassion, and sympathy did nothing but confuse both doctors and patients? These general terms, however, give little guidance about the behaviors pain medicine strives to cultivate. Occasionally, humanistic attributes of pain care will be listed in greater detail: respect, compassion, or empathy. Taught without critical discussion of what specific behaviors are meant by these concepts, these, too, fall short of the goals for authentic pain medicine (16).

What if the bridge we wanted to build by adopting the empathic attitude did not represent anything but an escape from a situation that seems to have gone out of control? What if identification with one's patients damages, rather than benefits, the medical relationship? If the notion of bodily empathy evokes the expectation that the doctor should experience more or less the same pain as the patient, this might lead to a depreciation of equanimity, and a loss of the good (14).

Such identification does not, in fact, seem to produce the opening that is characteristic of authentic questioning, but on the contrary, provokes the closure, a concentration on oneself, and an incorporation of the other. In this closure, that request for help that initiates the medical act remains unheard.

To be technically right, a decision must be objective; to be good, it must be compassionate. In the fusion of these attitudes, the ends of medicine are fulfilled. The physician must be able to feel something of the experience of the patient. The physician must apprehend something of the patient's pain, for that is what compassion literally means (17). But where and how can common ground between pain physician and pain patient be found? How can we balance objectivity and sharing of experience?

The road toward real integration must first involve a clarification of terms—empathy, sympathy, and compassion—and, at the same time, examine the crucial nature of the clinical meeting. Why do I define this clinical meeting as crucial? Because it places face-to-face not one, but two human beings who position for mutual acknowledgement: the physician wants to be acknowledged as expert and exercises his or her power through science and skill; the patient wants to be acknowledged in his or her need and, when the patient goes to the doctor, can decide whether to use his or her vulnerability as an instrument of power (e.g., to trust or not to trust? to tell all or to conceal?) or acknowledgement, whether to accept the other or accept the distance that separates him or her from the other. The subjectivity makes these separate strivings for acknowledgement all the more patent. The physician employs (if not exploits) knowledge and skills in the objective domain (e.g., science, techniques, technology) to exert power and bridge the gap of subjectivity. The patient presents those aspects of his or her subjective experience of pain in an attempt to overpower his or her own vulnerability and engage the help of the physician. Both engage subjectivity to assert objective means and ends.

EMPATHY, SYMPATHY, COMPASSION: ALL SYNONYMS?

The word *empathy* has assumed a wide variety of uses since its original presentation as *Einfühlung* (that was first used by Herder*) to *empathy* (coined by Edward Titchener, an English pupil of Wilhelm Wundt). Dictionaries trace the word to its Greek origins as the calque of *empàtheia*. The phonetic similarity, however, conceals a semantic abyss. As if this were not enough, the term is often used as a synonym of the word *sympathy*, which, in turn, is used as a synonym of compassion.† This is why many people talk of empathy without effectively defining. Some intend it as the technical ability to put yourself in someone else's shoes, an ability that both the competent physician and anyone who wants to communicate effectively must know how to use.‡ Even though the common term for this action is *empathy*, I should like to suggest calling it *trial identification*.

Others conceive empathy as an essential *predisposition* for compassion, thus glossing over the specific meaning of the term. Compassion is the virtue by which we have a sympathetic consciousness of sharing the distress or suffering of another person, and on that basis are inclined to offer assistance in alleviating or living through that suffering. Compassion, together with its related attitude of empathy, requires sharing in the suffering of another person (19). But then, in light of this ambiguity, what does it mean to feel empathy? When we feel empathic, do we really think that we are suffering, by putting ourselves in another person's place? Do we really imagine our reactions to be fused, in some mysterious way, with those of the sufferer? The same scholars who advance these theories at the same time advance doubts regarding the authentic possibility of assuming the first-hand experience of the other. Consequently, far from being resolved, the problem of empathy is further complicated. The need to clarify the issue (that empathy has an objective aspect that sympathy lacks) is the basis of the work of Robert Katz. Katz initially asks whether it is really possible to distinguish empathy from sympathy. It is true that in both sympathy and empathy, we permit our feelings for others to become involved. The purposes of the two activities are different, however (12). The conclusion Katz reaches is that, although both terms refer to situations in which feelings are involved, empathy focuses attention not on the subject feeling, but on the object/subject felt and, therefore, has an objective aspect.§ When we empathize, writes Katz, we focus our attention on the feelings and situation of another, and when we sympathize, we establish a parallelism between our feelings and those of the other, so that we are not concentrated on the objective reality and personal character of the other's situation.

* Together with the poet Novalis, an exponent of European Romanticism. In these two authors, the concept of empathy has an eminently aesthetic value and expresses the total identification of the subject with the life of nature, conceived as a spiritual living being. Through empathy, the object-nature has its aesthetic resonance inside the subject-poet. In the writings of these two authors, the word empathy is, however, used occasionally and does not have a theoretical meaning.

† Cfr. C.D. Batson (18). In this study, empathy is assimilated with compassion. Batson stated that he prefers the term empathy insofar as it is less moralistic and less misleading than sympathy, but he does not consider that it has an essentially different meaning.

‡ R. Fliess quoted in R.L. Katz (12).

§ Warren Reich declares that he was inspired by this argument.

In this case, analogy replaces attention, and by losing its objectivity, comprehension of the other is compromised.

The distinction is important. Unless it is sharply made, it will be difficult to appreciate empathy as a professional tool, in both the arts and sciences. Practitioners of empathy are committed to objective knowledge of other personalities. If we use our own feelings, it is for the purpose of learning more about what actually belongs to the other person. But we do not exercise our own feelings to gratify our needs. In contrast, when we sympathize, we are aware of and focused upon our own state of mind, and much of our attention is devoted to our own needs. When we empathize, we cannot fully escape our own needs, but we use our own feelings as instruments of cognitions (12). So, empathy, like sympathy, consists of feelings and implies an emotional involvement, but unlike sympathy, empathy does not establish that similarity which generally leads oneself to ask: what would I do if I found myself in his or her position? Katz seems to be on the right track: by so defining empathy, as attention, he asserts that it is the basis for building authentic relations of care. When we empathize, we may not ever entirely escape our own needs—thus stressing how it is impossible to become one with another, to cancel the distance and difference of experience. Yet, he maintains that during the empathic act, we completely abandon our self-awareness to lose ourselves, temporarily, in the object/subject we are dealing with (12).

But, I ask, how is it possible to establish such intimacy with the patient, how is it possible to feel at home with someone who is going through a deeply disturbing experience of pain? If pain, and particularly pain-as-illness, transforms not only the body, but also the life-world, if it is something that uproots the patient, what *home* are we talking about? I believe that the misunderstandings and incapacity to establish what empathy is—whether a going-inside, a sort of imitation, an emotional involvement, or something else—mainly reflects the fact that most scholars refer to the concept of empathy inherited from Theodore Lipps, which implies a movement made by the subject toward the object. It ignores the use that phenomenologists* make of the term which, in turn, refers to the Greek sense of the word, Εμπάθεια, that means "being exposed, being subject to"—in other words, a movement opposite to the type we have been dealing with so far. This is a movement from the outside inward. This is why the *em*, here, is not in a dynamic reference to another self, and does not have a projective or fusional sense with regard to another; rather, it means a reinforcement of the pathetic dimension that characterizes sensitivity.† According to the theory of imitation, elaborated by Lipps, the experience of the extraneous is realized in me, through my internal imitation of another's action (or his reaction to some occurrence). In the modern and contemporary meaning, empathy, *Ein-fühlung*, indicates an act of emotional participation and identification—understanding with regard to another human subject.

The feeling Lipps talks of can be considered as an interior movement that allows us to appropriate ourselves of what another is feeling—or at least to some extent, by relating it to our own experience. The recent work of Giacomo Rizzolati and

* In particular, I am referring to Scheler, Husserl, and Stein, as we shall see below.

† Cfr. On this subject, the observations of M.F. Basch (20).

coworkers appears to have provided some evidence for the neurobiological basis for such effects (21). The elucidation of so-called *mirroring neuron* systems in brain networks that are involved in extero/interoceptive cognition, and which are involved in self–other emotional identification, may subserve the phenomenon addressed by Lipps, at least in part. Still, understanding that there are putative neurological mechanisms for empathic action only establishes a biological predisposition or capacity. Engaging this capacity, and doing so in ways that enable, if nothing more, a sensitivity to another, may be the critical step, not in taking another in, but in realizing the necessity to reach out to those in pain.

Those who experience pain first-hand often stress that no one can understand how they feel. Perhaps this will always be true. But, perhaps what we must understand is that the request for help does not call for illusory identification, but rather is a need for contact, a desire for intimacy tempered by an awareness of the unbridgeable and unavoidable distance that the illness of pain can impart.

REFERENCES

1. Marcum, J.A. 2004. Biomechanical and phenomenological models of the body, the meaning of illness and quality of care. *Medicine, Health Care, and Philosophy* (7)3: 311.
2. Jonsen, A. 1998. *The birth of bioethics*. New York: Oxford University Press.
3. Reich, W.T. 1995. Introduction. In: *The encyclopaedia of bioethics*. ed. W.T. Reich, XXI. New York: Simon Schuster Macmillan.
4. Gracia, D. 1989. *Fundamentos de Bioetica*. Madrid: Eudema.
5. Pellegrino, E.D., and D.C. Thomasma. 1993. *The virtues in medical practice*. Oxford: Oxford University Press.
6. Lain-Entralgo, P. 1986. La palabra y el silencio del medico. *Ciencia, técnica y medicina*. Madrid: Alianza Editorial.
7. Lipps, T. 1906. Einfühlung und ästhetischer Genuß. *Die Zukunft* 54 (Jan.): 100–114.
8. Switankowsky, I. 2004. Empathy as a foundation for the biopsychosocial model of medicine. *Humane Health Care* 4(2): E5.
9. Glick, S.M. 1993. The empathic physician: Nature and nurture. In: *Empathy and the practice of medicine,* ed. H. Spiro, M. Curnen, E. Peschel et al., 90. New Haven: Yale University Press.
10. Carmel, S., and S.M. Glick. 1996. Compassionate-empathic physicians: Personality traits and social-organizational factors that enhance or inhibit this behaviour pattern. *Social Science and Medicine* 43(8): 1253–1261.
11. Levasseur, J., and D.R. Vance. 1993. Doctors, nurses, and empathy. In: *Empathy and the practice of medicine,* ed. H. Spiro, M. Curnen, E. Peschel et al., 77. New Haven: Yale University Press.
12. Katz, R.L. 1963. *Empathy. Its nature and use*. London: Free Press of Glencoe Collier-Macmillan.
13. Neuwirth, Z.E. 1997. Physician empathy—Should we care? *Lancet* 350(9078): 606.
14. Rudebeck, C.E. 2001. Grasping the existential anatomy. In: *Handbook of phenomenology and medicine,* ed. S.K. Toombs, 311. London: Kluwer Academic.
15. Engel, G.L. 1980. The clinical application of the biopsychosocial model. *American Journal of Psychiatry* 137: 535–544.
16. Giordano, J. 2009. *Pain: Mind, meaning, and medicine.* Glen Falls, PA: PPM Press.

17. Pellegrino, E.D. 1983. The healing relationship: The architectonics of clinical medicine. In: *The clinical encounter: The moral fabric of the patient–physician relationship*, ed. E.E. Shelp, 165. Dordrecht: D. Reidel.
18. Batson, C.D. 1991. *The altruism question: Toward a social–psychological answer*. Hillsdale, NJ: Lawrence Erlbaum Associates.
19. Reich, W. 1989. Speaking of suffering, a moral account on compassion. *Soundings: An Interdisciplinary Journal* 72(1): 83–108.
20. Basch, M.F. 1983. Empatic understanding: A review of the concept and some theoretical considerations. *Journal of the American Psychoanalytic Association* 31: 101–126.
21. Rizzolati, G., and C. Sinigaglia. 2008. *Mirrors in the brain: How we share our actions and emotions*. Oxford: Oxford University Press.

6 Spirituality, Suffering, and the Self*

James Giordano and Nikola Boris Kohls

CONTENTS

INTRODUCTION

Due to the fact that the phenomenon of pain defies clear classification in either of the Cartesian categories of *res cogitans* or *res extensa*, within medicine we have come to recognize pain as both a physiological event of the nervous system as well as a psychological phenomenon of consciousness (1,2). As such, while the objective properties of the sensation of pain may be quantifiable, the qualitative dimensions of the experience of pain are individually variable and, in many ways, are unique to the person who suffers—reflective of the ongoing interaction of hereditary, behavioral, and environmental interactions throughout the life span that are both predispositional to and affected by pain (and its broadly biopsychosocial manifestations). The event of pain is inextricable from the event of (self-) consciousness (2,3). As a conscious, self-referential sensory process, it manifests subjectivity and transparency (only) to self.

This first-person experience of the subjective *self* is grounded in personal circumstance and bounded by place and time. But what is this *self*? Polymath Douglas Hofstadter claims that "an 'I' is an abstract pattern that arises naturally…in human brains" (4). To be sure, as we previously noted, our *selves* arise from our brains, and our brains are nested within and are the stuff of our universe (5). Yet, try as we might, attempts to reduce consciousness or self-consciousness to the mechanisms of interacting molecules or even atoms have failed (at least to date). It may well be, as Hofstadter

* This chapter has been adapted, with permission, from Giordano, J., and N. Kohls, 2008, Spirituality, Suffering, and the Self, *Mind and Matter* 6(2): 179–191.

stated, that "life resides on a level…that no being could survive if it concentrated on that level" (4, p. 70). Hence, the experiences of body, brain, and world as well as culture and society organizing and structuring human relations are essential to the self, and these experiences assume meaning based upon circumstance and manifestation.

Although the self-conscious experience of a self cannot be argued away, the ontological status of the self has been extensively debated within cognitive neuroscience and neurophilosophy. For example, Metzinger recently argued that there are no such things as selves in the world, only conscious self-models (6). In this chapter, we use the term *self* as an epistemological neutral umbrella term for denoting the phenomenon of a self-conscious first-person experience.

Our *self* is what Damasio called "the feeling of what happens" (7), not simply on a sensory level, but on levels of cerebral function that involve concomitant temporal and intentional interpretation, as relevant to each person's history and anticipations of the future. However, consciousness is not only dependent on distinct brain states, as perception and cogitation also hinge upon schemes and representations that are created, shaped, reinforced, and changed by cultural matrices composed of symbolic and linguistic patterns. Thus, this abstract, physically embodied, and culturally embedded *self* allows metarepresentations of what we are, allows interactions with others, enables deciphering of societal signals, and becomes the basis of interacting with the world at large.

PAIN, THE "SELF," AND SPIRITUALITY

Given that the physical process of pain can affect the brain substrates of consciousness, and thereby affect this self, we may view pain as a phenomenal event that can trap the person within a lived body to which that person has become disattuned, and limits the capacity for other experiences of the inner and outer environment that constitute each person's life-world (8). Notably, Sigmund Freud, suffering from jaw cancer that caused the last 16 years of his life to be spent in ever-increasing and ceaseless pain, described his state in 1939 shortly before he died as follows: "my world is…a small island of pain floating on an ocean of indifference" (9). As in Freud's case, the existential reality of (chronic) pain can be both an intrinsic part of the person and a way of being in the world—it becomes a part of the self, may define the self, and can become something greater than the self.

As spiritual experiences are often associated with crises and suffering (10), one could also speculate that pain may not only evoke a spiritual experience but could potentially be regarded as a distinct category of spiritual experience—taken here in the strictest sense as a liminal or transliminal, sublime occurrence—based upon definitions of (1) the spiritual as the essence of being and the qualities that bring significance, purpose, or meaning to life; and (2) spiritual experiences as extraordinary conscious events that are profound, difficult to objectify, may be polyvalent, and are often ineffable (11). For example, Grof and Grof reported that spiritual experiences can occur during, or may be precipitated by, illness and suffering (12). We agree with Grof and Grof, and we would like to go one step farther and offer that the nature of illness and suffering might also directly elicit a spiritual experience, as a phenomenally relevant property of brain–mind, both in response to and to affect bodily states.

Although spiritual experiences are often contextualized within religious beliefs and practices, it is important to note that (1) these events can and do occur in nonreligious persons (e.g., during meditative states, environmental interactions, and epiphanic moments) (13,14); thus, (2) the constellation of phenomenal features and many of the putative physiologic effects have been described by and reported in both religious and nonreligious individuals. Clearly, such experiences are subjectively powerful, and accumulating evidence suggests that such impact reflects both the involvement of hierarchical neural and other physiological systems, and potentially health-promoting (salutogenic) effects that these experiences respectively entail and obtain (15).

SUPPORT FROM OUR EMPIRICAL RESEARCH

Our research has shown that individuals with regular spiritual practice not only report more spiritual experiences, experiences of ego loss, and visionary dream experiences than spiritually nonpracticing individuals, but they also evaluate these experiences as significantly more positive (13,14). Particularly, our data suggest that experiences of ego loss seem to be tied to positive spiritual experiences as their phenomenological correlate. This emotional ambiguity of spiritual experiences is well known in spiritual traditions as boundaries of the self change. It may be perceived either as a joyful and expansive enrichment of the self, or as a terrifying experience of ego loss, depending on the self concept and corresponding worldview of the individual (16). Interestingly, two of our main findings indicate that regular spiritual practice of any sort may actually alleviate the negatively emotional impact that arises from the perceived threat of such deconstructive experiences. Supporting this, our work has demonstrated the following:

1. A comparison of spiritually practicing and nonpracticing individuals by means of a linear regression analysis revealed different pathways from experiences of ego loss to psychological distress (14). Although spiritually practicing individuals reported more exceptional and spiritual experiences, they accounted only for 7% of psychological distress (as measured with the Brief Symptom Inventory, BSI) in the spiritually practicing sample, but for 36% of distress in individuals with lack of spiritual practice. It is thereby noteworthy that experiences of ego loss had no effect on psychological distress in the group of individuals with regular spiritual, contemplative, or meditative practice, while they exhibited significant impact on distress in individuals with lack of spiritual practice. Based on these findings, we suggested that spiritual practice could be considered to be a specific coping strategy for the distress caused by experiences of ego loss.
2. There was a significantly lower test–retest reliability for the BSI in the spiritually practicing sample after 6 months (17). Thus, individuals engaged in spiritual practice seem to perceive distress as temporary states rather than permanent traits. This finding argues against the employment of distress scales as a single criterion for assessing the effects of pain in spiritually practicing individuals, or, alternatively, one would need information about an individual's regular spiritual practice to interpret the results correctly.

NATURE AND BRIEF ANTHROPOLOGY OF SPIRITUAL AND RELIGIOUS PRACTICES

Given that spiritual experiences of some sort can occur in individuals irrespective of secular or religious orientations, it may be that their profound nature (and observed or recognized salutogenic effects) became progressively codified into oral and later written cultural traditions that sought not only to preserve their positive effects, but also to provide explanatory models in accordance with metaphysical interpretations of self-awareness and the environment (18). For a discussion of progressive capacity for understanding, and codification of salutary practices, see Marshack (19).

The evolution of the human brain afforded progressive development of humans as a symbolic and linguistic species, therefore, the preservation and culturally reinforced augmentation of spiritual experiences could be seen as providing both significant ecological and survival advantage (20). The organization of religious practices, in all their diversity, served the positive effects of unifying rituals and meanings so as to fortify the health practices and reduce existential trepidations within common groups of individuals in particular demographic circumstances (21). In this light, note that the word *religion* is based on the Latin *relego*, to gather or bind together. In other words, the beneficial psychological (cognitive, emotional, and motivational) and physiological effects of spiritual experiences were bound together by religious practices as fundamental forms of communitarian public health (15).

The positive effects of these experiences (and durability of the diverse practices used to induce or interpret them) may reflect activity in brain areas involved in the perceptual and emotional dimensions of consciousness, as well as the capacity of conscious process to alter brain-mediated changes in sensation, apperception, and, perhaps, physiological state that incurred increased hardiness and resistance (i.e., through enhanced immunologic and endocrine function) and which were hereditarily (i.e., both genetically and culturally) maintained and preserved. It has been hypothesized that spiritual experiences and their codification into religious practices afforded some type of benefit by reducing allostatic load; thus, spiritual practices may be seen as conferring some form of culturally codified salutary benefit and ecological facilitation (22).

PUTATIVE NEURAL SUBSTRATES OF SPIRITUAL EXPERIENCES

A considerable body of literature has been devoted to elucidating neural structures that might be involved in spiritual experiences and effects. Most notable are the studies of D'Aquili and Newberg (23) and Austin (24). Lee and Newberg summarized the possible relevance of neural substrates of spiritual experiences to those subserving pain and analgesia (25). Working with our group, Peter Moskovitz (26) proposed a neurobiological model of chronic pain and suffering in which activation of hierarchical networks conjoining reticular, thalamocortical, and thalamolimbic axes are also involved in other, more positively valent conscious states and emotions, such as empathy and altruism.

Our work has focused on a neurophysiological perspective to sustain the possibility that pain and spiritual experiences are events of hierarchical neural processing that mediate both an underlying *state* or *feeling* of the conscious *self*, and are concomitant to more attentive *awareness* or consciousness *of* experience that are directly self-referent (15,27). Thus, might it be that pain can be or evoke a spiritual (i.e., liminal or transliminal) experience? And if so, how might this occur?

In brief, the strong activation of brain stem pathways can engage ascending input to the (ventroposterolateral) thalamus to activate the sensory and associative cortices, as well as numerous linking pathways to the limbic system (including the insula, cingulate, parahippocampal gyrus, hippocampus, septal nuclei, and amygdala). Of note is that activation of the amygdala, particularly nuclei of the right amygdala, appears to evoke strongly negative emotions including fear, distress, and anxiety (28). Of interest is that reticular activation increases sympathetic output and concomitantly stimulates the (paraventricular nucleus of the) hypothalamus to secrete corticotropin-releasing hormone (CRH) to engage the pituitary–adrenal axis, resulting in increased production and release of cortisol, thereby augmenting the stress response and sensations of apprehension (29).

However, continued activation of the right amygdala may induce a spillover effect that engages the nuclei of the left amygdala to alter patterns of limbically dependent brain–mind networks and evoke positively connoted feelings of *wellness* (28). In addition, it has been suggested that the right–left amygdalar shift can decrease hypothalamic release of CRH (and, therefore, diminish hypothalamic-pituitary-adrenal output) and disinhibit activity of the right lateral prefrontal cortex, altering thalamocortical and corticolimbic activity to produce a sense of expectational gain, and stimulating the arcuate hypothalamic nucleus to release beta-endorphin at subcortical, limbic areas, and into the systemic circulation (25,30).

Even though the neural effects of beta-endorphin are well-known (e.g., engagement of dopaminergic substrates to evoke feelings of euphoria, satiety, and reinforcement, as well as analgesic effects in concert with other opioid and nonopioid, centrifugal mechanisms) (31), endorphinergic effects on immune organs and cells and on smooth muscle of several tissues have been increasingly validated (32,33). Thus, we believe that pain can *be* a spiritual experience and *evoke* spiritual experiences, and that such spiritual experiences may affect brain state to affect pain (and suffering). In this model, brain state affects the properties of consciousness; this incurs *top-down* effects such that these properties can *feed back* to affect the activity of brain networks; and these top-down effects can, at least within limits, alter conditions of brain and engage neuroendocrine and neuroimmunological systems to alter physiologic state; this reengages afferent neuraxes subserving the internal state; and, ultimately, evokes sensation and perception of the internal and external dimension.

PATIENTS' SPIRITUALITY AND ITS IMPORTANCE FOR THE CLINICAL PRACTICE OF MEDICINE

For some persons, pain—particularly intense or chronic pain—may activate pathways that lead to less ego-centered sensory bodily perception, consequently producing profoundly spiritual effects. This has been contextually referred to as spiritual emergency (12). For others, the evocation of spiritual experience through religious or secular rituals and practices may serve to engage brain–mind mechanisms that attenuate or modify the perception of the nocuous nature of pain (15,25,34). In light of this, we have argued that spiritual experiences and practices are important and potentially relevant to clinical medicine (15), in general, and to the management of chronic conditions, as focal to pain and suffering, more specifically. Spiritual experiences and the rituals and practices to induce them were historically preserved and codified in ways that reflected the mimetic and mythic (i.e., oral and written) nature of human communicative tradition (35). The contemporary understanding of self and world in theoretical terms is such that health practice and medicine have become almost exclusively applications of scientific knowledge, and the explanatory paradigms of medicine reflect this value-ladeness (36). Thus, any attempt to validate the potential benefit of spiritual experiences and practices within the medical model requires a theoretical understanding to initiate realistic appraisal and evaluate possible utility.

However, we believe that while science is critical to medicine, medical practice is not simply applied science but is a humanitarian tradition and endeavor grounded in beneficence (similar to most forms of spiritual practice). We support Pellegrino's view of beneficence as conditionally fourfold: certainly, the biomedical good must be ensured, but the humanitarian dimensions of medicine as a healing profession rely upon consideration and sustenance of the good that upholds the patients' choices, their inherent (human) quality, and acknowledgment of persons as spiritual beings (37).

On the first condition, we maintain that it is important that clinicians acknowledge that the biomedical dimension reflects unique, ongoing interactions of several physiologic systems, and pain specifically reflects a biological process that is expressed and affected by psychological and social domains of the patient's life-world. Thus, it is crucial to develop and implement technically competent medical care that accommodates specific needs in these domains, as required and viable within the clinical encounter. In this regard, it is equally important that clinicians recognize the putative physiologic mechanisms, and psychological and social effects of spiritual experiences, beliefs, and practices, and how these are affected by, and affect pain and suffering. This relates to the second and third conditions—namely, that patients are human, and as such have choice preferences that reflect the particularities of their life circumstances and experiences; personal dignity is very often dependent upon the ability to maintain control of idiosyncratic practices and attributes that are of value to the self, and there is intrinsic dignity to patients as sentient, self-conscious living beings, who can and often do manifest a spiritual sense or a belief in a transcendental principle. Patients may wish to engage their spirituality both as salutary practice and as means to maintain or enhance their perceived locus of control and preservation of self-integrity. Obviously, this speaks to the fourth condition—that the spiritual good of the patient must be a fundamental element of beneficent care.

SPIRITUALITY AND PRACTICAL PAIN MANAGEMENT

But clinicians' approach to patients' spirituality remains a contentious issue, and there are equivocal opinions ranging from advocating clinicians' complete separation from patients' spiritual issues (38) to supporting clinicians' direct involvement in patients' spiritual practices (39). A recent study, for instance, examined a stratified cross-sectional random sample of over 1,000 U.S.-physicians' attitudes toward religious and spiritual topics in the clinical encounter (40). While 91% of the physicians deemed it appropriate to discuss issues related to spirituality and religiosity if brought up by the patient, 45% of the physicians believed that it is usually or always inappropriate for physicians to inquire about these topics of their own accord. In contrast, we have opined that given the potential influence that spiritual experiences may exert over physiologic and psychosocial aspects of patients' pain (and response to care), it is important to assess each patient's spiritual enfranchisement and the expression and scope of this spirituality in secular or religious practices (15). Moreover, as there is evidence that spirituality may change the way stress and pain are perceived, we believe that questions about spirituality should be included in standard clinical assessment procedures (i.e., a *spiritual history*). Byrum and Materson provided an excellent overview of such assessment as relevant to pain patients (41). This is important not only toward evaluating patients' spiritual needs, but also to evaluating whether particular beliefs or practices may exert potentially negative influences, such as diverting patients from urgently needed care or generating emotional anguish (through guilt or fear) (25,42).

This brings up a number of ethical issues, some obvious, and others a bit more implicit. Of course, the most obvious are those situations in which patients' religious orientations and beliefs proscribe against particular medical treatments (e.g., the classical ethical scenarios of the Jehovah's Witness refusing blood products despite a life-threatening hemorrhage, the Christian Scientist eschewing medical care, etc.) that may place patients' wishes in apparent conflict with the beneficent intentions of the clinician. At first, a respect for patient autonomy—in the true sense, as the negative right of a rational, competent person to refuse treatment—would seem to prevail. Yet, the notion that one person's autonomy cannot trump another must also be considered, particularly in light of the moral affirmations, obligations, and responsibilities of medical practitioners. As well, even though morals and ethics are not equivalent to law, the medicolegal aspects of these decisions must be regarded in the circumstantial analysis that constitutes prudent decision making. Often, a casuistic approach is used in such cases, together with other ethical approaches (such as the use of midlevel principles, deontologic ethics, or agent-based approaches, such as virtue ethics).

A more implicit moral and ethical issue involves the question of the existential loss versus existential gain involved in the experience of pain, suffering, or relief from these conditions. With the rise of modern medicine, spiritual approaches to coping with and understanding distress have been largely abandoned, perhaps with the exception of psycho-oncology and nursing of terminally ill patients. Instead, distress has been defined by mainstream conceptualizations as a negative phenomenon, consisting of a physical and a psychological component only. However, for some, the burden of pain and suffering may be interpreted (in secular or religious terms) as enhancing the capacity to "know" life, and viewed as an enriching event. For others, current developments in medical science and technology offer means to reduce or relieve pain and suffering on a variety of levels, and sustain claims that alleviation of any suffering is a means to existential gain, by enabling a "fuller" ability to flourish in diverse areas of life (for example, activities, communication, etc.).

Obviously, this issue is beyond the scope of the present discussion, yet these considerations are important so as to prevent a myopic view of this subject and its implications. For more discussion, see Kass (43) and Fuchs (44).

We previously stated (15), and reiterate here, that even though assessment and respect for patients' spirituality are certainly obligations of medical care, the direct involvement in patients' spiritual practices is beyond (most) clinicians' professional expertise, and such fusion of personal beliefs and practices with the professional role may diminish the equanimity (i.e., compassionate competence and discretionary spacing) that defines the clinical encounter, and the medical relationship at large.

Of course, there are dually qualified professionals who are clinicians as well as members of the clergy. Still, professional capability in two domains does not necessarily mandate simultaneously acting in these roles. Rather, the dually qualified professional should act within the boundaries and expert knowledge of their role as requested by the patient, and at all times act in ways that are consistent with the patient's best interests.

To be sure, we live in a spiritually (and religiously) polyglot society. Physician-philosopher Daniel Sulmasy has offered guidelines on how clinician–patient interactions may occur in circumstances of similar or dissimilar spiritual and religious beliefs (45). Summarily, these may be described as exercising prudence, reverence, some degree of personal restraint, and effacement of self-interest. As with any medical decision or act, recognition of patients' needs that are beyond the scope of clinical practice mandate referral to and conjoinment of appropriate resources that are collaborative in the patients' best interests. This reinforces the need for multidisciplinary pain management that is inclusive of clergy to address patients' religious needs, and other practitioners and resources that allow patients to gain more secularly expressed spiritual experiences and enrichment (e.g., optimized healing environments, provision of meditative or contemplative spaces, etc.).

CONCLUSION

The assessment of patients' spirituality, acknowledgment of the effects of and effects upon pain, and utilization of pluralist resources to accommodate patients' spiritual needs reflect our most current understanding of the physiology, psychological, and sociocultural aspects of spirituality and spiritual experiences (regardless of religious or secular expression).

We assumed a somewhat secular stance based upon a natural philosophy, but this in no way dispels the notion of the transcendent, the power of belief—whether secular or religious—and the importance of these to the human condition. As Christian Smith noted,

> One…continually places one's faith in premises, assumptions and suppositions that cannot be objectively substantiated or justified without recourse to other believed-in premises, assumptions and presuppositions. Everyone—the secularist and nonreligious included—is a believing animal, ultimately a person of faith. (21)

Perhaps the power of medicine resides in such faith, whether it is in the abilities of modern science and technology to deliver new and novel means of therapeutics, in the trust that the clinician will act for the good of the patient, or in the capacity for practical pain care to be scientific, yet nonetheless morally humanitarian.

REFERENCES

1. Giordano, J. 2006. Understanding pain as disease and illness: Part 1. *Prac Pain Manage* 6(6): 70–73.
2. Giordano, J. 2006. Pain as disease and illness: Part 2—Structure and function of the ethics of pain medicine. *Prac Pain Manage* 6(7): 65–68.
3. Hardcastle, V. 1999. *The myth of pain.* Cambridge: MIT Press.
4. Hofstadter, D. 2007. I am a strange loop. *Seed* 2(9): 70–72.
5. Giordano, J. 2007. A big picture: Neurogenesis, pain and the reality and ethics of pain medicine. *Prac Pain Manage* 7(2): 37–52.
6. Metzinger, T. 2004. *Being no one: The self-model theory of subjectivity.* Cambridge, MA: MIT Press.
7. Damasio, A. 1999. *The feeling of what happens: Body and emotion in the making of consciousness.* New York: Harcourt.
8. Resnick, D.B., Rehm, M., and R.B. Minard. 2001. The understanding of pain: Scientific, clinical, cultural and philosophical factors. *Med, Healthcare Philos* 4: 277–288.
9. Schur, M. 1972. *Freud: Living and dying.* New York: International Universities Press.
10. Wardell, D.W., and J.C. Engebretson. 2006. Taxonomy of spiritual experiences. *J Relig Health* 45(2): 215–233.
11. Koenig, H., McCullough, M.E., and D.B. Larson. 2001. *Handbook of religion and health.* Oxford: Oxford University Press.
12. Grof, S., and C. Grof. 1989. *Spiritual emergency: When personal transformation becomes a crisis.* Los Angeles: Jeremy Tarcher.
13. Kohls, N. 2004. *Aussergewöhnliche Erfahrungen—Blinder Fleck der Psychologie? Eine Auseinandersetzung mit aussergewöhnlichen Erfahrungen und ihrem Zusammenhang mit geistiger Gesundheit.* Münster: Lit-Verlag.
14. Kohls, N., and H. Walach. 2006. Exceptional experiences and spiritual practice—a new measurement approach. *Spirituality and Health Int* 7(3): 125–150.
15. Giordano, J., and J. Engebretson. 2006. Neural and cognitive basis of spiritual experience: Biopsychosocial and ethical implications for clinical medicine. *Explore* 2: 216–225.
16. Otto, R. 1958. *The idea of the holy—An inquiry into the non-rational factor in the idea of the divine and its relation to the rational.* Oxford: Oxford University Press.
17. Kohls, N., and H. Walach. 2008 Psychological distress, experiences of ego loss and spirituality: Exploring the effects of spiritual practice. *Soc Behav Pers* 36: 2.
18. Boyer, P. 1994. *The naturalness of religious ideas: A cognitive theory of religion.* Berkeley: University of California Press.
19. Marshack, A. 1990. Early hominid symbolism and the evolution of human capacity. In: *The Emergence of Modern Humans*, ed. P. Mellars, 457–498. Edinburgh: Edinburgh University Press.
20. Knight, C., Powers, C., and I. Watts. 1995. The human symbolic revolution: A Darwinian account. *Cambridge Archeolog J* 5: 75–114.
21. Smith, C. 2003. *Moral, believing animals.* Oxford: Oxford University Press.
22. Atran, S. 2006. Religion's innate origins and evolutionary background. In: *The innate mind, vol 2: Culture and cognition.* eds. P. Carruthers, S. Laurence, and S. Stich, 302–318. Oxford: Oxford University Press.

23. D'Aquili, E.G., and A.B. Newberg. 1999. *The mystical mind: Probing the biology of religious experience*. Minneapolis: Fortress Press.
24. Austin, J.H. 2001. *Zen and the brain*. Cambridge: MIT Press.
25. Lee, B.Y., and A.B. Newberg. 2005. Religion and spirituality in pain management. In: *Weiner's pain management: A guide for clinicians,* 7th. ed., eds. M.V. Boswell and E.B. Cole, 1473–1490. Boca Raton, FL: Taylor & Francis.
26. Moskovitz, P.A. 2006. A theory of suffering. *Pain Pract* 16(1): 74–81.
27. Prinz, J. 2004. A neurofunctional theory of consciousness. In: *Philosophy and neuroscience*, ed. A. Brook and K. Akins, 381–396. Cambridge: Cambridge University Press.
28. Halgren, E. 1992. Emotional neurophysiology of the amygdala within the context of human cognition. In: *The amygdala,* ed. J.P. Aggelton, 191–228. New York: Wiley.
29. Ziegler, D.R., Cass, W.A., and J.P. Herman. 1999. Excitatory influence of the locus ceruleuas in hypothalamic-pituitary-adrenocortical axis responses to stress. *J Neuroendocrinol* 11(5): 361.
30. Kiss, J., et al. 1997. Metabotropic glutamate receptor in GHRH and beta-endorphin neurons of the hypothalamic arcuate nucleus. *Neuroreport* 8(17): 3703–3706.
31. Giordano, J. 2005. Neurobiology of nociceptive and anti-nociceptive systems. *Pain Physician* 8(3): 277–291.
32. Stefano, G.B. 1989. Role of opioid neuropeptides in immunoregulation. *Prog Neurobiol* 33(2): 149–159.
33. Illes, P., and C. Farsang (eds.). 1988. *Regulatory roles of opioid peptides*. Weinheim: VCH Verlagsgesselschaft.
34. Wachholtz, A.B., and K.I. Pargament. 2005. Is spirituality a critical ingredient of meditation? Comparing the effects of spiritual meditation, secular meditation, and relaxation on spiritual, psychological, cardiac, and pain outcomes. *J Behav Med* 28(4): 369–384.
35. Donald, M. 2001. *A mind so rare: The evolution of human consciousness*. New York: Norton.
36. Hofman, B. 2001. On the value-ladenness of technology in medicine. *Med, Healthcare and Philos* 4(3): 335–345.
37. Pellegrino, E.D. 1987. *For the patient's good: The restoration of beneficence in health care*. Oxford: Oxford University Press.
38. Vandecreek, L. 1999. Should physicians discuss spiritual concerns with patients? *J Religion Health* 38: 193–202.
39. Dossey, B.M., L. Keegan, and C.E. Guzzetta. 2003. *Holistic nursing: A handbook for practice*. 4th ed. Boston: Jones & Bartlett.
40. Curlin, F., M. Chin, S. Sellergren et al. 2006. The association of physicians' religious characteristics with their attitudes and self-reported behaviors regarding religion and spirituality in the clinical encounter. *Med Care* 44(5): 446–453.
41. Byrum, C.S., and R.S. Materson. 2005. Assessing patient spirituality: A compelling avenue for discovery. In: *Weiner's pain management: A guide for clinicians,* 7th ed., eds. M.V. Boswell and E.B. Cole, 1465–1471. Boca Raton, FL: Taylor & Francis.
42. Satterly, L. 2001. Guilt, shame, and religious and spiritual pain. *Holistic Nurs Pract* 15(2): 30–33.
43. Kass, L. (ed.). 2003. Beyond therapy: Biotechnology and the pursuit of happiness. *President's Council on Bioethics Report*. Washington, DC: President's Council on Bioethics.
44. Fuchs, T. 2006. Ethical issues in neuroscience: History and philosophy. *Curr Opin Psychiatry* 19(6): 600–607.
45. Sulmasy, D.P. 1997. *The healer's calling: A spirituality for physicians and other health care professionals*. New York: Paulist Press.

7 Expressions of Chronic Pain and Suffering in Western Art

Scott L. Karakas

CONTENTS

Art is the creation of forms symbolic of human feeling.

Susanne K. Langer, *Feeling and Form* (1953)

INTRODUCTION

As suggested in the above quote by a prominent philosopher of art in the 1950s, each of the arts represents a type of communicative activity between artist and audience, with the forms of expression varying from medium to medium. In the visual arts, this communication is accomplished through the creation and manipulation of visible and sometimes tactile forms, such as line, color, texture, space, mass, and volume. The artist selects and applies these visual forms to create patterns, which often portray meaning through the use of simple or complex symbols representing objects, ideas, and emotions. Visual forms, patterns, and symbols tend to be culturally specific, and allow art historians to identify when, where, and by whom an individual work was made. Because the arts consume human and material resources that could otherwise have been used for acquiring necessities such as food and shelter, artists have also tended to occupy specialized niches within those cultures, either as persons of relative wealth and authority, or through the patronage or purchases of such persons.

Chronic pain is a powerful human feeling, and yet relatively few depictions of the condition appear in the history of the visual arts. One reason for this may be that, at least up until the last 100 years or so, average human life spans have been rather brief, and pain has tended to be a short-term phenomenon, leading quickly to either recovery or death. Another may be that chronic pain is internal, and therefore invisible to others (1,2).* Still another may be that the depiction of such pain has not often seemed to meet the needs or desires of those with the financial resources to serve either artists, or patrons of the arts. Whatever the underlying reasons, it appears that the social, cultural, and economic conditions necessary for the production of art with this subject matter have until recently been relatively uncommon. Even so, a comprehensive examination of such works would be far beyond the scope of this single chapter. Instead, I hope to provide a small but representative selection of works of sculpture, painting, and other visual arts from the Western world which depict aspects of chronic pain and suffering, together with very brief discussions of the historical and cultural contexts within which each work was made.

The earliest representations of chronic human suffering appear in art of the Mediterranean world during the Hellenistic period, c. 300–30 BCE. During this time, patrons for large-scale artworks consisted of royals and nobles in successor kingdoms to the conquests of Alexander the Great, each comprising a tiny Greek-speaking elite ruling a large native subject population. While earlier Greek art had featured images of generalized youthful beauty based on mathematical ideals,† art of the Hellenistic period demonstrates an ever-increasing diversity of subject matter, with an emphasis on naturalism, exotic subject matter, and highly dramatic effects, including representations of extremes of youth and old age (3).‡ One well-known example is a sculptural type representing an old fisherman. This type is known from several Roman copies, including one in the collection of the Louvre in Paris.§ This particular figure is portrayed as if wading in shallow water, shuffling, stooped, and emaciated, with thinning hair, heavily lined face, and prominent varicose veins. Another figure, now in the collection of the Metropolitan Museum of Art in New York, depicts an old woman carrying a basket of fruits and vegetables (Figure 7.1).

The figure is depicted wearing fine clothing and an ivy wreath, indicating that she is on her way to a religious festival, possibly in honor of Dionysos, the Greek god of wine, vegetation, and theater. Like the old fisherman, the woman's stooped posture, wrinkled face and neck, and sagging breasts highlight her age and infirmity. In both images, the notion of suffering is visually conveyed through posture and expression—the figures' curved spines, stooped shoulders, and strained expressions appear to have been designed to render pain and infirmity visible and palpable, thereby allowing them to be perceived by the viewer. Images like these would have stood in sharp contrast to the idealized physical perfection found in representations of gods, athletes, and the nobility.

Reasons for the development of this type of representation in Hellenistic art are obscure and much debated, as are the contexts and conditions under which they were

* For a more comprehensive discussion of historical cultural responses to pain in the Western world, see Morris (1). For the most complete treatise to date on the expression of pain in Western art, see Spivey, N.J., 2001, *Enduring Creation*, Berkeley, University of California Press (2).

† See Doryphoros by Polykleitos, Roman copy of a bronze Greek original, c. 450–440 BCE. Marble, ht. 2.12 m. Museo Archeològico Nazionale, Naples. (An Internet search for the image title, artist, and museum information will provide a selection of online images for this work.)

‡ For further discussion of these figures and the issues involved, see Pollitt, J.J., 1986, *Art in the Hellenistic Age*, Cambridge, Cambridge University Press (3).

§ See Old Fisherman, Roman copy of a Hellenistic original, c. 200–150 BCE. Black marble and alabaster, ht. 1.22 m. Musée du Louvre, Paris. (An Internet search for the image title, artist, and museum information will provide a selection of online images for this work.)

FIGURE 7.1 *Statue of an old market woman*. Roman. Early Imperial, Julio-Claudian, 14–68 AD. Marble, Pentelic, h. 49 5/8 in. (125.98 cm), Rogers Fund, 1909 (09.39). The Metropolitan Museum of Art, New York. (Image Copyright © The Metropolitan Museum of Art/Art Resource, New York.)

originally presented. Some Roman copies appear to have served as decoration for villas and baths, where they may have served to help create a bucolic rural theme, or as reminders of the rustic virtues held by followers of Stoic and Cynic philosophy. The Hellenistic originals may have also served private ends, or possibly as votive offerings for religious festivals. Some scholars posit that the figures were meant to be viewed with sympathy, and that representation in monumental sculpture was intended to confer a degree of nobility on their suffering. Others have pointed out the dangers of projecting modern cultural assumptions and responses on people living in a very different place and time, and utilize contemporary literary sources to suggest that the figures were instead created to provide aristocratic patrons with images of human types to which they could feel physical, intellectual, and moral superiority. It is also possible that the figures could have conveyed both messages at once; their ambiguity and openness to personal interpretation are certainly in keeping with the diverse moral and intellectual climate of the time. In any case, these figures clearly represent early attempts to convey aspects of long-term physical infirmity and suffering in a visually compelling way.

A later flowering of this dramatic Hellenistic style appears in the first century AD in works commissioned by Roman emperors, most famously the *Laocoön* group from Rome.* Excavated in 1506 from a structure that had once been part of the *Domus Aurea* of Nero and later used as a residence by the Emperor Titus, the sculpture is described

* See *Laocoön and His Sons*, probably early first century AD. Attributed by Pliny to Athanodoros, Hagesandros, and Polydoros of Rhodes. Marble, ht. 1.84 m. Vatican Museums, Rome. (An Internet search for the image title, artist, and museum information will provide a selection of online images for this work.)

by Pliny the Elder, who attributed them to the artists Athanodoros, Hagesandros, and Polydoros of Rhodes (*Natural History* 36.37) (2,3).* The style is so similar to works from the second century BCE that many scholars have assumed that the *Laocoön* group also dated to the Hellenistic period. However, the discovery in 1957 of several sculptural groups signed by the same three artists in a grotto that functioned as a dining room in a villa for the Emperor Tiberius has led other scholars to conclude that the Rhodian artists' works were all created during the first century AD. According to the Roman poet Virgil, Laocoön was the priest who warned the Trojans against taking the huge wooden horse of the Greeks into the city, and who was killed, along with his sons, by a huge sea serpent sent by the sea god Poseidon (*Aeneid* 2.199–277). In the *Laocoön* group, the title figure is represented in muscular, heroic nudity, his body dramatically twisted as he and his sons attempt to fight off the murderous serpents. Most strikingly, his head is dramatically turned back and tilted upward; his face is contorted, with upturned eyes and deeply furrowed brows above a mouth contorted in an agonized scream. As Livy and other Romans viewed the Trojan Aeneas as the founder of their civilization, it appears that the dramatic suffering of Laocoön and his sons was intended to be viewed sympathetically by the viewers. Although depicting a short-term event, from the time of the sculpture's rediscovery, it has been viewed as a universal expression of human pain, suffering, and despair (4,5).†

Subsequent representations of sympathetic suffering appeared in Western art only after the Roman Empire officially adopted Christianity as a state religion in the early fourth century, and then only very gradually. Stories of the passion of Christ and the martyrdom of saints played an important role in the theological basis for the early Church, promoting the notion of suffering for one's faith as an important spiritual virtue. However, the earliest images of the Crucifixion and martyred saints typically depicted their subjects as if they did not feel any noticeable effects from their painful and gruesome ends. This may reflect a degree of discomfort among early Christians with the means of death, particularly crucifixion, which in the Roman world was reserved for slaves, who were considered the lowest of the low (2). At the same time, it also indicates the dominance of spiritual over physical values in the theology of the early Church, and may reflect a lingering tradition from classical Greek and Roman art, in conflating the concept of nobility with the representation of impassive and idealized physical beauty. A good example can be seen on a plaque from an ivory casket dating to c. 420, and now in the collection of the British Museum (Figure 7.2).

In this scene, the body of the crucified Christ is muscular and alert, his eyes open and a youthful, beardless face uplifted and calm. This forms a sharp contrast with the limp, lifeless form of Judas hanging from a tree, the 30 pieces of silver spilling from a sack on the ground at his feet. The symbolic message is clear: sinners and betrayers suffer death, while Christ and his followers triumph over it, raised to noble status by virtue of their superior spiritual nature.

---

* See also Spivey, N.J., 2001, *Enduring Creation*, Berkeley, University of California Press, pp. 25–34, 120–126, with notes (2).

† For discussion of artistic and literary responses to the *Laocoön* group, see Bieber, M., 1967, *Laocoön*, University of California Press (4); Brilliant, R., 2000, *My Laocoön*, University of California Press (5).

FIGURE 7.2 *Death of Judas and Crucifixion*, c. 420. Plaque from an ivory casket, 7.62 × 9.84 cm. British Museum, London. (Image Copyright © Trustees of the British Museum.)

Changes in the Christian representations of virtuous suffering began to appear during the tenth century in the Holy Roman Empire in north-central Europe. There, around the year 970, the Archbishop Gero commissioned a crucifix for the cathedral of Cologne (6). Carved from wood and painted, the crucifix was hung behind the altar.

In contrast with the early heroic representations, this image of the crucified Christ was depicted with a hanging head and slumped posture, his sagging weight indicated by wrinkles of skin stretching from his shoulders and chest down to his distended and sagging belly. Looking up into the face, the viewer was confronted with closed, deep-set eyes and a contorted, downturned mouth, intended to convey a new emotional intensity. This *expressionistic* figure appears to have been designed to elicit both sympathy and empathy, presumably to create a better sense of spiritual and emotional connection with the viewer. Earlier images of Christ were portrayed as supernaturally triumphing over death; this one emphasized the fact that Christ suffered as we would, suggesting a new focus on the redemptive (or at least human) quality of such suffering.

The increasing popularity of this new image of a suffering Christ on the cross is attested by its gradual spread over both the Latin and Eastern Christian worlds, as well as the appearance of a new, related form of image in northern Europe during the second half of the thirteenth century, conflating Christ with the Old Testament Man of Sorrows, as described in *Isaiah* 53.

Modeled after an icon from the city of Rome, images of the Man of Sorrows (see Figure 7.3) from Italy depicted Christ immediately following the Crucifixion, arms folded across his chest with blood flowing from the wounds in his hands, eyes closed, and head downcast in a manner reminiscent of the *Gero Crucifix* (7). The figure was designed to remind viewers of the relevant passages from both the Old and New

FIGURE 7.3 School, Umbrian. *The Man of Sorrows,* c. 1260. Part of a diptych. Egg tempera on wood, 32.4 × 22.8 cm. Bought, 1999 (NG6573). National Gallery, London, Great Britain. (Image Copyright © National Gallery, London/Art Resource, New York.)

Testaments, while also creating a heightened sense of identification and empathy, by bringing the figure of the suffering Christ down from the cross to confront the viewer directly. This and other such images were often set in diptychs for private devotion, paired with representations of the Virgin and Child.

In the diptych's facing leaf, the Virgin gazes through downcast eyes at both the playful infant in her arms and the crucified adult in the opposite panel, linking the two and creating a sense of compassion and foreboding. Through Mary, the viewer was encouraged to feel both sympathy and empathy with the suffering of her son.

Some later images of the Man of Sorrows appear to elaborate on this message, as in the panel painted in c. 1420 by Master Franck, a Dominican monk in Hamburg* (7). This image includes the cross in the background, bloody nails still in place. In the foreground, the figure of Christ is supported by a trio of angels with sad eyes and mournful expressions, who also bear miniaturized versions of the spear, sponge, and scourge. In this case, they stand in for the Virgin in creating a sense of sympathy for the figure of Christ, who is depicted still bleeding from the crown of thorns as his emaciated form hangs limply in the angels' arms, with one hand fingering the wound in his side while the other holds a corded whip. Most strikingly, Christ's sorrowful eyes are represented as open to catch the viewer's gaze. This appears to have been intended to symbolize the theological message of the resurrection to follow the crucifixion, the spiritual return of

* See Master Francke, *Man of Sorrows*, c. 1420. Tempera on wood panel, 42.5 × 31.5 cm. Museum der bildenden Künste, Leipzig. (An Internet search for the image title, artist, and museum information will provide a selection of online images for this work.)

life after death. At the same time, establishing eye contact was also designed to further enhance the sympathetic bond between viewer and subject.

An interesting related development is the eventual recognition of a saint specifically dedicated to those suffering prolonged afflictions, named Lidwina (or Lydwina) of Schiedam. According to *The Lives or the Fathers, Martyrs and Other Principal Saints* by Alban Butler, Lidwina was born in the Dutch town of Schiedam in 1380 (8), and from an early age evidenced a very strong religious devotion (9).* At the age of 15, she broke a rib while ice skating. Rather than healing, an abscess formed in the wound, placing her in severe and chronic pain, and leaving her bedridden for 38 years. During that time, Lidwina suffered a series of additional symptoms, including headaches, vomiting, fever, thirst, bedsores, toothaches, muscle spasms, and blindness, until she finally died on April 14, 1433. Lidwina's devotion and pious behavior during her lengthy period of suffering brought her to the attention of local citizens and Church officials, and her grave became a popular site of pilgrimage for people suffering from pain and sickness. In 1890, she was formally confirmed by Pope Leo XIII as a patron saint for both ice skating and chronic sickness, with her feast day held on April 14, the day on which she had died. In addition, some of the symptoms suffered by Lidwina have caused her to be identified as representing the earliest documented case of multiple sclerosis (10).

Lidwina is visually represented as a young girl falling while ice skating, or as an idealized bedridden female figure receiving a lily or rose branch from an angel.† In Christian iconography, both the rose and the lily are associated with the Virgin Mary as symbols of purity, chastity, and virtue, although both had also been appropriated from the earlier pagan goddess Aphrodite. The pose of the bedridden Lidwina also appears to be adapted from images of the Dormition of the Virgin, the earliest examples of which are found on Byzantine icons of the ninth century.‡ Although Lidwina is represented with eyes open and hands reaching to grasp the rose branch being offered to her by the angel, her prone pose and mantled cloak closely emulate those of the sleeping Virgin, suggesting a parallel between the forms of suffering experienced by the two figures. Through her suffering, Lidwina's faith is portrayed as bringing her closer to the divine suffering of Mary and her son.

Although a spiritually based iconographic tradition focusing on the redemptive nature of suffering continued in Western art for the next several hundred years, secular versions of the theme also begin to appear in the aftermath of the French Revolution of 1789. There, the new republican government raised social and popular concerns to the fore, and new generations of artists sought to create secular forms for themes that previously had been the domain of religious artists and patrons. The most famous example of this is *The Raft of the Medusa,* by the Romantic painter Théodore

* See also Verkerk, D., and H. Luttikhuizen, 2006, *Snyder's Medieval Art*, 2nd ed., Upper Saddle River, NJ, Pearson Prentice-Hall (9).

† See Koimesis (Dormition) of the *Virgin Mary*, late tenth century. Ivory, 18.4 × 14.6 cm. Metropolitan Museum of Art, New York. (An Internet search for the image title, artist, and museum information will provide a selection of online images for this work.)

‡ See Lidwina of Schiedam, from the *Vita Alme Virginis Lidwine* by Jan Brugman, 1498. Woodcut, Koninklijke Bibliotheek, The Hague, Netherlands. (An Internet search for the image title, artist, and museum information will provide a selection of online images for this work.)

Géricault (11,12).* From the beginning of the Renaissance in the late fifteenth century, artists had shown renewed interest in using classical Greco-Roman forms and techniques to monumentalize their subjects. In his huge painting, Géricault sought to combine these with intensely dramatic effects to draw attention to the events surrounding a shipwreck that had occurred 3 years earlier.

On July 2, 1816, the frigate *Medusa* had been run aground off the coast of North Africa, largely due to the incompetence of its captain, a political appointee under the newly reconstituted monarchy. When the ship began to break up, it was found that there were only enough lifeboats to hold 250 of the ship's 400 passengers and crew. The rest were relegated to a hastily built makeshift raft, which was soon abandoned by the lifeboats and left to drift in the open ocean. With no means of steering and limited food and water, the raft's occupants soon descended into armed conflict and cannibalism. When the survivors were finally rescued 12 days later, only 15 of the original 150 remained alive. Accounts by two of the survivors were eventually published, creating a huge scandal and popular outcry against the royalist Bourbon government.

For his work, finished in 1819, Géricault chose to create a large-scale history painting, depicting the dramatic moment when the survivors were attempting to signal the rescue ship on the distant horizon.† The scene is composed on a series of diagonal lines formed by the planks, mast, and sail of the raft, causing it to appear to jut into the viewer's space and bringing the viewer into the scene. The color palette is dark and subdued, with greens, browns, and tans predominating. The drama of the scene is intensified by the use of theatrical lighting, highlighted by sharp contrasts between intense light and deep shadow. The figures are arranged as a tangled jumble of bodies in a variety of dramatic poses, including a bearded male whose chin rests in his hand in a gesture of despair. Although Géricault had done extensive research on the shipwreck, including interviewing survivors and visiting hospitals and morgues to sketch corpses, he chose to minimize the physical signs of starvation, injury, disease, and death, preferring to heroize his figures with dramatic gestures and heavily muscled bodies. In this, he was emulating earlier classicizing artists such as Raphael, Michelangelo, and Caravaggio, and in this way, evoking Hellenistic depictions of suffering like the *Laocoön,* which had similarly idealized the suffering of its title character. What is new is the application of this monumental iconic elevation of the suffering of ordinary people, a pointed criticism against the perceived incompetence of the ruling political authority.

Another example of the dramatic portrayal of classicized suffering may be seen in a sculpture by Jean-Baptiste Carpeaux, entitled *Ugolino and His Sons* (Figure 7.4) (13).

Carved between 1856 and 1867, the work is based on Canto 33 from Dante's *Inferno,* in which the poet encounters Count Ugolino Ugolino dell Gheradesca, a nobleman who had been the chief magistrate of Pisa in 1284. Accused of having betrayed his city through negligence in battle, he was condemned and locked in a tower and left to starve with his two sons and two grandsons. The children offered their own bodies as food to Ugolino so that he might remain alive, but ultimately

* See also Spivey, N.J., 2001, *Enduring Creation*, Berkeley, University of California Press (2).

† See Géricault, Théodore (1791–1824) *Raft of the Medusa*, 1818–1819. Oil on canvas, 4.91 × 7.16 m. Musée du Louvre, Paris. (An Internet search for the image title, artist, and museum information will provide a selection of online images for this work.)

FIGURE 7.4 Carpeaux, Jean Baptiste (1827–1875). *Ugolino and His Sons*. 1867. Saint-Béat marble, h. 77 in. (1.95.6 m). Purchase, Josephine Bay Paul and C. Michael Paul Foundation Inc. Gift, and Charles Ulrick and Josephine Bay Foundation Inc. Gift, and Fletcher Fund, 1967 (67.250). The Metropolitan Museum of Art, New York. (Image Copyright © The Metropolitan Museum of Art/Art Resource, New York.)

all five starved to death. In Carpeaux's marble sculpture, Ugolino's compressed and contorted pose follows the established conventions for portraying despair, as also seen in the elderly figure from Géricault's painting (see above). The Count's agonized expression and deeply undercut eyes create an intensely dramatic portrait of mental anguish and physical suffering.

At the same time, the muscular bodies, dramatic gestures, and intense expressions of the figures of the children show clear influence from the works of Michelangelo, and also from Hellenistic models like the *Laocoön* group. Carpeaux heroized his figures with the idealized muscular nudity favored by the artistic establishment, rather than depict the unpleasant realities of their actual physical condition. As in Géricault's painting, this appears to have been done in order to engender the greatest degree of sympathy from the educated aristocrats who were the intended audience for the work.

During this period, artists of Social Realism used the medium of caricature to portray a wide variety of physical and mental diseases, in inexpensive graphic works created on printing presses, thus making the images and associated ideas available to a broader spectrum of increasingly urban European populations (14).* Among these

* See also Morris, D.B., 1991, *The Culture of Pain*, Berkeley, University of California Press (1).

FIGURE 7.5 Daumier, Honoré (1808–1879). *La Colique.* Lithograph, published in: *Le Charivari,* February 19, 1833. Robert D. Farber University Archives and Special Collections Department, Brandeis University, Waltham, Massachusetts. (Image Copyright © Robert D. Farber University Archives and Special Collections Department, Brandeis University.)

are a number of depictions of intense or chronic pain, such as Honoré Daumier's *La Colique* (Figure 7.5).

In this work, severe abdominal pain is depicted not only by the facial expression of the subject as he doubles over, clutching his stomach, but also by metaphorical imps and demons who pull at ropes and saws that go through his body, creating a concrete and readily identifiable visual expression of his experience of gnawing, searing pain, rendering visible the suffering that would otherwise have been hidden from external view. Devoid of pretensions to classical idealism, such images nevertheless utilize ancient and familiar religious notions of pain being caused by unseen supernatural forces, in order to illustrate these invisible ills in a manner that would be readily understood by the broadest possible audience.

In the later nineteenth century, Realist painters sought to depict the suffering of ordinary people in a relatively straightforward, naturalistic manner, devoid of any religious iconography or classicizing idealism. By this time in Europe, the traditional aristocratic and religious hierarchies were losing power and influence, and in addition to work in popular print media, some artists were utilizing more traditional and prestigious media such as oil paintings in the service of revolutionary political movements espousing the rights of the poor. Socialists, anarchists, and civil reformers cited epidemics of tuberculosis and other degenerative diseases as examples of failures and inequities in the existing social order. One striking example is a painting entitled *Sick Girl,* created in 1880–1881 by the Norwegian painter Christian Krohg (Figure 7.6).

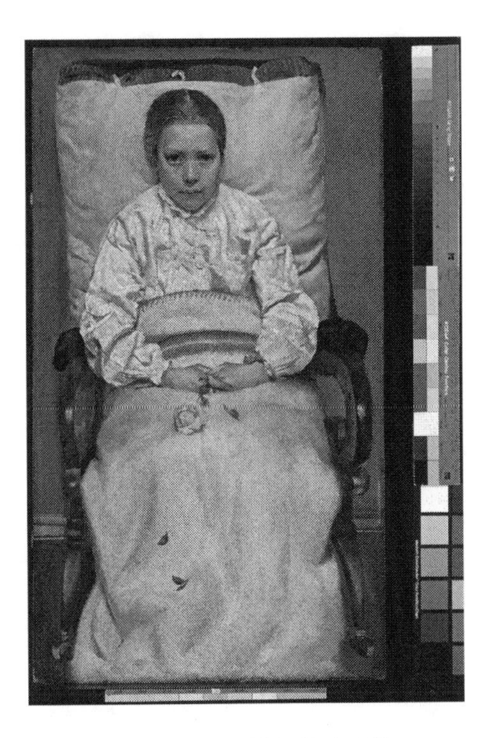

FIGURE 7.6 Christian Krohg, *Sick Girl,* 1880–1881. Oil on canvas, 103.5 × 51.4 cm. NG.M.00805. Nasjonalmuseet for kunst, arkitektur og design/The National Museum of Art, Architecture and Design, Oslo. (Image Copyright © The Munch Museum/The Munch-Ellingsen Group/ARS 2010.)

An author and journalist as well as a painter, Krohg was interested in depicting scenes from the lives of everyday people, as well as the darker aspects of contemporary urban life (15).*

In *Sick Girl,* the subject is a young patient in a tubercular ward, portrayed in the Realist style. Seated in a hospital chair, the girl's suffering is indicated by her fragile form, sallow complexion, and sad and sunken eyes that look directly at the viewer. Her frailty is highlighted by her small size in comparison with the rocking chair in which she sits, as well as the symbolic dying rose she holds, its leaves scattered across the blanket covering her legs, and its pallid color matching her own. Other symbolic features include the girl's blanket, pillow, and blouse, whose stark whiteness indicates not only her antiseptic hospital setting, but also provides reference to traditional images of purity and innocence. This contrasts sharply with the blood-red color of the blanket's trim decoration, which together with the rose, suggests her life draining away. The dull brown color of the wall in the background enhances the somber mood of the painting. Krohg's composition is designed to ennoble the figure of the girl, while also establishing a strong empathic connection with the viewer. He also idealizes his subject, to the extent

* See also Norwegian Realism (16).

FIGURE 7.7 Edvard Munch, *Sick Child,* 1885–1886. Oil on canvas, 119.4 × 119 cm. NG.M.00839. Nasjonalmuseet for kunst, arkitektur og design/The National Museum of Art, Architecture and Design, Oslo. (Image Copyright © The Munch Museum/The Munch-Ellingsen Group/ARS 2010.)

that he has chosen an attractive and engaging figure upon which to focus his work. This is designed to generate the maximum degree of sympathy for the condition of the subject, and by extension, all others suffering under similar circumstances, in order to effect social change.

This theme was engaged by Krohg's protégé Edvard Munch, in his painting entitled *Sick Child* (Figure 7.7).

Munch lost his mother to tuberculosis when he was only 5 years old, and his sister Sophie also died from the disease when he was 14 (17). As a result, Munch painted and sketched the same subject numerous times in a variety of media during his early career. Munch places the pallid young girl in her sickbed, in a pose reminiscent of Krohg's earlier work. Unlike Krohg, Munch adds the figure of the girl's mother, her head bowed so low in grief that her face is hidden from the viewer. The communication in the scene is between the girl and her mother, leaving the viewer as a voyeuristic interloper to an intensely personal and emotional moment. Here the sympathetic connection with the suffering of the subject is indirect, conveyed through her mother's expression of sorrow. Munch also sought to convey emotion by deliberately avoiding a naturalistic representation, instead breaking up the surface of the painting with a sharply drawn series of crosshatched lines in contrasting colors, using the formal techniques of line and color to viscerally convey the emotional intensity of the scene (12).

The Expressionistic use of line and color to depict suffering is carried even further in Munch's most famous work, *The Scream,* from 1893 (Figure 7.8).

Munch would eventually create nearly 50 different versions of this work, in a variety of media. Set on the road from Ekeberg to Oslo, overlooking the Oslo fjord, the painting depicts a contorted figure clasping his hands over his ears, eyes wide,

FIGURE 7.8 Edvard Munch, *The Scream,* 1893. Oil, tempera, and pastel on cardboard, 91 × 73.5 cm. NG.M. 00939. Nasjonalmuseet for kunst, arkitektur og design/The National Museum of Art, Architecture and Design, Oslo. (Image Copyright © The Munch Museum/ The Munch-Ellingsen Group/ARS 2010.)

and mouth open in horror, set against a churning red and orange sky.* According to Munch,

> I was walking along a path with two friends—the sun was setting—suddenly the sky turned blood red—I paused, feeling exhausted, and leaned on the fence—there was blood and tongues of fire above the blue-black fjord and the city—my friends walked on, and I stood there trembling with anxiety—and I sensed an infinite scream passing through nature. (18)

Munch utilizes intense contrasts in bright and dark color, as well as the swirling lines of the figure in the foreground and fjord in the background, to heighten the expression of horror on the face of the generalized and anonymous subject. This Expressionistic use of line and color also generate strong feelings of claustrophobia and anxiety, while the painting's lack of specific references have caused it to serve as a universal expression of modern angst and psychological suffering. As a representation of generalized horror, it has become one of the most widely recognized and copied paintings in the modern Western world.

By the mid-twentieth century, in response to Expressionistic works by Munch and others, some artists' work began to directly express their own physical pain and suffering. Most famous among these is the Mexican artist Frida Kahlo. Born in 1907, Kahlo suffered horrific injuries at age 18, as the result of a collision involving the

* For a thorough recent treatment of the painting and the artist, see Prideaux, S., 2005, *Edvard Munch*, New Haven, Yale University Press (18).

bus in which she was riding. These included serious fractures to her collarbone, ribs, pelvis, and spine, as well as being impaled on an iron handrail, severely damaging her uterus. From then until her death in 1954, Kahlo underwent 32 separate medical operations and suffered frequent periodic episodes of severe pain. It was reportedly during one of her frequent hospital stays that she took up painting.*

A work famously illustrating the artist's perception of her own suffering is *The Broken Column,* painted in 1944. Surrealist in its psychological style, this self-portrait depicts the artist as seminude, the lower half of her body hidden by a white drapery, and a body brace framing her torso. The body is represented as torn open, revealing a broken classical column in place of her fractured spine, which may also represent her disillusionment with traditional society in light of the Mexican Revolution. Her pain is represented by dozens of nails piercing her flesh, and tears pouring down her cheeks. At the same time, the artist depicts herself as standing proudly and looking the viewer squarely in the eye. The painting has been interpreted as expressing the duality of outer pride and inner pain, both in her physical condition and in her response to numerous affairs had by her husband and fellow artist, Diego Rivera. In Kahlo's own words, she is

> Waiting with anguish hidden away, the broken column, and the immense glance, footless through the vast path...carrying on my life enclosed in steel...If only I had his caresses upon me as the air touches the earth. (20)

In her painting, Kahlo reveals on canvas the pain that she attempted to conceal behind her outward countenance. In directly expressing the experience of physical and mental pain, Kahlo's autobiographical work marks a turning point in the visual representation of chronic pain and suffering.

In recent years, some chronic pain sufferers have begun to explore the possible therapeutic and educational benefits to creating art that visually represents the pain (21).† Studies suggest that various forms of art therapy can serve as effective communication tools between patient and practitioner, and can also reduce the symptoms of pain and anxiety associated with cancer and other prolonged illnesses (23,24).‡ A prominent online site for this is PainExhibit.com, established in 2001 by chronic pain sufferer Mark Collen as an educational forum and space in which fellow pain sufferers could express themselves through art (25). The inaugural image in the collection was submitted by Collen and is entitled *Chronic Pain.*§ The piece consists of a torn sheet of white paper over a black background. In the upper left corner is a small photographic portrait of the artist, his entire face covered by brown paper, representing his sense of pain and isolation. Only one of the artist's eyes is shown through a hole in the covering, making direct contact between the disturbing image and the viewer.

---

* For more on the artist, see Prignitz-Poda, H., 2004, *Frida Kahlo*, New York, Schirmer/Mosel (19). See also Watt, G., 2005, *British Journal of General Practice* 55(517): 646–647 (20).

† See also Flynn, B., 2006, Pain Monsters, *PainFoundation.com* (22); Art Therapy Can Reduce Pain and Anxiety in Cancer Patients, *ScienceDaily.com* (23).

‡ Examples of recent study results include Luzzatto, P., V. Sereno, and R. Capps, 2003, *Palliative and Supportive Care* 1: 135–142 (24). See also Collen, M.R., 2006, *PainExhibit.com* (25).

§ See Mark Collen, Chronic Pain, 2001, Image: http://www.painexhibit.com/about (accessed March 16, 2010). (Image Copyright © Chronic Pain Visual Arts Project.)

Below the photograph is an inscription that fills the lower half of the composition:

> *Falling*
> I think I have reached the bottom but I am standing on the wall
> Falling again into the abyss
> The taste of suicide is delicious in my brain
> Fantasies of death—
> Shooting myself in the head, hanging from the bedroom door
> There are moments when a ray of hope filters through the darkness
> But it is too black to make a difference
> I escaped before, but it found me again

The image and text personalize and *deprofessionalize* Collen's artwork, allowing the sufferer of chronic pain to express his subjective experience directly, without the mediating influence of a third party.

According to the Web site, Collen has received more than 500 entries from chronic pain sufferers throughout the world. From these, 70 works, in wide variety of artistic styles and media are included in the online exhibit, divided into thematic categories such as "Portraits of Pain," "But You Look So Normal," and "Hope and Transformation." The individual works range from relatively naturalistic representations heavy with metaphorical symbols, such as *Things Could Be Worse (Self Portrait with Pain)* by Gregory Maskwa* to highly abstract pieces like Georgia Davidson's *Tower of Pain/Tower of Strength*.†

Each of the Pain Exhibit artworks is highly expressionistic, seeking to visualize extremely personal and deeply felt emotions and experiences, with the dual purpose of therapeutic self-disclosure and public education, a combination for which the communicative medium of the Internet seems particularly effective. Many of the works of art also employ stylistic and symbolic techniques explored by artists discussed earlier in this chapter, such as Francke, Daumier, Munch, and Kahlo. In doing so, they represent the most recent developments in the long history of interpretations of chronic pain and suffering in Western art. Thus, it is clear that art, like pain, remains a durable aspect of the human condition, and that depictions of pain in art will be iterative and illustrate interpretations and expressions of pain, and in this way continue to serve as a bridge between subjective suffering and subjective apprehension—by others and society at large.

REFERENCES

1. Morris, D.B. 1991. *The culture of pain*. Berkeley: University of California Press.
2. Spivey, N.J. 2001. *Enduring creation: Art, pain, and fortitude*. Berkeley: University of California Press.

---

* See Gregory Maskwa, *Things Could Be Worse* (Self Portrait with Pain). Acrylic on masonite, 101.6 × 76.2 cm. Image: http://www.painexhibit.com/ag108_Maskwa (accessed March 16, 2010). (Image Copyright © Chronic Pain Visual Arts Project.)

† See Georgia Davidson, *Tower of Pain/Tower of Strength*. Paper and glue, 34.3 × 19 cm. Image: http://www.painexhibit.com/ag907_Davidson (accessed March 16, 2010). (Image Copyright © Chronic Pain Visual Arts Project.)

3. Pollitt, J.J. 1986. *Art in the Hellenistic Age.* Cambridge: Cambridge University Press.
4. Bieber M. 1967. *Laocoön: The influence of the group since its rediscovery* (rev. ed.). Detroit: Wayne University Press.
5. Brilliant, R. 2000. *My Laocoön: Alternative claims in the interpretation of artworks.* Berkeley: University of California Press.
6. Verkerk, D., and H. Luttikhuizen. 2006. *Snyder's Medieval Art* (2nd ed.). Upper Saddle River, NJ: Pearson Prentice-Hall.
7. Camille, M. 1996. *Gothic art: Glorious visions.* New York: Harry N Abrams, pp. 116–119.
8. Butler, A. 1999. *Butler's lives of the saints: April* (rev. by P. Doyle). Collegeville, MN: Liturgical Press.
9. B. Lidwina, commonly called Lydwid, V. 1380–1433. 2006. Accessed at http://www.ewtn.com/library/MARY/LIDWINA.htm, March 16, 2010.
10. Medaer, R. 1979. Does the history of MS go back as far as the 14th century? *Acta Neurologica Scandinavica* 60: 189–192.
11. Eitner, L. 1972. *Géricault's raft of the Medusa.* London: Phaidon.
12. Eisenman, S. et al. 2002. *Nineteenth century art: A critical history.* London: Thames & Hudson.
13. *Ugolino and His Sons.* 2010. Accessed at http://www.metmuseum.org/toah/hd/carp/hod_67.250.htm, March 16, 2010.
14. Stafford, M.S. 1985. From "Brilliant Ideas" to "Fitful Thoughts:" Conjecturing the unseen in late eighteenth-century art. *Zeitschrift für Kunstgeschichte* 1985; 48 Bd., H. 3: 329–363, esp. 337.
15. Christian Krohg (1852–1925). 2000. Accessed at http://www.artnet.com/library/04/0480/T048038.asp, March 16, 2010.
16. Norwegian Realism. 1999. Accessed at http://www.norway.org/aboutnorway/culture/painting/realism/, March 16, 2010.
17. Beller, G.A. 2006. Edvard Munch's renditions of illness and dying: Their message to contemporary physicians. *Journal of Nuclear Cardiology* 13(3): 309–310.
18. Prideaux, S. 2005. *Edvard Munch: Behind the scream.* New Haven: Yale University Press.
19. Prignitz-Poda, H. 2004. *Frida Kahlo: The painter and her work.* New York: Schirmer/Mosel.
20. Watt, G. 2005. Frida Kahlo. *British Journal of General Practice* 55(517): 646–647. Accessed at http://www.pubmedcentral.nih.gov/articlerender.fcgi?artid=1463226, March 16, 2010.
21. Sullivan, M. 2004. Subjective made object: Pain in contemporary art. *APS Bulletin* [serial online]. July/August, 14(4). Accessed at http://www.ampainsoc.org/pub/bulletin/jul04/path1.htm, March 16, 2010.
22. Flynn, B. 2006. Pain monsters: Art therapy and chronic pain. Accessed at http://www.painfoundation.org/learn/library/pain-topics/complementary-medicine/art-therapy/pain-monsters-art-therapy.html, March 16, 2010.
23. Art therapy can reduce pain and anxiety in cancer patients. Accessed at http://www.sciencedaily.com/releases/2006/01/060102104539.htm, March 16, 2010.
24. Luzzatto, P., V. Sereno, and R. Capps. 2003. A communication tool for cancer patients with pain: The art therapy technique of the Body Outline. *Palliative and Supportive Care* 1: 135–142.
25. Collen, M.R. 2006. *PainExhibit.com.* Accessed at http://www.painexhibit.com/aboutexhibit.html, March 16, 2010.

8 Maldynia as Muse
A Recent Experiment in the Visual Arts and Medical Humanities

Nathan Carlin and Thomas Cole

CONTENTS

Every picture is a picture of the body.

James Elkins, *Pictures of the Body*, 1999 (1)

A minister is not a doctor whose primary task is to take away pain. Rather, he deepens the pain to a level where it can be shared.

Henri Nouwen, The Wounded Healer,
***Images of Pastoral Care*, 2005 (2)**

INTRODUCTION

The invitation to write on the topic of pain and art immediately brought to mind the multifaceted nature of pain and suffering, as well as the various media of art. What aspect of pain should we consider, and what kind of art? After a few weeks of reading and discussion, we began to see profound limits of language in communicating about pain. Pain resists language even as it demands interpretation and expression. Virginia Woolf puts it this way: "English which can express the thoughts of Hamlet and the tragedy of Lear has no words for the shiver or the headache.… The merest schoolgirl when she falls in love has Shakespeare or Keats to speak her mind for her, but let the sufferer try to describe a pain in his head to his doctor and language at once runs dry" (3, p. 194). Severe pain, Elaine Scarry points out, "not only resists pain but actively destroys it, bringing about an immediate reversion to a state anterior to language, to the sounds and cries a human being makes before language is learned" (4, p. 4). These reflections led us to focus on the visual arts, hoping to learn about the potential of images to help span the seemingly unbridgeable gap between the one who suffers pain and the one who hears about pain.

Then there was the matter of choosing an aspect of pain to consider, and also the matter of defining pain. In keeping with the theme of this volume, we chose chronic, persistent pain, using James Giordano's reformulation of *maldynia*—understood as an *illness* (the experience of one who suffers) "of seemingly idiopathic, intractable, chronic pain and subjective suffering within bio-psychosocial contexts." Maldynia, according to Giordano, has emerged in the postmodern era because medical technology now permits an extended life span. But technocentric medicine is ill-prepared to address the biocultural consequences (e.g., maldynia) that have accompanied the epidemiological transition from acute to chronic disease.

What about the definition of pain? The International Association for the Study of Pain defines pain this way: "An unpleasant sensory and emotional experience associated with actual or potential tissue damage, or described in terms of such damage" (5). It was striking for one author (Carlin), who is a newcomer to the field of medical humanities, that pain here is understood as a sensory *and* emotional experience, which challenges the classic distinction between pain and suffering which is prominent in medicine and the humanities. The most prominent exposition of this distinction lies in the work of Eric Cassell, often cited for his 1982 *Journal of the American Medical Association* article "The Nature of Suffering and the Goals of Medicine," in a book by the same title published in 1991. Cassell argues that while medicine has been attending to pain and to bodies, it has not attended to suffering and to persons. Bodies feel pain, but *people* suffer, Cassell suggests. In one sense, Cassell is on the mark. Mainstream medicine still generally operates within the old Cartesian *mind/body* dualism. And attention to the experience of suffering remains all too rare. But within the specialty of pain medicine, late twentieth-century theories began to integrate psychological experience and neurological substrates in their definitions. Here, for example, is Merskey's 1976 comment: "Pain is a disagreeable experience which we originally associate with a bodily lesion, or describe in terms of tissue damage, or both simultaneously" (6, p. 332). As noted, the term *maldynia* here refers to the persistence of pain in face of the biotechnological advances in medicine. In some cases, the biotechnological advances of medicine have exacerbated pain and suffering.

If seeing is believing, one wonders how maldynia might be represented in art—that is, seen and made knowable. We are not experts in pain research, and we are not experts in the history of art. We bring different perspectives to this topic, perspectives we hope will enrich the understanding of pain in the medical humanities. One of the authors (Cole) is a trained historian, but he comes to this topic from the perspective of a medical humanist. He is currently the Director of the McGovern Center for Health, Humanities, and the Human Spirit at the University of Texas Health Science Center in Houston, Texas, and he has had a long interest in the arts. Cole is also a practicing Jew. The other author (Carlin) comes from a different perspective. He is trained as a Presbyterian pastor, and he received his Ph.D. in religion and psychology from Rice University, Houston, Texas. Carlin is a practicing Presbyterian, and his interest in the arts was sparked by Donald Capps's "The Lessons of Art Theory for Pastoral Theology" (7). So even though we come from different perspectives, our interests converge under the theme of *caring*—Cole from a humanistic perspective and Carlin from a pastoral perspective.

In this chapter, we are concerned with the care of sufferers of chronic pain. We interpret and analyze a recent book in the medical humanities, Deborah Padfield's *Perceptions of Pain*. Padfield makes the case for *objectifying* pain by means of artistic representation so that sufferers can disassociate the pain from their being. The photographs she produces in consultation with chronic pain patients are striking, and her efforts represent a new moment in the history of the visual representation of pain (though we suggest that her images recall the theme of *vanitas* in art history) and also a new moment in patient care. Padfield suggests that her efforts could be used in a number of ways, and she makes several recommendations. We pick up on one of them—namely, her call for an exploration of psychological issues—and we make recommendations of our own. We call for greater attention to the visual arts in the ongoing humanistic education of medical doctors, especially in the United States, as the British seem to be quite ahead of us in this respect. *Perceptions of Pain* is a case in point, which is one reason we decided to focus on it. Another reason is that we argue that Padfield's efforts might be understood as maldynia playing the muse and that Padfield's art can aid in the communication process between the doctor and the patient, especially in the case of maldynia.

PERCEPTIONS OF PAIN: THE ESSAYS

In her *Perceptions of Pain*, Deborah Padfield, who is an artist and a sufferer of chronic pain, offers a collection of photographs from the art exhibition *Perceptions of Pain*, as well as scholarly and reflexive writings that address such topics as the nature of pain, the representation of pain, the experience of pain, the doctor and patient relationship, the relationship between the photographer and the subject, and the relationship between the arts and medicine. This exhibition appeared in the Sheridan Russell Gallery in London and has since been shown widely in London, including at Guy's and Thomas' Hospitals and the Royal College of Physicians, and it also made various tours that were funded by the Arts Council of England. It is now owned by the Napp Educational Foundation (8).

Padfield's photographs mark a radical disjuncture in the history of representations of pain. As far as we know, in Western art all prior paintings, prints, woodcuts, drawings, lithographs, posters, photographs, sculptures, and so forth, have represented pain *within* a particular human or animal body or bodies. From antiquity, the most famous icon of pain is that of Laocoön and his sons, taken to the extremes of pain by a monstrous sea serpent attacking them (9). There are countless paintings of the sufferings of the saints and the Crucifixion of Christ from the first century to Grunewald's *Issenheim Alter* piece of 1919. Images of suffering and pain in the punishment of criminals are a major genre in the Middle Ages and the Renaissance (10). Images depicting the pain and suffering of war, of poverty, and of deprivation become more prominent in the modern era. And then there are the depictions of pain in modern artists such as Vincent Van Gogh, Kathe Kollwitz, Edvard Munch, and Frida Kahlo. The point here is that all these images represent pain and suffering *in a particular person's body*. The fresh and unprecedented move made by Deborah Padfield is to create images of pain itself, outside of the body, for therapeutic purposes within the health care setting.

Before discussing and interpreting some of the images from the exhibit, it is appropriate to consider the essays in the book so that the reader will understand what Padfield is trying to do.

Looking at Pain

The first essay, "Looking at Pain," is by Brian Hurwitz, Professor of Medicine and the Arts at King's College, London. He begins by noting that pain "has driven people into the hands of doctors more often than any other symptom in human history" (8, p. 7). Therefore, Hurwitz suggests, an understanding of pain is "foundational to medicine" (8, p. 7). Hurwitz then notes that pain can have a manifold of bodily expressions, everything from shivers and winces to sighs and screams. And although pain can sometimes be beneficial—because it allows biological dysfunction to be fairly precisely pinpointed and treated—some pain, particularly chronic pain, is maladaptive, because, in Hurwitz's view, such pain does not seem to be chemically related to the process of healing (8). Pain that disturbs sleep, emotions, relationships, and work—all of life, really—pain that has no easily discernable solution or explanation persisting for months and years, "This sort of pain," Hurwitz writes, "is termed chronic pain" (8, p. 8).

Hurwitz recognizes that there is no objective measure of pain and that pain is always experienced subjectively. So when fixed descriptions of pain are taken to be authoritative or objective, and when the subjective experience of pain does not match these descriptions, the result is that "sufferers [often] feel disbelieved and invalidated" (8, p. 8). When considering the subjective—that is, the idiosyncratic and deeply personal—nature of pain, the arts become all the more compelling. Strikingly, though, historically, "the subjective nature of pain seems largely to have been ignored in artistic works." Hurwitz writes: "Visual portrayals of pain…[have] often focused on painful situations rather than on the nature and quality of pain experienced" (8, p. 8). He also notes that sufferers have rarely expressed their pain artistically. Hurwitz identifies Gillray's *The Gout* "as one of the most arresting depictions of pain in western imagery." Yet, this is not a depiction of pain itself (8, p. 9). In *The Gout,* we have an image that is a demon or devil inflicting pain on the body. We have, in other words, a body in pain—not pain itself. We would also add that what we have in *The Gout* is an agent attacking the body who is depicted as ontologically other to the human host—thus, a dissociative representational schema.* If the reader would like to see an image of *The Gout*, a quick search on the Internet will turn up several good Web sites.

Padfield, Hurwitz points out, attempted to remedy this lack of representation of pain. For 8 months, she worked with sufferers "to create photographs that represented and expressed as nearly as possible each sufferer's particular experience of pain" (8, p. 9). Through many conversations and workshops, "pain sufferers were able to project the private sensations and experiences of pain—its associations, intensities, qualities, and significance—outward on to the publicly accessible surfaces of canvasses and photographic plates" (8, p. 10). The upshot is that "The resulting images do not so much depict pain as they express it; they help, thereby, to objectify pain" (8, p. 10). In so doing, "*Perceptions of Pain* reminds us that visualizations can play a role in medicine beyond that of illustration, to include expression and treatment as well" (8, p. 11).

A Visual Language for Pain

In her "In Search of a Visual Language for Pain," the second essay in *Perceptions of Pain*, Deborah Padfield draws heavily from the work of Elaine Scarry, the Walter M. Cabot Professor of Aesthetics and the General Theory of Value at Harvard University. In *The Body in Pain*, Scarry argues that

> Physical pain has no voice, but when it at last finds a voice, it begins to tell a story, and the story it tells is about the inseparability of...the difficulty of expressing physical pain; ...the political and perceptual complications that arise as a result of that difficulty; and...the nature of both material and verbal expressibility or, more simply, the nature of human creation. (4, p. 3)

Padfield elaborates: "The challenge that pain makes on language is inextricably bound up with the demands it makes on our ability to move beyond our individual experience and empathise with that of another" (8, p. 17). When patients are trying to describe pain, the problem with language is in defining "it" (that is, the patient's particular experience of pain). And if there is no understanding of "it," how can there be empathy? As one patient put it:

> You can't see pain, so people don't believe it. I had that even more so with doctors. One doctor sat there and said, "You cannot be in the pain that you say that you are in," I [responded], "What do you want me to do to show you that I am in pain?" (8, p. 17)

And so Padfield suggests that a visual language for pain is needed because pain "requires a language which works on a more instinctual and primal level than words"

* This is Marcia Brennan's insight.

(8). Not incidentally, the epigraph for her essay is taken from Elaine Scarry's *The Body in Pain*: "To have pain is to have *certainty*; to hear about pain is to have *doubt*" (4, p. 13). One would be justified in saying that the major objective of Padfield's book is to combat this doubt, as visual modes of expression can be instrumental in building a bridge of empathy and understanding between the sufferer and the world.

Perhaps the force of *Perceptions of Pain* is that Padfield is no stranger to pain. After surgery, she became disabled by chronic pain, and the world that she knew previously no longer existed. Her first response was to run from the pain, but to no avail. Her doctor suggested that she face her pain, that she confront it by thinking about it, by writing about it, and by drawing it. This helped. Padfield says that because she was able to externalize her pain, she no longer felt personally connected to it, which alleviated some of it. She also felt that she was able to think more objectively about herself and her situation. She talked with a second surgeon and felt understood (unlike her experience with her first surgeon), and she eventually gained enough trust in the second surgeon to let her reoperate (8).

Before the first surgery, Padfield had a career in the theater, but she was unable to continue that work afterward. She retrained in the fine arts and was able to explore issues of the body and pain in creative ways. She found that "every sculpture, drawing, and painting I produced, succeeded better as a photograph" (8, p. 19). She continues:

> I began to realise that having spent so long on the passive receiving end of the medical gaze, it was through photography and an active control of the lens, that I was regaining a sense of control over my world. I was using photography as a way of regaining ownership of my own body and its experiences. Was it possible that a similar process could be useful to other people with chronic pain? (8, p. 19)

If one does not feel any sense of control, it is unlikely that they will be able to experience the empathy of others. Art, therefore, aids empathy in at least two respects—namely, it helps the chronic pain sufferer communicate his or her pain, which enables another to empathize, and it also helps the sufferer experience the empathy of others, because in producing the art, one has a sense of agency and a stronger sense of self.

A failure of empathy by the doctor—which Padfield experienced with her first surgeon—can be devastating to the patient: "I believe the need to prove its existence increases a person's experience of pain, trapping them within one of the few certainties left to them" (8, p. 20). Padfield elaborates:

> If pain can be trapped within an external representation such as a photograph, instead of within the body, then it passes from a private hidden individual space into a visible, public and collective space. Once the photograph has been seen and accepted as "real" by others, the sufferer no longer has to prove the existence of the pain and the pain is [then] reduced. (8, p. 20)

The doctor and patient then have a common reference point, Padfield notes—namely, the photograph. And Padfield did not pick the photograph as the primary medium for expression haphazardly. Quite the contrary, the photograph, Padfield knew quite rightly, often carries with it a certain weight of objectivity and truth. Padfield is not naïve to the hermeneutical issues here (e.g., the selection and construction of the objects to be viewed, as well as the personal interpretive location and politics of the viewer). In any case, the illusion of objectivity that the photograph carries works to the advantage of the sufferer of chronic pain in that it validates his or her experience by trading in various visible representations rather than invisible sensations. If seeing is believing, then the patient's fear that the doctor will not be empathetic might be ameliorated, and it is likely that such a process will help create an environment for sharing (8).

With all of this talk of empathy, one might wonder if Padfield's efforts were therapeutic in nature, and whether Padfield might regard herself as an art therapist of sorts. This is not exactly the case, as Padfield made it clear to her volunteers that she was *not* a therapist. And even though the whole process had therapeutic effects, Padfield says that this was not due to "any attempted psycho-dynamic relationship," though the therapeutic benefits, Padfield speculates, were probably due to "the freedom [the sufferers had in] exploring their own narratives visually, the released experienced in making visible what had hitherto been invisible, and the inherent therapeutic effect of the creative process" (8, p. 22). Padfield is well aware of Susan Sontag's warning that the aesthetic representation of pain might produce an emotional detachment from pain, but Padfield suggests that this is precisely the point (8). The emotional disconnection from the pain by the sufferer could very well lead to an emotional connection between the doctor and the patient—that is, empathy, or precisely the opposite, at first. As one doctor reported, "Viewing the pictures was a numbing and uncomfortable experience for me....[F]aced with something too hard to understand and to which I felt I had no answer, I wanted to shut off" (8, p. 21). Paradoxically, then, the visual arts can produce an intimate estrangement between the doctor and the patient, one that later enables the doctor to give empathy and the patient to receive empathy precisely because the patient has taken agency in producing or choosing the photograph. The paradox of intimate estrangement here is that, even though the doctor may feel himself of herself "shutting off," the doctor is probably in a better position to be empathetic because awareness of one's own feelings is the beginning of connection to another's.

Unspeakable Pain

In his "Unspeakable Pain," the final essay and epilogue in *Perceptions of Pain*, Charles Pither, a consultant pain specialist at St. Thomas's Hospital in London and Padfield's own pain consultant, begins by noting that both patients and doctors want certainty, treatment, and closure. The doctor listens attentively to the symptoms and sometimes orders tests. And once the doctor is certain of the diagnosis, he or she then often stops listening and only hears. Pither writes, "The problem has changed from a diagnostic puzzle to a case of X or Y. The physiology and pathology now take precedence over the subjective experience: the person takes second place to the disease" (8, p. 125). But sometimes the doctor does not find out what is wrong. And sometimes, the inexplicable pain intensifies. The doctor often understands this inexplicable pain as unresolved emotional issues manifested psychosomatically in the patient. We do believe that emotional problems can and do affect the body, but Pither notes that such a response from doctors often leaves the sufferer feeling rejected and not taken seriously (8). And what to do, Pither asks, when one finds oneself in pain for 10 years and still disbelieved? Pither writes:

> It is now clear that chronic pain causes degrees of suffering and limitation of quality of life worse than many better-defined diseases. Nor is it the case that there is nothing physically causing the pain within the body. The experience of pain is virtually always caused by processes within the body, albeit that such processes are not identifiable or definable within current medical or psychological understanding. The concept of imaginary pain is untenable. One either has pain or one doesn't: pain is always real to the sufferer. (8, p. 126)

Pither concludes by suggesting that pain cannot be confined to words, that pain is unspeakable and yet is able to be communicated—just not with words alone (8).

PERCEPTIONS OF PAIN: THE ART

The Process

Padfield got the idea for this project through her own experience. She had a long interest in the arts, and, as noted, she also experienced a great deal of suffering at the hands of the medical profession—and since she has become an advocate for patients. When she initially discussed her idea of using photographs with her pain consultant (Pither), he supported her interests because he was interested in improving the work that has been done with pain drawings. She discussed her ideas further with Stephen Dwoskin and Nell Keddie, artists and pain sufferers, and, in Padfield's words, "a project emerged" (8, p. 19).

Padfield sent letters requesting chronic pain sufferers to volunteer to participate in her study at St. Thomas's Hospital. Those interested met with Padfield, and she explained to them that they would be working with her to "co-create photographs which represented as nearly as possible their individual experience of pain, and that they would be offered the opportunity of taking a selection of the photographs produced to a follow-up consultation with Dr. Pither" (8, p. 23). All who initially expressed interest actually volunteered, perhaps because these sufferers had been in pain for many years (between nine and 42 years), without resolution, and perhaps also because they were seeking community, a community in which they could craft a language together (8).

The program at St. Thomas lasted 4 weeks. The patients began by describing their pain to Padfield, and they brainstormed together on how their pain—its elements and its effect—might be represented in photographs. Padfield then created photographs from these sessions to consider with the patients. But this was just the beginning: "Sufferers intervened in the process through stitching into, drawing onto, tearing or cutting into the photographs or at times returning with their own" (8, p. 24). Padfield met with the patients once a week for the rest of the 4 weeks, and they "continued to deconstruct and reconstruct the images" (8, p. 24), discussing angles, colors, textures, and whatever else the patient identified. The patient then was able to discuss these images with a medical specialist, in this case, Pither (8).

As already indicated, these photographs were used to create an exhibit on art and chronic pain, and many doctors at various levels in their careers studied these images. Some of these doctors asked to use these images with other patients in their own practice, and although these patients were not involved with the creation of the photographs, the doctors reported that the images proved to be helpful to them. Perhaps what is most striking is that while pain seems to deconstruct verbal language, its visual representation seems to (re)construct a *verbal* language for pain that the doctor and patient cocreate while interpreting the images. Diagnosis, then, truly becomes a collaborative effort.

In the history of medicine, such a collaborative effort recalls the significance of the Belmont Report. It is impossible to overestimate the importance of this foundational document, because it lays out the three pillars or central principles of bioethics—namely, autonomy and respect for persons, beneficence, and justice. At the close of her essay, Padfield reflects on the experience of handing one's body over to doctors, which is an experience that inherently deals with the question of autonomy. In interpreting the images with the doctor, as Padfield is advocating, the patient is likely to regain a sense of autonomy. Pain drawings, of course, have functioned in a similar way for patients, but this kind of art provides a new avenue for doctors and patients to explore (8).

THREE PHOTOGRAPHS: INTERPRETATIONS

Padfield's project of objectifying pain in art becomes all the more compelling when the body is *not* disfigured but nevertheless is in severe pain. One chronic pain sufferer illustrates this point particularly well: "To make your friends and family understand is even harder when you look physically fine" (8, p. 50). How might a sufferer of back pain communicate his or her pain, when his or her back looks fine? Perhaps the photograph shown in Figure 8.1 could be of use.

These cards, falling and about to collapse, express imagery of extreme fragility and vulnerability—the foundation of one's "house" is literally collapsing before our eyes. The blurring, due to the fact that the camera is not fast enough to produce a still life image, seems to call to mind a blurring of consciousness, or perhaps a loss of consciousness. In Padfield's book, the sufferer, for whom and by whom the images were cocreated (but, in this case, the patient wished more to advise Padfield in creating the image rather than collaborating with her), describes his pain as "red hot wire being pulled around the joints" (8, p. 72). His fits of pain are sudden but, fortunately, do not last for very long. However, he has been suffering for 11 years. During exercises, sometimes his pain builds up slowly and he visualizes a bathtub being filled with water, with the water symbolizing the pain. When the tub overflows, he loses consciousness. Yet, he also describes the water as himself:

> The overflow thing is "me" totally. It is not just my body that is overwhelmed, but it [is] my consciousness. The consciousness instead of being in the bath is on the floor. It

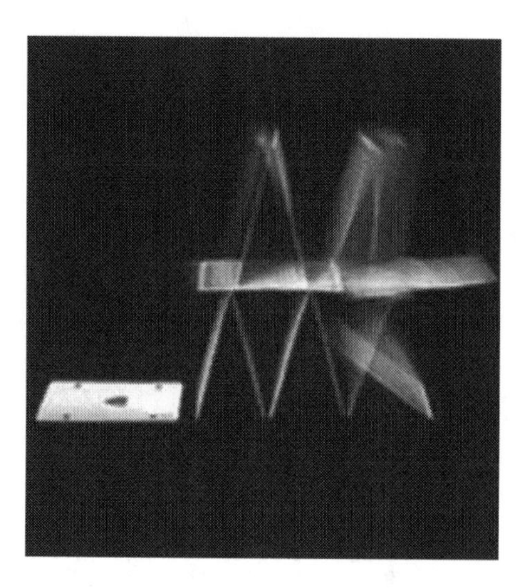

FIGURE 8.1 *Pack of Cards.* (From *Perceptions of Pain*, pp. 78–79. Reproduced with the permission of Deborah Padfield © 2003 and her colleagues [Deborah Padfield with Robert Ziman-Bright].)

> is my brain that is actually being swamped, and that is the part I resent—that I totally lose control of my ability to think rationally. (8, p. 72)

And so the patient describes these exercises as "a very fine balancing point," which is what the cards in Padfield's book represent—*pain as a very fine balancing point* (8, p. 72).

Another patient, when describing her chronic pain, said that "At its worst it feels like rusty, hot barbed wire. There is almost a taste of iron. It goes from my heel up my leg and into my back—a throbbing length of hot barbed wire. It is wound round itself, twisted up with hot sharp points like the red hot part of blown glass" (8, p. 30). She also said, "To help deal with the pain and control it, I tried to perceive it as a tangible thing, I had to be able to describe it" (8, p. 30). This patient *needed* to objectify her pain, and the photograph in Figure 8.2 is the result.

In this image, the spikes on the barbed wire are the hottest, and a flame seems to be ascending from the wire. The image of barbed wire recalls a prison—perhaps pain is locked into the body, or perhaps pain locks us in. Along the lines of prison imagery with demarcations of forced boundaries, the photograph suggests historical practices of torture devices, such as branding. The barbed wire could also recall a property fence, intended to keep others out, or perhaps also a fence for cattle, intended to keep them in. Perhaps pain locks the sufferer in and keeps others out. But in communicating pain in this manner, maybe the fence can be jumped.

The image shown in Figure 8.3, unlike the other images included here and unlike many of the other images in *Perceptions of Pain*, depicts a person. In the others, pain is objectified and externalized, but here pain is located in the body.

In creating this image, the sufferer tried to convey his pain as "ants crawling all over and underneath my skin but there is no way I can scratch, push or wipe them

FIGURE 8.2 *Red Hot Wire*. (From *Perceptions of Pain*, p. 31. Reproduced with the permission of Deborah Padfield © 2003 and her colleagues [Deborah Padfield with Linda Sinfield].)

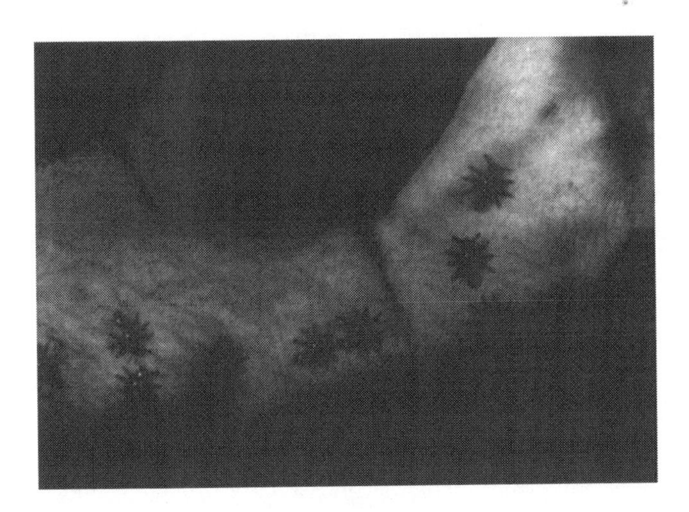

FIGURE 8.3 *Arms and Ants*. (From *Perceptions of Pain*, pp. 56–57. Reproduced with the permission of Deborah Padfield © 2003 and her colleagues [Deborah Padfield with Rob Lomax].)

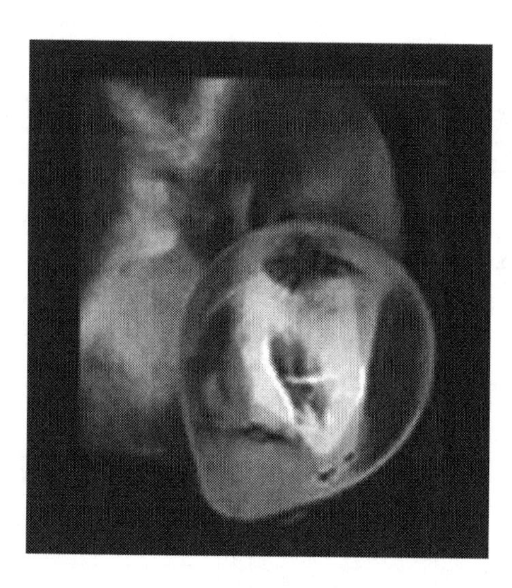

FIGURE 8.4 *Face and Balloon Off Side.* (From *Perceptions of Pain*, p. 58. Reproduced with the permission of Deborah Padfield © 2003 and her colleagues [Deborah Padfield with Rob Lomax].)

off until the cycle of pain passes. At this stage my facial expressions change with my cheeks and eyes puffing up and it feels like it is going to burst my skin open and release all the pressure" (8, p. 50).

In Figure 8.4, we have a close-up shot of a face, so close, in fact, that the clarity of the image is poor. In the bubble, though, there is more clarity. Does pain produce its own kind of clarity? Or is it that pain produces a skewed perspective? Rays of light extend from the eye, but they do not make it outside of the bubble. Does pain confine clarity? The bubble is not perfectly round, as one would expect with swelling. The only clarity we have in this image is inside the bubble, and it is of the eye and the mouth, both of which are downcast.

These images give the reader a sense of Padfield's book, though there are many more photographs and testimonies in the book. The last image seems to run counter to Padfield's objective of objectifying pain and projecting it onto another object, but this particular image illustrates her larger project: to convey the idiosyncratic and subjective experience of pain, and *all* of the images are readable as metaphorical self-portraits. The final image just incorporates a different mode of stylized literalism—an anamorphic appearance of the image, viewing life from a skewed perspective.

CONTEXTUALIZING PERCEPTIONS OF PAIN

From the perspective of art therapy, Padfield's efforts may be read as an extension of the practice of pain drawings. Pain drawings are often used to help patients convey the level of pain that they are feeling. In Figure 8.5, for example, children who suffer from migraine headaches depicted their pain as "pounding pain." These drawings were made by the children (11).

FIGURE 8.5 Pain drawings from *The Usefulness of Children's Drawings in the Diagnosis of Headache*. (Reproduced with permission from *Paediatrics*, 109, 463, http://pediatrics.aappublications.org/cgi/content/full/109/3/460. Copyright © 2002 by the AAP.)

In these drawings, the sufferers created the artwork. In Padfield's project, however, the sufferer and artist cocreated. An advantage of Padfield's approach could be that her collaborative method seems to be more conducive to the experience of empathy. As the sufferer and the artist negotiate the image, the sufferer will likely feel that at least the artist understands his or her pain. In the case of these pain drawings, it is possible that the sufferer might feel that no one understands them or believes them. Note, too, that the pain is depicted in the body, that the bat and hammer inflict pain (but are not pain itself). And one cannot help but wonder if these images were conjured up from actual physical abuse. In any case, the experience of being disbelieved about the experience of having a migraine may be a form of abuse in its own right, and the children may very well have felt that they were being battered around by the experts.

How else might we contextualize Padfield's *Perceptions of Pain*? As noted, Padfield draws from the work of Elaine Scarry. Scarry writes: "The mental habit of *recognizing* pain *in* the weapon (despite the fact that an inanimate object cannot 'have pain' or any other sentient experience) is both an ancient and an enduring one" (4, p. 16). Padfield participates in this ancient and enduring tradition. One patient, when describing the experience of chronic pain lasting over 40 years, said that "There are two kinds of pain: the mental pain and the physical pain. I am constantly battling with the physical pain. You could possibly describe it as swords on fire. It is as if they are ripping down my leg all the time....If I could [just] get my hand in, I could lift it out" (8, p. 60). Her pain is represented in Figure 8.6, which, like the barbed wire, recalls torture.

Like the image of the barbed wire, again the theme of heat is emphasized, as the tip of the sword is white-hot. The sword, of course, is phallic, but it is also curved like a woman's body. Because the sufferer describes the pain inside her body, it is striking that this image seems to represent the body. So, not only is pain objectified here, but perhaps the body is as well (a theme intimated in one of our epigraphs, with more on this theme to follow below). The imagery of the sword recalls medieval

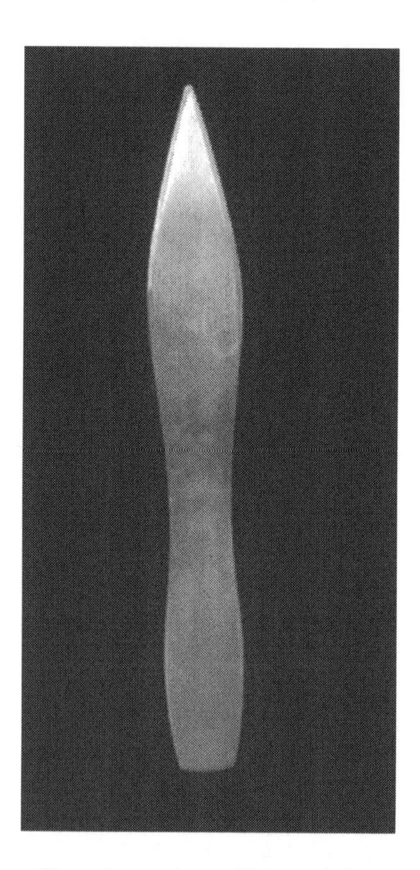

FIGURE 8.6 *Red Dagger.* (From *Perceptions of Pain*, p. 61. Reproduced with the permission of Deborah Padfield © 2003 and her colleagues [Deborah Padfield with Frances Tenbeth].)

or ancient battles, so perhaps what is being conveyed, in part, is an "old"—that is, chronic—battle with pain.

This old battle with pain also recalls the theme of *vanitas* in art history. One well-known representation of this motif is Pieter Claesz's *Vanitas*, and, again, this work can be readily and easily found online, and many works like this one, so we did not feel compelled to reproduce this image. Deriving from Latin, *Vanitas* in art reflects the vanity of life. Just as ink dries out, candles melt away, and flames extinguish, so life ends in death—all is vanity, as Ecclesiastes instructs: "All is vanity" (1:2b). *Vanitas* relates to the fleeting nature, temporality, and fragility of all things, and in art, such fragility and temporality is often depicted with an iconographic topos. Rising in prominence in seventeenth-century Europe, particularly in regions influenced by Calvinism, *vanitas* imagery spread to New England by the 1690s. John Martin writes: "The secret of many a seventeenth-century still life painting lay in the fact that it offered to the observer a stimulating and life-enriching illusion of reality, while at the same time inviting him—paradoxically—to reflect on the brevity of man's existence and the insubstantiality of all worldly things" (12, p. 134). Playing cards often appear in *vanitas* motifs, symbolizing the "illusionary pleasures of this life" (12, p. 136), as well as the vulnerability and fragility of life. We wondered if Padfield meant to recall the *vanitas* theme in her photograph of playing cards.

Here we are interested in pointing out that Figure 8.4 may reflect the *vanitas* theme. Perhaps the nose in Figure 8.4 is intentionally missing or, if it is there, hard to decipher—that is to say, the very ambiguity of seeing the (missing) nose in Figure 8.4 recalls the image of a skull (and also the theme of *vanitas*). *Could it be for chronic pain sufferers that their suffering is in vain, and that all efforts to alleviate it are as well?* Also of interest here is that the realization of the vanity of human life, of life and existence, can be a form of psychological suffering. If existential suffering can be objectified in a decomposing book, why not pain in a pack of cards? In correspondence with her later, we learned that this motif has in fact influenced her a great deal, as she reads it in a positive light, because, if nothing is permanent, then pain, too, will not last forever.

In any case, our point here is simply that Padfield's project did not emerge out of a vacuum. Quite the contrary, it seems to be a logical extension of the practice of pain drawings, and, more generally, her project can be seen to have emerged from the history of the representation of pain in art, with particular affinities to the theme of *vanitas*. In other words, while clearly all representations are produced in specific historical, cultural, and subjective contexts, at the same time, the evidence for pictorial archetypes in visual representations of pain is striking and suggestive.

CONCLUSIONS AND RECOMMENDATIONS

Padfield's project is commendable. At first glance, though, we found her pictures in *Perceptions of Pain* to be a bit curious, because they only make sense within the patients' testimony. There is no reason to assume, for example, that some playing cards could represent pain. But in context of a particular patient's description of pain, the images become powerful, and even chilling and disturbing. It is striking that Padfield is making the case that pain destroys language, because her images need words to be understood. Initially, we also wondered about questions of *dis*-embodiment: *By objectifying pain and locating it symbolically in an object, might the sufferer be somehow dehumanized in the process, just as the object is implicitly humanized or anthropomorphized?* But because the sufferers cocreated the images, we figured that it is unlikely that they would have experienced these images as dehumanizing. Further, in researching for this chapter, we came across the work of James Elkins and his intriguing thesis that "Every picture is a picture of the body." He elaborates: "To say this is to say that we see bodies, even where there are none, and that the creation of a form is to some degree also the creation of a body" (1, p. 1). So, rather than dehumanizing sufferers by objectifying their pain, the world becomes humanized, and perhaps more hospitable. This was enough to put our objections to rest.

Padfield makes several recommendations, but we want to focus on one of them—namely, her recommendation for the use of psychological methods for understanding issues raised by looking at photographs such as the ones she cocreated. Because Padfield employs psychoanalytic concepts (e.g., projection), one would expect that psychoanalysis would complement her efforts.

Padfield's Call for Psychological Interpretations

In her essay in *Perceptions of Pain*, Padfield writes: "These responses [i.e., the responses of doctors who looked at her exhibition] open up possibilities for future work, such as involving a therapist or psychologist to explore the issues raised by the images" (8, p. 26). We think that this is right, and we are particularly interested in the possibilities that psychoanalysis could open up. Discussion of psychoanalysis in the medical humanities is not at all new. In *The Silent World of Doctor and Patient*—a classic text in the medical humanities—Jay Katz explores the issues of transference and countertransference in the doctor and patient relationship. With regard to Padfield's photographs, we think that the recent work of Christopher Bollas, who might be considered a *transformational* psychoanalyst, is an especially useful way of exploring these images further. Art historian Marcia Brennan is currently convincingly making the case for understanding art, both its creation and its interpretation, as a transformative practice, so it makes sense to ask in what sense transformational psychoanalysis might be related to such an enterprise.*

In his *Being a Character: Psychoanalysis and Self Experience*, Bollas argues that "we consecrate the world with our own subjectivity" (13, p. 3). "Very often," Bollas writes, "we select and use objects in ways unconsciously intended to bring up such imprints; indeed, we do this many times each day, sort of thinking ourself out, by evoking constellations of inner experience" (13, pp. 3–4). Bollas, of course, is drawing from object relations theory, but finds

> It rather surprising that in "object relations theory" very little thought is really given to the distinct structure of the object which is usually seen as a container of the

* For the term *transformational object*, see Bollas, Christopher, 1987, *The Shadow of the Object: Psychoanalysis of the Unknown Thought*, New York, Columbia University Press.

> individual's projections. Certainly objects bear us. But ironically enough, it is precisely *because* they hold our projections that the structural feature of any one object becomes even more important, because we also put [ourselves] into a container that upon re-experiencing will process us according to its natural integrity. For example, if I put a feeling of joy derived from early adolescent skills in baseball into a piece of music—such as Schubert's C Major Symphony—and if that same week I project an erotic response to my girlfriend into Salinger's *The Catcher in the Rye*, then encountering these objects in adult life may elicit the self experiences stored in the objects; but equally, the musical experience and the literary process are different types of object, each with its own "processional potential," by which I mean that employing one or the other will involve me in a different form of subjective transformation, deriving from the object's structure. (13, pp. 4–5)

In other words, Bollas is calling for closer attention to the *form* of the objects in which we store ourselves, and, it follows, closer attention to our capacity to reshape our responses to these forms, thereby transforming our own consciousness.

We suggest that this insight from Bollas is especially useful for the type of exploration that Padfield is recommending. This is so because, as noted, Padfield worked with chronic pain sufferers and revised her images a number of times to depict their pain; that is, the patients are already concerned with *form*. The patients gave reasons for picking particular objects, and these reasons, of course, would be conscious. But what might be the unconscious motivations in these choices? Bollas mentions storing joy in objects, and he is interested in how various forms of objects work in our adult lives. Padfield is interested in storing pain in objects, also with close attention to form. So we could ask, with Bollas, how can and do these objects transform the lives of pain sufferers today, or we could ask, with Brennan, how does this artistic enterprise potentially operate as a transformative practice?* That is, how can and do these transformational artistic objects work for particular individuals? Why is it, for example, that one patient insisted that his back pain is more like a deck of cards about to fall rather than a cement jacket (8)? Why such emotion surrounding form? A psychoanalyst might conjecture that he was once cheated in a card game and experienced this as being "stabbed in the back" and therefore "stored" his back pain in playing cards. In any case, our point here is that psychoanalysis, in all its varieties, but particularly in the work of Bollas, could be a useful tool for deeper explanation. Attention to *form* would very likely yield significant insights.

MALDYNIA AS MUSE

Giordano, the editor of this volume, writes: "[T]he physician must rely upon subjective descriptions to gain access into the life world of the patient as affected by pain. However, as can be seen, while explanation may be inherently difficult, this narrative is critical to establish the nature, meaning and impact of pain upon a particular person" (14, p. 6). He advocates what he calls a "phenomenological approach to pain,"

* What we learned from Marcia Brennan has come from a number of conversations. For a fuller understanding of her thought and work, see her work *Curating Consciousness: Mysticism and the Modern Museum*, Cambridge, MA, MIT Press, 2010.

and he suggests that diagnosis should be "both a technical and moral act...based upon both scientific skill and humanitarian art" (14, p. 8). Now this, of course, is a tall order, and it would require more attention by medical schools to the humanistic education of doctors. We believe that Padfield's efforts could aid in this regard, as well as more attention to the visual arts in general in the medical humanities. But whatever the *form* we choose to use to convey pain—be it, say, the visual arts, poetry, or music—there is also a need for a greater attention to what moves us, what repels us, and what we feel represents our experiences, so that we—as doctors and patients and artists and scholars and teachers and caregivers—might understand ourselves, and our pain, better. The *forms* of our objects matter—and our ability to create these forms, and hence, to create *ourselves* symbolically—for our self-experience, Bollas teaches us.

When we were invited to write on the topic of art and chronic pain, Giordano sent us a synopsis of his anticipated book. In it, he wrote:

> Modern medicine has assumed a technocentric, process-orientation and has utilized technologic advances to improve public health by reducing the mortality from disease through both environmental hygiene and implementation of advanced diagnostics and therapeutics. Yet, these very means have resulted in an increased prevalence of chronic morbidity, the effects of which are multi-factorial. As we move into the post-modern era, this has become strongly represented by *maldynia*: the "illness" of seemingly idiopathic, intractable, chronic pain and subjective suffering within biopsychosocial contexts.

If medicine cannot defeat pain, if the persistence of pain in the face of all of biotechnological advances, that is, if "maldynia" means that pain is here to stay, then maybe we would do well to disassociate from our pain by objectifying it and locating it outside of the body, thereby making it dialogical, making it a conversational partner that is also a symbolic representation of ourselves. If medicine cannot defeat pain, this leads us to suggest, *What science cannot conquer, maybe art can deal with.* And here maldynia could play the muse, inspiring new kinds of art. In Padfield's book, we find precisely this: We see a new kind of pain, a pain that is unspeakable and unanswerable even in light of, *especially* in light of, our biotechnological advances, and this pain gave rise to a new kind of art, an art that objectifies pain and locates it outside of the body. But the irony, if Elkins is right, is that all pictures are of the human body. So, it would seem that all of our attempts at defeating pain and suffering once and for all are in vain, but perhaps there is potential for understanding pain's message that possibly yields a kind of transformative wisdom. This, of course, is what many psychoanalysts and artists have been saying all along, and perhaps what Ecclesiastes said even longer ago. Concerning maldynia, Giordano's fear is that "The clinician may retreat to an increasingly paternalistic position, which can instigate conflicts in autonomy, raise questions about the nature and meaning of beneficence and promote intersection between several ethical and legal issues in patient care." But if Maldynia can play the muse, maybe she will bring Hermes along as well, so there will be no need for paternalism, even in the face of unexplainable, incurable, unspeakable, and persistent pain, because, if the photographs are successful, there will be greater understanding and new opportunities for self-creation.

ACKNOWLEDGMENTS

We wish to thank Marcia Brennan and Jeffrey Kripal for their assistance with this essay.

REFERENCES

1. Elkins, J. 1999. *Pictures of the body: Pain and metamorphosis.* Stanford: Stanford University Press, p. 1.
2. Nouwen, H. 2005. The wounded healer. In: *Images of pastoral care: Classic readings*, ed. R. Dykstra, 77. St. Louis, MO: Chalice Press.
3. Woolf, V. (ed.). 1967. On being ill. In: *Collected essays, Vol. 4,* 194. New York: Harcourt.
4. Scarry, E. 1985. *The body in pain: The making and unmaking of the world.* New York: Oxford.
5. International Association for the Study of Pain, http://www.iasp-pain.org/terms-p.html#Pain.
6. Rey, R. 1995. *The history of pain.* Cambridge: Harvard University Press.
7. Capps, D. 1999. The lessons of art theory for pastoral theology. *Pastoral Psychology* 47: 321–346.
8. Padfield, D. 2003. *Perceptions of pain.* Stockport, UK: Dewi Lewis.
9. Spivey, N. 2001. *Enduring creation.* Berkeley: University of California Press.
10. Merback, M.B. 1998. *The thief, the cross, and the wheel.* Chicago: University of Chicago Press.
11. Stafstrom, C.E., R. Kevin, and A. Minster. 2002. The Usefulness of children's drawings in the diagnosis of headache. *Paediatrics* 109: 460–472.
12. Martin, J. 1977. *Baroque.* New York: Harper and Row.
13. Bollas, C. 1992. *Being a character: Psychoanalysis and self experience.* New York: Hill and Wang.
14. Giordano, J. 2007. Pain, the patient and the practice of pain medicine: The importance of a core philosophy and virtue-based ethics. In: *Ethics of chronic pain management,* ed. M. Schatman, 1–15. New York: Taylor & Francis.

9 Maldynic Pain in Image and Experience
Engraving Meaning through Subtraction

Rosemary Feit Covey

CONTENTS

> There is always a well known solution to every human problem—neat, plausible, and wrong.
>
> **H.L. Mencken**

INTRODUCTION

Art is a means of obtaining knowledge. This knowledge cannot be acquired through rote learning or an analytical process. Rather, it must be absorbed through one's senses and imagination. Aesthetic experiences move us beyond our logical frame of reference, combining psychological symbols and cultural references to form a language that resonates in each viewer's subconscious. This is not knowledge that leads to a neat and easy solution; instead, it raises the most fundamental questions of existence.

When one reads a book or studies a visual artist's work, one gets a true snapshot of that artist's experience. A picture, set in place and time and providing the proper context, may be more intimate than any actual encounter. This glimpse into an interior world organizes one's mind and allows an understanding of places, people, and situations, that although never encountered, seem eerily familiar. A work of art acts as a source of understanding the ineffable, subtle, hidden, and historic, in ways that reverberate within our complex human psychology. Art evokes emotion and carries us beyond our own experiences, while simultaneously linking us to our thoughts. It hones our imaginations and increases our abilities to access new lexicons of images. As a viewer, one gains the ability to come as close as any mortal to entering the mind of another. Like meditation, artworks aid our ability to move into a transcendent or reflective state. Art can express the indescribable—for example, individual feelings of pain or euphoria. A deep absorption of art is not a leisurely activity that can be relegated to a plane flight's good read or an occasional visit to a museum. Instead, it is an interactive experience that provides a conduit to creative thought of all types.

The artist must rely on all of his or her senses, intellect, and powers of focused concentration to create meaningful work. The artist must also study art as a discipline, learning technique, composition and history. The artist must examine the world directly and find a means to express those observations. This includes what is commonly referred to as *voice*. Artistic voice is an integration of a technique and a style that uniquely expresses a personal point of view. Many artists struggle for years to find their true path. In my case, a personal style emerged early in childhood and is as natural to me as my signature. Assembling all of these elements, studied or obtained through conscious effort, artists must eventually allow all to fall away, allowing his or her unconscious mind and hand to lead during the act of creation.

Drawing, science, and medicine are learned as disciplines that must eventually be practiced as an art. Deep knowledge comes from the combination of the learned and the intuited. Without years of study, the knowledge that comes without conscious thought will not occur. The artist, scientist, or physician will not be able to close the gap between learning and mastery of his or her discipline. Really knowing means allowing the brain to work and combine the many disparate elements that are a mixture of accumulated facts, observation, and instinct into a diagnosis or picture that goes beyond even one's own understanding of how an insight was obtained. Years of accrued knowledge must be quickly and assuredly accessed to provide conclusions that will result, at their most accomplished, in alchemy that is unobtainable through the pathways of pure reason.

Solitude is the fertile soil for most creative thought. This applies to the sciences at their most creative level, as well as to the arts. In a relationship-based world, where love is equated with happiness, and being a good person means being a social creature, the need for a creative person to spend the necessary time alone is often viewed as selfishness. Anthony Storr, the noted British psychiatrist and philosopher, deals with this misunderstood phenomenon in his book *Solitude, A Return to the Self*: "The creative person is constantly seeking to discover himself, to remodel his own identity, and to find meaning in the universe through what he creates. He finds this his most valuable integrating process which like meditation or prayer, has little to do with other people, but which has its own separate validity" (1, p. xiv). In his book, *How Doctors Think*, Jerome Groopman deals at length with the need to apply both knowledge and intuition creatively in diagnosis. In our current educational system, in which students are taught to rely on algorithms and substitute statistics for thoughtful contemplation, he observes that, "complicated problems cannot be solved in a rush. The inescapable truth is that good thinking takes time. Working in haste and cutting corners is the quickest route to cognitive errors" (2, p. 268).

When I decide to create an engraving, I commit myself to endless hours of sustained work that require solitude both during and surrounding the creative act. Initially, I rely on an almost subconscious sense of pull toward my subject. A visual image usually appears, in black and white, dancing in my mind; dream-like it reappears on a regular basis. Its constancy alerts me to its importance. Often, I receive additional cues from unexpected or unlikely sources, perhaps due to a heightened state of awareness. Eventually, I sit down to work, although I am nervous and afraid the act of drawing may destroy this fragile mind picture. I begin the process by sketching. It is the hardest part of a difficult progression—this teasing out of the mental picture, forcing it onto a cold and empty page. Usually, the first day of drawing is a form of hell. The image dies and seems useless. I try it again and again. Covered in ink and lead from my pencils, I cannot get it to work or conversely let it go. Hours pass in fruitless searching for the mind picture that seems now illusive and hopeless. It is not a good time to spend in my company.

The second day is better. A strange metamorphosis has occurred overnight. The first day's work is not wasted. It has acted not as a conclusive event but, as a springboard to entering the second phase of a subconscious process. By the second or third day, my brain has combined the mind picture and the deliberate act of drawing, creating the first stage of the work on paper. This is usually a very rough depiction of the image. To my eye, it is readable, almost finished; to anyone else, it is scratches and dashes and means next to nothing.

After I start the actual engraving, I aim to work in a state of flow. A great gift in my early 20s was reading the work of Mihaly Csikszentmihalyi, a psychologist who first described the circumstances in which mastery of a subject can lead to a state of relaxed awareness that facilitates inspiration. What struck me about Csikszentmihalyi's theory was that he equated flow with happiness (3). His work was seminal in cementing my drive to create art; my single greatest pleasure—as well as a means to communicate and earn a living. By understanding the principle of flow (similar to what is described as the *zone* in sports), I have been able to develop my ability to achieve this state almost at will. It has become my greatest tool and the only time the work becomes a source of renewal, not anxiety.

Where do these images come from? This, in many ways, remains a mystery. Raw emotion, often pain—recently experienced, remembered, or accessed through the power of empathy—is the basis for the strongest artwork, mine and that of others. Experiencing pain, illness, death, or any of life's vicissitudes and turning these emotions into prints and drawings acts as my salve. For me and many artists, it's not just what we experience, but how we access and apply the raw material of life, both lived and imagined, that enables us to create forms that symbolically communicate our mental images.

Chronic pain—or any illness for that matter—has the ability to reduce us to our most basic selves. We feel diminished and removed from the trappings of our lives that can often define us. Sometimes there can even be a sense of relief to having so many extraneous physical elements removed. We are base, a body, a specific pain; our mortality suddenly in relief. It is somewhat the same producing artwork. Art is supposed to sort and sift, reducing an idea or emotion to its core elements. The best art does not rely on artifice or manipulation. Just like pain, it is a visceral experience beyond words. Art conveys humanity in the way the touch of the hand conveys comfort when we are sick. It means solitary work for the artist and a willingness to reconnect to past pain, both mental and physical.

I work whether well or sick. Over many years, I have lived with the chronic pain of *ankylosing spondylitis*. This pain as well as every other experience becomes the fodder for my work. This is not intellectual knowledge but absorbed knowledge. Because I struggle to convey my experiences through my artwork, ultimately selling the work, I am considered a professional. I have been lucky to find an audience for my particular vision and to be awarded commissioned work that is compatible with my style and interests. For example, 2 years ago, I worked on a commission that involved a year following a young man with a brain tumor and reflecting his experience artistically. Although the experience was entirely vicarious, I was able to effectively absorb his emotional state. However, in order to do so, I also had to read about and research his disease extensively, accompany him on visits to the hospital, be with him when glimmers of hope shone, and conversely when the disease and treatments took their highest tolls. In other words, there was an extensive amount of work that took place prior to starting the *artistic* phase of the work.

The artwork included in this chapter (see Figures 9.1 through 9.6) was created using the technique of woodcut, specifically wood engraving. The artist incises a white line into the dark surface of the wood, using simple knives called gravers. In wood engraving, unlike woodcut, the gravers are not gouges but knives; the wood is hard, not soft as in woodcut. After the painstaking engraving is completed, the inked block is impressed on a piece of paper using a specially designed press. I use a Vandercook, a relic from the 1920s. (In its heyday, it was used for proofing at the Bureau of Engraving in Washington, D.C.) As with other printmaking techniques, the design on the block is a mirror image of the printed composition.

Everything in my work is built from this seemingly simple construct: facial expression, form, and content, line by line without any modern contrivances. Using a technique that dates back to the Middle Ages, all meaning is conveyed by the hand, the tool, and the combination of lines and dots filtered through my imagination. Engraving is produced by subtraction, carving out the image, the way a sculptor must find the figure when he chisels into marble.

In some sense, this is maldynic pain—it engraves meaning by subtraction—of the lived body, the life-world, and relationships. To intuit greater meaning, the most obvious solutions to either an artistic problem or a medical one may not be correct. The answer lies in the subtle gray tones or less obvious diagnoses, coaxed from beneath the surface of skin or wood. An engraving relies on line but also on the areas of the wood block left uncut; these spaces print as black. Similarly, in science, the best insights cannot be drawn from information found only through scientific formulae. To succeed in finding the less obvious but correct solution requires an imagination willing to look at these spaces in between the words or lines. It is impossible to practice either fine art or the medical arts well, without the knowledge that emerges through inspiration and concentration, as well as through skill.

> The human spirit will not even begin to try to surrender self will as long as all seems to be well with it.
>
> **C.S. Lewis**

FIGURE 9.1 *Bone Pain* (drawing): Bone pain feels alive—suddenly aware of our own skeleton, we can become its prisoner. Shackled to it, it owns us. Our usually quiet bones have made an almost physical appearance. No longer silent, they request and demand our attention, hugging us to them. Pain, while experienced, has no beginning or end. This sense of entrapment arouses an acute sense of our own frailty. (© Rosemary Feit Covey. This image is the sole property of the artist and is reprinted here with permission.)

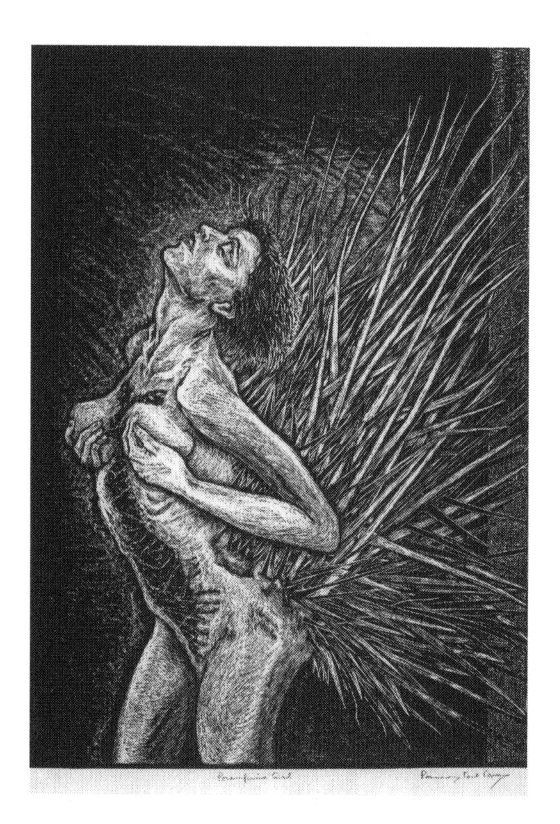

FIGURE 9.2 *Porcupine Girl* (standing): The experience of pain causes the outer body and will to be subsumed, all concentration centered on the afflicted area. With a toothache, our head becomes our only appendage. Internal reality, our thoughts, fantasies, desire for pleasure, or plans for the future, is obliterated. Focus is present-oriented. The core of our sense of self is replaced almost exclusively by the physical sensations we experience. Although it can seem paradoxical, extreme pain can also be liberating and make us feel very much alive. Pain both alienates us from others and connects us back to our most intensely personal selves; not our sense of our place in the world, but our most physical sense of our body as a known organism. When pain or suffering is chronic, the mere absence or abatement of it (however fleeting) can cause moments of ecstasy; lack of pain is experienced as a sensation. This sensation is not available to the totally well body. What can be normally taken for granted suddenly rushes to the forefront. Pleasure and pain become intensified by this ebb and flow. The porcupine woman's face does not depict agony. Despite her opened belly and twisted and thorny back, she is experiencing intense pleasure, in spite of her predicament. (© Rosemary Feit Covey. This image is the sole property of the artist and is reprinted here with permission.)

FIGURE 9.3 *Disconnect*: Pain forces the sufferer to develop a new sense of self and can manifest illness as depression. This can change the balance of power in a relationship and alter the fabric that connects a couple in the delicate dance of responsibilities and problems that make up a normal daily routine. In this print, the man's illness has affected this balance to the extent that the woman feels she is carrying the weight of the man's former strength. He is diminished, sliding out of the picture, his face glazed. She is strong but unhappy carrying this new weight. They both fear the present as represented by the monkeys and the future obscured, but not looking hopeful in the person of a seated figure, half hidden by the tent. The tent, while decorated, is not benign but suggests claustrophobia and an enclosure that threatens to envelop all. The disease process has changed both the sense of self and the connections and threads of the relationship. (© Rosemary Feit Covey. This image is the sole property of the artist and is reprinted here with permission.)

FIGURE 9.4 *Porcupine Girl* (quills off): The figure in this print is opened up—unzipped from the back; her spine is exposed. Her pain has left her vulnerable and alone. The quills that have fallen all around her represent the vulnerability of chronic illness, wherein sense of dignity and the trappings we use to preserve our status in the world are stripped away. Illness always involves a series of losses, whether it is the loss of energy, loss of work, or the more fundamental losses of health, mobility, and ultimately of the life-world. A porcupine's quills act to protect the animal. Its underbelly is its weakness; when in a defensive posture, it is invulnerable to attack. In pain, we are turned on our backs, attacked from within, and we keenly feel the multiple losses of what used to constitute our sense of self. (© Rosemary Feit Covey. This image is the sole property of the artist and is reprinted here with permission.)

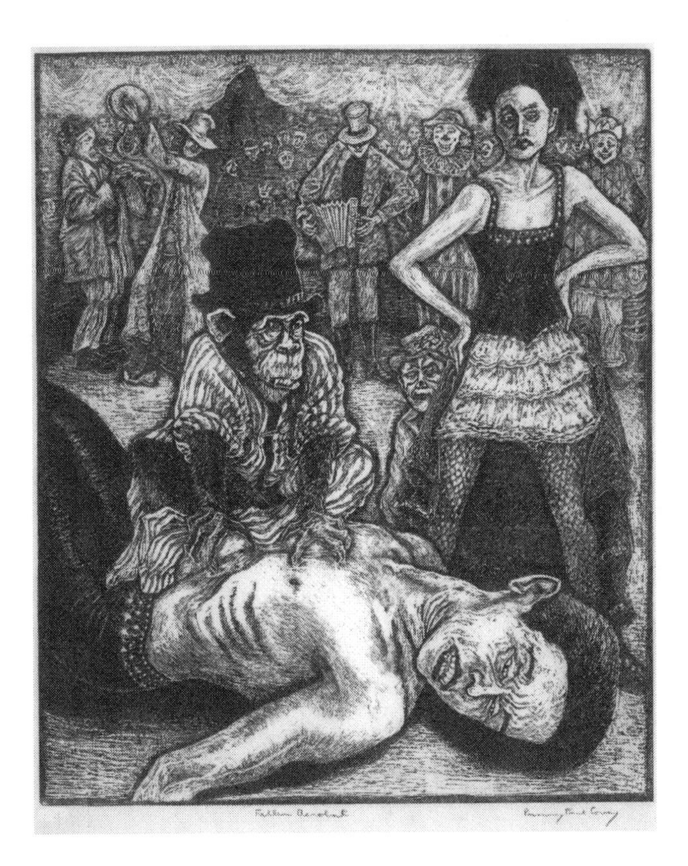

FIGURE 9.5 *Fallen Acrobat*: As in the print *Disconnect*, the subtle balance in a relationship is dramatically shifted in pain as illness. Using the circus as a metaphor, the fall of an acrobat is symbolic of pain and its chronicity as illness. The female performer, behind the acrobat, is the acrobat's caregiver, who is rarely the sainted figure others imagine. Like a conversation or a life-plan broken in midstream, the acrobat's fall has destroyed the performance for both him and his female partner. She is temporarily unsympathetic to his plight, as the consequences so heavily impact her own rhythm and routine. The monkey suggests pain and illness, sitting atop the man's chest crushing and suffocating him. The background figures form a distant audience, watching, playing, but ultimately distanced from the intimate drama occurring in the ring. The small clown looking out from behind the woman's dress represents the black humor and sometimes mockery that resides even in the worst of circumstances. (© Rosemary Feit Covey. This image is the sole property of the artist and is reprinted here with permission.)

FIGURE 9.6 *Good-Bye to the Elephant*: Loss is a part of pain and illness. It damages our sense of equilibrium. The elephant represents this loss in its entire perceived enormity. The woman touches the elephant gently, as she has reached a stage of acceptance of her situation; time has worked its soothing effect to lessen the most painful elements of her struggle. She has come to terms with her new level of capability, reached an understanding with her illness, and has realized that both growth and calm can extend from what was earlier experienced as shock and calamity. (© Rosemary Feit Covey. This image is the sole property of the artist and is reprinted here with permission.)

REFERENCES

1. Storr, A. 1988. *Solitude: A return to the self.* New York: Ballantine, p. xiv.
2. Groopman, J. 2007. *How doctors think.* New York: Houghton Mifflin, p. 268.
3. Csikszentmihalyi, M. 1975. *Beyond boredom and anxiety: Experiencing flow in work and play.* San Francisco, CA: Jossey-Bass.

10 Musical Representations of Physical Pain

Elaine Peterson

CONTENTS

INTRODUCTION

We are coming close to achieving an understanding of the molecular and physiological mechanisms by which pain is experienced, but we still cannot answer all of our questions about the complex experience of pain (1–6). Individual responses to pain vary greatly from person to person and across cultures. Pain tolerance, reactions to pain, and meaning found in pain are as much learned as they are biologically determined (5).

The ways in which we interpret and express pain are as relevant as the physiology and chemistry of the process. Pain is difficult to communicate, and it can be difficult to recognize and interpret in others. Music is often recognized as a tool to soothe pain but may also be effective as a means by which to directly express its experience. In examining this possibility, it becomes necessary to sort out what pain is, and if representations of physical pain can be separated from its psychological dimensions. Several musical examples are presented here that relate to physical pain. The interpretation of music is subjective, and the labeling of specific musical examples as types of represented pain may be challenged as a manifestation of my own system of representation (7). Yet, because we build up shared meanings within our communities, this allows us a basis for understanding.

PAIN AND LANGUAGE

Elaine Scarry (8) describes pain as resisting verbal representation. There is no direct language to express the experience of acute pain. Physical pain does one of two things: it reduces the sufferer of pain to a preverbal state in which only grunts, screams, or silence are available means of audible communication, or it requires the use of metaphor and symbol for verbal communication. We recognize these metaphors, but only because they have contextual meaning that is culturally learned. The afflicted speak of, for example, feeling "as if" they are being burned and stabbed ("it feels as if my arm is on fire"), when they are not actually on fire or being cut. In clinical settings, we rely on artificial numerical scales (e.g., "rate your pain on a scale of 1 to 10") to communicate degrees of pain. This, too, is an example of using an imposed symbolism to interpret the pain experience.

These metaphorical descriptions of bodily state are attached to the actual experience of pain, for which we do not have direct language. Physical pain is not the same as damage or agency, and can exist without it, yet these referential images are often called upon to convey the experience of pain. As Mark Johnson explains in *The Body in the Mind* (1990), metaphor plays a central role in human experiences of meaningfulness. Metaphor is not simply a poetic or literary device, but a process through which we generate meaning, coming to understand one area of experience in the framework of another. So it is that the use of metaphor is central to our understanding of pain.

When people are in the presence of those who are in physical pain, it is difficult for them to recognize the pain of the other (8). Even when one becomes aware of another's pain, it is difficult to relate to the actual sensations they are experiencing, unless the experience is in some way mutual. It is difficult to remember with detail one's own past experiences of pain. One can remember that a certain experience was, for example, the most painful moment they ever endured, but full retrieval of the specific sensations remains tangential, unless those sensations are being endured in the present. To be in pain is to be fully in the present.

PAIN AND ITS OBJECT

Scarry (8) describes states of consciousness other than pain, such as hunger or desire, that will, if deprived of an object, begin to approach the sensation of pain, as, in this case, unsatisfied hunger or objectless longing. Pain is an objectless state of consciousness. People in excruciating physical pain are stripped of language and reduced to cries and moans. It is only when the afflicted regains his or her power of speech that he or she also regains powers of self-objectification.

Gerald Edelman (10) designates painfulness as a type of quale. Qualia are the experience of a property—for example, of blueness, warmth, or painfulness. We experience warmth and pain in degrees of intensity, as we also perceive the blueness of a shade of blue. Edelman stated, "Qualia are high-order discriminations that constitute consciousness" (10, p. 10). He argues that qualia are always experienced as part of the integrated whole of consciousness and are never experienced as a single quale. An example used by Edelman is that one cannot experience "red" in isolation.

According to him, it needs an object and other reference points to be part of normal conscious process. Is it possible that the objectless, preverbal state of acute pain to which Scarry refers is actually the experience of the quale of pain in isolation? Consciousness functions as a system and is organized to incorporate multiple qualia to create a sense of time, place, and reference. The suspension of language and object orientation that occurs with pain may be due to the experience of an isolated quale. The intensity with which one experiences pain could be correlated to the solidarity of one's experience of the quale of pain. As pain increases, the recognition of other qualia drops away.

Alternatively, it may be that Scarry and Edelman are referring to different types of pain that are processed in different ways. Scarry's writing emphasizes experiences of acute physical pain. This type of pain may not be a type of qualia at all. Even though it takes the afflicted to a preverbal state, it may be doing so because it is operating on a preconscious plane. Acute pain may occur outside the borders of consciousness. The pain that Edelman refers to as a quale may be functioning as an extension of sensation, which would include chronic pain and the experience of pain when the consciousness returns to a verbal state.

MUSIC AND MEANING

Music can communicate tremendous meaning. It can touch the passions and ignite our souls. It can move people to ecstasy, influence them, open up new worlds to them, and reinforce current identity. The power of music has been argued from the Greeks (e.g., Plato and Aristotle) to Tipper Gore (e.g., the Parents Music Resource Center, PMRC, of 1985). Music is a highly organized human activity that engages abstract thinking and is capable of communicating emotions and perceptions that are difficult or impossible to express through words (or through words alone, in the case of song). We cannot express pain through language without the use of metaphor, but with music we may be able to come closer to direct communication of the experience.

Theoretically, anything can be represented musically. In my exploration of musical examples that may be interpreted as representations of physical pain, I have come across many that are fitting but could also be interpreted as musical examples of violence or emotional turmoil, such as anger. It is likely that some types of physical pain can be represented by music more readily than others. Care must be taken to distinguish between representations of physical pain, emotional pain, violence, and illness.

To interpret music is to make a personal statement, not about what the music means, but what the music means to the individual making the judgment. As Robert Walser stated, "Underpinning all semiotic analysis is, recognized or not, a set of assumptions about cultural practice, for ultimately music doesn't have meanings; people do" (11, pp. 135–136). Although meaning in music is different for different individuals, this does not suggest that meaning in music is arbitrary. As Edelman (10) pointed out, languages can be powerful tools *because* of their ambiguity. Metaphor leaves room for interpretation, which is one of its great strengths.

MUSIC AND TIMBRE

There has been relatively little research done in the realm of timbre, partly because there is no single, widely accepted definition of this multidimensional perceptual attribute of music (12). Dowling and Harwood (13) describe timbre as everything that is left over once pitch, loudness, and duration have been accounted for. They go on to explain that timbre, or tone-color, incorporates the psychological properties of sounds that make them qualitatively distinguishable from each other even if they have the same pitch and loudness. All musical sounds originate from the vibration of some physical medium (i.e., a string on a guitar or one's vocal chords). Physical objects do not vibrate in simple patterns but in several modes at once (6). Combinations of different vibrations create different tone *colors*. Attack (how a sound is initiated) and flux (how a sound changes after it begins) are also important components of timbre. Timbre is how one distinguishes the sound of a flute from a clarinet or how one is able to instantly recognize the voice of his or her mother on the phone—her voice has a distinct timbre. Levitin (6) pointed out that human perception of timbre is so acute that we can even tell when someone close to us is happy or sad or coming down with a cold, all based on the timbre of that person's voice.

MUSIC AND LYRICS: MASOCHISM TANGO

I am quite fond of the work of musical satirist Tom Lehrer (b. 1923), and the "Masochism Tango" (14) is one of his most popular tunes. It is a cheery, upbeat song that describes the sadomasochistic use of physical pain for pleasure. The lyrics of the last four verses are as follows:

> Your eyes cast a spell that bewitches.
> The last time I needed twenty stitches
> To sew up the gash
> That you made with your lash,
> As we danced to the Masochism Tango.
> Bash in my brain,
> And make me scream with pain,
> Then kick me once again,
> And say we'll never part.
> I know too well
> I'm underneath your spell,
> So, darling, if you smell
> Something burning, it's my heart... [hiccup]
> 'Scuse me!
> Take your cigarette from its holder,
> And burn your initials in my shoulder.
> Fracture my spine,
> And swear that you're mine,
> As we dance to the Masochism Tango.*

* Lyrics to the song "Masochism Tango" by Tom Lehrer.

Lehrer's lyrics focus almost exclusively on agents of pain. The words explain how pain will be achieved, but descriptions of the pain are lacking. This is consistent with the ways in which we communicate experiences of physical pain (8). The lyrics of "Masochism Tango" are representational of pain, but they have been coupled with music that sets the subject matter ironically. The pleasant, sexually playful music deemphasizes the brutal nature of the violence depicted in the lyrics. Without the consideration of the lyrics, the music cannot stand on its own as depicting pain.

Daniel Levitin's *The World in Six Songs* (15) focuses study upon music with lyrics. In his examination, he considers the emotional impact and meaning of the text alongside the meaning and impact of the music. There are many song lyrics about painful experiences and violence, but they deal with agents and consequences of pain, rather than the sensation of pain. My examination of meaning in music does not consider lyrics, but focuses solely on musical representation through the elements of melody, rhythm, harmony, and timbre. In essence, I seek to examine those musical elements that go beyond words.

MUSIC AS ABSOLUTE: PACIFIC 231

In spoken language, we can refer to specific ideas, for example, "I am enjoying the banana pudding. Thank you." On the other hand, music is a means of abstract and emotional communication. Melody and rhythm are self-referential, meaning that music, without text, can be about the music and not about anything external. A Mozart symphony is not intended to paint a picture, set a scene, or be descriptive. Musicologists describe this as absolute music. Absolute music, such as a Carlos Santana guitar solo or a Mozart symphony, communicates powerful human emotion and creates social connection.

Some music does intend to paint specific pictures of external objects. Music that attempts to do this, not through lyrics but through pure musical elements, is called program music. There are symphonic pieces about battles and Don Quixote and the countryside. There are songs that use the instrumental accompaniment to imitate spinning wheels and heartbeats and foreign places. Music is used again and again throughout time to represent external objects and ideas, with varying degrees of aesthetic success.

Pacific 231 (1923) by Arthur Honegger is a programmatic piece for orchestra that is known for depicting a steam locomotive. Train buffs will recognize *231* as a reference to wheel arrangement and thus recognize the Pacific 231 as a specific type of locomotive built for heavy loads and high speeds. It is a popular piece often cited as a quality example of program music. However, when Honegger wrote the piece, he did not have a train in mind. He wrote it as an abstract exercise to give the impression of a mathematical acceleration of rhythm, while the pace music actually slows through the movement (it is part of a three-movement collection). He later gave it the title simply because the original title *Movement Symphonique* was bland. In his autobiography (16), Honegger notes that a critic writing about the piece had confused *Pacific* with the Pacific Ocean, and wrote about how the music evoked smells of the open sea. This suggests just how easily one can be swayed by external labels and imagery in the interpretation of musical meaning.

It is easy to argue for the effectiveness of the depiction of a train in *Pacific 231*, but this imagery relies on the listener already possessing knowledge of what sounds are imitations of the subject in order to work as a programmatic piece. Will someone listening to the piece without knowing the title or popular interpretation identify it as a train? Is the music just as aesthetically pleasing without the programmatic meaning? If a listener has never been exposed to trains in the past, the music can still be enjoyed as an exercise in the building of momentum and power, but it has a different meaning. The music does not clearly communicate the concept suggested by the title without the addition of that external context in which to place it. The emotional impact of the music may still be present, but the representation is not. Returning to the issue of pain, this raises the question: Can pain be represented as a direct emotional or psychological component of music, or does it rely upon an external story to be conveyed? Can the communication of physical pain fall under the rubric of absolute music, or is it reliant on program music and lyrics to be conveyed?

I have chosen to examine examples from three broad genres of music familiar to Western audiences: film music, classical music, and popular music. Do these examples effectively communicate the concept of pain? What musical elements are being used to represent pain? Is there a reliance on song texts, or can the concept be carried by musical timbre and structure alone? Is the musical example aesthetically pleasing within its genre, style, and culture?

FILM MUSIC: PSYCHO

One of the most famous musical moments in Hollywood film history is the shower scene in Alfred Hitchcock's 1960 thriller *Psycho*, the film score composed by Bernard Herrmann. The scene begins in silence as the character Marion Crane (played by Janet Leigh) showers. From the viewpoint inside the shower, the audience sees the shadow of a knife-wielding figure appear on the curtain. As Marion is attacked and stabbed repeatedly with the knife, the action is accompanied by a high-pitched, dissonant, repetitive violin motive, which mimics the slashing of the knife. The music at this point is often described as razor-sharp, piercing, knife-screeching, and violent (17–19). Originally, Hitchcock wanted the entire shower scene to play without music, but Hermann had written music for "The Murder" and insisted that Hitchcock give it a try. The director agreed that the music made the scene more effective (17). What was Herrmann intending to depict with the music here? The music may represent the knife; the music is sharp and pointed. It may represent the violence of the act; it is stabbing and forceful. It may represent the psychological panic of the situation; it is without melody, stark and focused. It may represent the victim's pain as the stabbing knife cuts her body; the dissonance and tension in the strings scream. The violins start on the same pitch, add the dissonant major seventh, each successive bar adding more discord to the tonality. Upon repeat of the music, glissandos are added to the timbre. Bruce describes the sound traveling through the orchestra as being that of a scream. Herrmann suggested that the music here is the shriek of birds, associated with the imagery of and musical references to birds used throughout the film (17).

"The Murder" music in *Psycho* may be interpreted in a variety of the aforementioned metaphors. If the music is depicting the weapon, which is the cause of pain, we may interpret that aspect of the music as also representing the pain of Marion Crane. The screaming nature of the music takes on the preverbal qualities of pain discussed by Elaine Scarry (8).

CLASSICAL MUSIC: THRENODY FOR THE VICTIMS OF HIROSHIMA

In the later 1950s, there was a trend in classical composition toward shaping music through organization of timbre and texture over the combination of such things as melody, harmony, and rhythm. Polish composer Krzysztof Penderecki (b. 1933) was a leader in this style, his works being clear in intent and emotional impact and, consequently, well received.

His most-widely known work is his *Threnody for the Victims of Hiroshima*, for 52 stringed instruments, completed in 1960. The work was originally called *8'37"*, referring to the length of performance time. Penderecki had conceived of the work in an abstract sense, giving rise to the abstract title, but upon hearing the work performed, he was struck with the emotional charge of the work. In searching for an association, he arrived at the decision to dedicate the work to the victims of Hiroshima (20).

Threnody is a "tone cluster" composition in which segments of music are built around simultaneously sounding blocks of pitches that are very closely spaced, incorporating both chromatic and microtonal pitches that are not part of the traditional Western scale. There is both large- and small-scale structure built into the *Threnody*, with sections purposely giving the impression of chaos (though many of these chaotic sections are actually highly organized through serial techniques). The result is a total loss of tonality. Individual pitches are lost within the mass of sound in which they are embedded, leaving the listener with no point on which to focus. The effect is that of noise, which at times in the composition can be likened to screaming and moaning. It is dissonant and humanoid to the point of being disturbing and disconcerting. It does not emanate from a single discernable source. It is this lack of identification that lends the work to possible interpretations as a representation of direct physical pain. Scarry's (8) theories on pain as an objectless state of consciousness are reflected in the objectless nature of the musical timbre.

Even though it was not Penderecki's original intention when he was composing *Threnody*, it comes potentially close to representing physical pain without reliance on context, because it creates and manipulates an objectless timbre. Listening to it can be difficult, uncomfortable, terrifying, and cathartic, yet also a pleasure to experience as it communicates directly and effectively. It is a well-constructed masterpiece because of its delicate organization and emotional impact.

POPULAR MUSIC: BACK TO BACK

Heavy metal music emerged in the late 1960s and early 1970s in the working-class culture of Birmingham, England, exemplified by Ozzy Osbourne, Black Sabbath, and Judas Priest (11). The genre was defined and codified in the early 1970s with the arrival of bands such as Led Zeppelin and Grand Funk Railroad. These bands laid out the sound that would be identified as heavy metal, with Led Zeppelin becoming the most celebrated heavy metal band of all time (21). By the 1980s, glam metal and speed metal were established subgenres of heavy metal. During this time, mainstream bands such as Iron Maiden, Mötley Crüe, Ratt, Twisted Sister, and Scorpions played arenas, while a subculture of metal developed in club venues. Underground styles blended together elements from both heavy metal and punk, creating music that was more violent, aggressive, and noisy than the heavy metal with which most people had come into contact. Fast, adrenaline-pumped tempos and lyrics dealing with social issues in blunt and sarcastic ways were taken from the hardcore punk genre, while emphasis on guitar virtuosity was taken from heavy metal, and then applied to the whole band. Fusions of these styles were dubbed thrash, speed, and death metal, through which metal reached a new commercial peak in the 1980s and early 1990s (21). Commercially successful thrash metal style can be seen in the music of the bands Metallica, Slayer, Anthrax, and Megadeth.

As one traces the development of heavy metal, or for that matter any musical genre in Western culture, a trend toward greater dissonance and a pushing of boundaries can be found. For example, the general trend in classical music has been a slow push toward greater acceptance of dissonance and more incorporation of timbres and harmonies that would have previously been classified as noise. Similarly, rock and roll, which was becoming increasingly gritty over the course of the 1960s and 1970s, gave rise to heavy metal that placed increased emphasis on harsh distorted timbres and loud volumes, pushing music closer toward the realm of noise. Harris Berger (22) explained how the history of heavy metal can be summed up as a progressive quest for ever-heavier music. The rise of thrash metal gave angry, young working-class listeners a sound that was more relevant to their punk-influenced tastes than the MTV-embraced pop metal being produced by Van Halen and Def Leppard (23).

According to Robert Walser (11), the overarching aesthetic behind heavy metal is the concept of power. The musical gestures, social conventions, and visual images that are part of heavy metal all point toward power and empowerment. The music is dominating and aggressive. No apologies are made. Nothing is held back. The sheer volume and intensity at concerts empower the audience, as evidenced by their shouting and headbanging. The performers display their sexual, vocal, and virtuosic power and command.

The most important musical element found across styles of heavy metal is the use of heavily distorted electric guitar, which not only adds noise to the timbre, but also boosts upper and lower harmonics, creating greater sustain and power (24). This distortion of sound is created when the electronic components carrying the sound are pushed beyond the limit of the amount of power they can optimally process. Amplifiers are designed to take a signal and increase the volume without distortion, but stressing this equipment to the point of noise (or what was previously defined as noise, but is now acceptable musical timbre) is an act of power (11).

Heavy metal vocal timbre is also distorted and overextended. Screaming is distortion, caused by an excessive use of power, as the capacity of the vocal chords is pushed beyond their normal limit. Typical metal singing is rough and guttural, and the music often incorporates screaming and other sounds of physical and emotional exertion. Distortion in both the guitar and human voice are signs of extreme power and intense expression, as both push the limits of their physical media. Historically, this overload of power may be viewed as equipment failure, but because in this aesthetic it is intentional, it becomes an expression of power, and becomes music instead of noise (11).

A current term associated with the thrash metal and the fusion of punk and metal styles is metalcore, which is taken from heavy *metal* and hard*core* punk, though tempos in metalcore tend to be slower than those found in thrash metal (25). The Chariot is an Atlanta-based band that considers itself part of the metalcore genre. "Back to Back" from their 2007 album *The Fiancée* may be interpreted as including representations of physical pain (26). The lyrics are as follows:

> This is the last chance you got, open wide.
> We both know we're both going to die, but there's a difference with you and I.
> You want Peace but refuse the fight.
> So shake hands with change tonight. Bathe in armor, for death feeds.
> O' death, don't bother me tonight.
> Be grace, my God, and stand still.
> Be grace, my God, and send more minutes.
> For churches have nuns, cowboys got guns and everyone's waiting to die.*

The lyrics do not point specifically to experiences of physical pain, but the music suggests otherwise. The song begins with multiple people screaming at full force, with frenetic fast-as-possible drumming underneath. There is also distorted guitar, which slowly, almost imperceptibly rises in pitch as the screaming continues, building the tension of the introductory section. Additional screaming is added as the intensity increases. High, rhythmic power chords played in unison with the high-hat cymbals mark the beginning of a new section (11 seconds into the song), after which all sound drops out save for high-pitched distortion (16 seconds into the song). At 20 seconds, the high-pitched rhythmic power chords are now combined with a driving bass line and drums as the lyrics are sung/screamed, while the song goes through several shifts in tempo, becoming slower and more aggressive until the very end where high feedback is again the primary timbre accompanying the final line of text. The song features multiple dramatic shifts in tempo and texture and highly virtuosic, carefully synchronized playing, all of which are important aesthetic elements of thrash metal and metalcore.

SCREAMING AS MUSIC

All three examples make use of dissonant timbre that replicates screaming, and it is this stylized musical use of screaming that I find most salient. Elizabeth Tolbert (27)

* Lyrics to the song "Back to Back" by Josh Scogin.

notes that the presence of stylized crying is present across a wide range of intercultural contexts, which are found throughout the world. She has written on the Finnish-Karelian Ritual Lament, which incorporates ritualized crying, sobbing, and wailing into the music. She points to the significant power of the human voice to arouse attention and emotion and to make immediate connection. The use of the human voice as an instrument allows the music to create an immediate social bond between singers and listeners. In a discussion of Tolbert's research, Nicholas Wolterstorff (28) commented on how important the timbral use of weeping in the Finnish-Karelian Ritual Lament is to the power of the music. Similarly, the affective power of heavy metal, avant-garde classical, and descriptive film music stems from the incorporation of screaming timbres. The intensity and emotional impact of the sound is multiplied because of the human quality of sound, whether produced through an instrument or vocally. The timbre creates immediate, intense social connection.

Laurence Kirmayer (29) relates an anecdote about his brother who plays very intense avant-garde jazz, which family members dutifully go to watch, though they cannot stand it, saying that it causes them physical pain. He notes that this raises the issue of training one's ear or learning a vocabulary so that the assaulting sounds cease to be noise and can be recognized as music. All three musical examples discussed here are quality representatives of their respective genres. Each beautifully handles their understood constraints and aesthetics, coming to creative, elegant, and emotionally intense solutions. Listeners being exposed to these genres for the first time may perceive only noise and chaos. Each of these examples push the current system of tonality to the edge of its defined parameters. The timbres of screaming may be reflective of anger or emotional pain. However, there are circumstances where interpretation can point convincingly toward a representation of physical pain.

CONCLUSION: PUSHING BOUNDARIES

Pain can occur when boundaries of either body or consciousness are pushed or broken. Common touch can become painful through increased intensity and repetition or through the perception of an increased intensity (5). The musical examples I have chosen as coming close to representing pain do so because they push aesthetic experience outside the normal parameters of earlier definitions of music. As timbre edges closer to the point of noise, it not only approaches the sound of screaming, which has close associations with pain, but becomes objectless, similar to the immediate objectlessness of pain. This pushing of boundaries in both musical and neurological perception moves one closer to the experience of pain. This experience of physical pain can be communicated through absolute music, but only through extreme timbres that reflect an extreme state of consciousness, or a preverbal, preconscious state.

REFERENCES

1. Bennett, G.J. 2000. Update on the neurophysiology of pain transmission and modulation: Focus on the NMDA-receptor. *Journal of Pain and Symptom Management* 19(1) (suppl.): s2–s6.

2. Clark, M.R., and G.J. Treisman. 2004. Neurology of pain. In *Pain and depression: An interdisciplinary patient-centered approach*, eds. M.R. Clark and G.J. Treisman, 78–88. Basel, Switzerland: Karger.
3. Coakley, S., and K.K. Shelemay (eds.). 2007. *Pain and its transformations: The interface of biology and culture,* Cambridge, MA: Harvard University Press.
4. Cross, I. 2003. Music as a biocultural phenomenon, In: *Annals of the New York Academy of Sciences: The neurosciences and music. Vol. 99*, eds. G. Avanzini, C. Faienza, D. Minciacchi et al., 106–111. New York: New York Academy of Sciences.
5. Horn, S., and M. Munafò. 1997. *Pain: Theory, research and intervention.* Buckingham: Open University Press.
6. Levitin, D.J. 2006. *This is your brain on music: The science of a human obsession.* New York: Plume.
7. Vendix, P. 1997. Cognitive sciences and historical sciences in music: Ways towards conciliation. In: *Perception and cognition of music*, eds. I. Deliege and J. Sloboda, 69–79. East Sussex, UK: Psychology Press.
8. Scarry, E. 1985. *The body in pain: The making and unmaking of the world.* New York: Oxford University Press.
9. Johnson, M. 1990. *The body in the mind: The bodily basis of meaning, imagination, and reason.* Chicago: University of Chicago Press.
10. Edelman, G.M. 2004. *Wider than the sky: The phenomenal gift of consciousness.* New Haven: Yale University Press.
11. Walser, R. 1993. *Running with the devil: Power, gender, and madness in heavy metal music.* Hanover, NH: Wesleyan University Press.
12. Deliege, I., and J. Sloboda (eds.). 1997. *Perception and cognition of music*, East Sussex, UK: Psychology Press.
13. Dowling, J.W., and D.L. Harwood. 1986. *Music cognition.* Orlando: Academic Press.
14. Lehrer, T. 1959. Masochism tango. *More of Tom Lehrer.* Lehrer Records: TL-102/102S, vinyl record.
15. Levitin, D.J. 2008. *The world in six songs: How the musical brain created human nature.* New York: Dutton.
16. Honegger, A. 1966. [W.O. Clough trans.]. *I am a composer.* New York: St. Martin's Press.
17. Bruce, G. 1985. *Bernard Herrmann: Film music and narrative.* Ann Arbor, MI: UMI Research Press.
18. Hickman, R. 2006. *Reel music: Exploring 100 years of film music.* New York: W.W. Norton.
19. Timm, L.M. 2003. *The soul of cinema: An appreciation of film music.* Upper Saddle River, NJ: Prentice Hall.
20. Jacobson, B. 1996. *A Polish renaissance.* London: Phaidon.
21. Romanowski, P., and H. George-Warren (eds.). 1995. *The new Rolling Stone encyclopedia of rock & roll*, New York: Fireside.
22. Berger, H.M. 1999. *Metal, rock, and jazz: Perception and the phenomenology of musical experience.* Hanover, NH: Wesleyan University Press.
23. Garofalo, R. 1997. *Rockin' out: Popular music in the USA.* Boston: Allyn and Bacon.
24. Walser, R. 1991. The body in the music: Epistemology and musical semiotics. *College Music Symposium* 31: 117–124.
25. Mudrian, A., and J. Peel. 2004. *Choosing death: The improbable history of death metal and grindcore.* Los Angeles: Feral House.
26. Chariot, The. 2007. Back to back. *The fiancée.* Tooth & Nail: TNL49015B.2, compact disc.

27. Tolbert, E. 2007. Voice, metaphysics, and community: Pain and transformations in the Finnish-Karelian ritual lament. In: *Pain and its transformations: The interface of biology and culture*, eds. S. Coakley and K. Kaufman Shelemay, 147–165. Cambridge, MA: Harvard University Press.
28. Tolbert, E., and N. Wolterstorff. 2007. Discussion: The presentation and representation of emotion in music. In: *Pain and its transformations: The interface of biology and culture*, eds. S. Coakley and K. Kaufman Shelemay, 208–209. Cambridge, MA: Harvard University Press.
29. Brust, J., S. Coakley, H. Fields, J. Jackson, L. Kirmager, E. Tolbert, T. Weiming, R. Wolf, and C. Woolf. 2007. Discussion: Neurobiological views of music, emotion, and the body. In: *Pain and its transformations: The interface of biology and culture*, eds. S. Coakley and K. Kaufman Shelemay, 210–216. Cambridge, MA: Harvard University Press.

11 Beyond Technology
Narrative in Pain Medicine

Lucia Galvagni

CONTENTS

INTRODUCTION

This chapter aims at providing an overview of the contribution that narration and narrative medical ethics may offer to pain medicine in the attempt to approach the experience of pain with resources and modalities that may allow forms of communication around this experience and, hence, providing better care. Setting off from initial considerations on narration and communication in the experience of illness, narrative medical ethics is analyzed; a phenomenology of the experience is charted, considering its impact on the three different dimensions of embodiment, temporality, and subjectivity; and the voices and narratives of some patients who describe their pain in figurative language are reported. The chapter concludes with an exploration of the possible meaning of metaphors and images in the experience of pain-as-illness. Communication plays an essential role in pain medicine, as it helps to develop more significant care relationships. Through them, the pain patient feels listened to, looked after, and cared for in the complexity of his or her own experience.

NARRATIVE AND COMMUNICATION IN THE EXPERIENCE OF PAIN-AS-ILLNESS

Illness, the body, and pain are often represented using figurative language, particularly by the patients. Such language is characterized by the presence of narratives, metaphors, and answers composed by means of images. Many different narratives can be generated around an illness, and voiced by patients, but also recounted by physicians, nursing staff, and caregivers; literary fiction and poetry provide notable accounts of the experience of pain—and the illness it evokes—and these appear especially rich in images and meanings: compared to the narratives recounted by patients, the narrative accounts provided by the caregivers often reflect a different language and other ways of perceiving and talking about the same situation.

Illness inserts itself into a life process as a moment in which the ordinariness of a personal history is interrupted, giving rise to a crisis, a caesura, a cut in the lived experience, regardless of the evolution that particular life will have and of the consequences that are going to follow. This life history, while remaining that of the same individual, is now read and narrated from a different perspective, sometimes even by using new categories.

Intractable pain in its own way can constitute an extreme situation for the individual on an existential level: the situation is particularly significant because it also sheds light on normality and informs us about the human condition (1).* In this way, the physical event of pain becomes the experiential illness of pain. While, or perhaps because this is a subjective experience of the inner dimensions, when faced with the extreme, one perceives the need to report about what is being experienced and lived through.

The urge to confer unity to one's personal history, while always present, makes itself particularly felt during illness. Within this urge, one can trace a narrative dimension: the illness becomes the subject of a narrative, thus constituting the beginning of a communication aimed at the construction and definition of an identity, which is also a life profile. Even the reporting of a case history gives rise to a narrative and hence to the narration of a lived experience or of a life. In fact, self-narration is never performed merely for oneself, nor is it ever simply about oneself: the act of engaging in a relationship and hence the recognition of the other and the constitution of a relationality are integral to the act of narration.† As Byron Good wrote: "At the same time that it offers suffering, pain and misery, when illness is transformed as narrative it has the potential to awaken us to conventionality and its finitude, provoking a creative response and revitalizing language and experience" (3, p. 165).

Whereas illness may be told through narrative, the suffering that goes along with it often remains unvoiced, unable to find a means of expressing itself or addressing language. Simone Weil affirmed that the type of thinking that is characteristic of suffering is not discursive in that it does not display the features of dialectic and deductive reason. Therefore, the experience of pain develops and strengthens our

* Todorov's reflections on the extreme condition and upon being exposed, which he articulates starting from an analysis of dramatic historical situations of the twentieth century, may also be extended, with due differentiation, to the experience of illness, which in its own way is characterized as an extreme situation (1).

† There are many interesting and valuable contributions on the significance of narration in medicine. Brody underscores the importance of medical ethics acquiring awareness of the narrative dimension that always characterizes illness (2). See also the above-mentioned work by B.J. Good (3). Different types of narratives of an illness may be produced, such as the story, which draws upon metaphors, allusions, and symbolic representation, or the aetiopathological explanation, which describes the causes and course of an illness in an objective and univocal fashion, or again through a clinical anamnesis in which the illness is reconstructed through knowledge and medical observation. See Cattorini, P., 1994, *Malattia e Alleanza*, Pontecorboli, Firenze (4).

intuition, allowing us to draw upon different modes of representing, symbolizing, and narrating our personal history and, within it, illness. The silenced component of suffering pain—silenced often for the very subject who is experiencing it—could perhaps be helped to emerge through such modalities.

It is through a narrative, although not exclusively so, that the patient's attitude toward his or her pain qua illness can emerge, giving voice to the patient's personal way of living and perceiving his or her condition. Putting such a lived experience into words is a first step on the path of change and perhaps recovery. In this narrative process, the metaphors through which each person represents his or her own lived experience and life history emerge, images that reach beyond the limits of conceptuality and, in so doing, concentrate and display a true and proper *life-world*.*

By listening to the narration of illness, we can understand what is at stake for the patient, and so help to chart the direction along which an intervention is to take place,† as specific to the vulnerability and weakness of persons as patients.

Pain can be described and defined as an emotional experience, but it can also be interpreted within a biocultural framework (5). Whereas in the modern era the dominant models where characterized by dualistic, mechanistic, and reductionist thinking, in postmodernity the interpretation of pain as disease and illness is inseparably linked to the cultural contexts within which it appears and is experienced. Illness is not just the result of organic and bodily dynamics, as much as it is an experience constructed at the crossroads of biology and culture. In such a context, the perception and interpretation of pain also undergo a change (5). Pain, chronic pain in particular, cannot be explained by anachronistic dualisms. A closer look at the cultural approach to pain shows that ethnicity, gender, and beliefs all have significant influence upon pain, its perception, its experience, and its expression.

How then, could one try to interpret pain? Are there measurements and technologies that could acknowledge its existence? A narrative approach could help to shed light on the conditionalities, dynamics, and experiences of pain that are difficult to objectify and for this reason are at risk of remaining unheeded. Rita Charon and Martha Montello observed that, pain-as-illness

> Is a biological and material phenomenon, the human response to it is neither biologically determined nor arithmetical. In extending help to a sick person, one not only determines what the matter might be; one also by the necessity of illness determines what its meaning might be. Such a search requires the narrative competence to follow the patient's narrative thread, to make sense of his or her figural language, to grasp the significance of stories told, and to imagine the illness from its conflicting perspectives.

---

* Husserl employs this term (in German: *lebenswelt*) to indicate the world of our common and immediate lived experiences.

† The modern-day difficulty in managing a therapeutic exchange aimed at the healing of an individual in a particular and unique situation could perhaps be related to the difficulty in establishing and consolidating a significant relationship between the person who offers care and the person who will receive care. Lacking the encounter and the privileged time of listening, the real possibility of an effective, complete, and respectful treatment is undermined. This should be a further reason to rethink the relationship between the clinician—the person who performs the act of caring—and the person who receives treatment, and to understand what kind of meaning should or could be endowed.

> Narrative approaches to ethics recognize that the singular case emerges only in the act of narrating it and that duties are incurred in the act of hearing it. (6, p. ix)

There is a narrative structure that permeates and characterizes our lives; therefore, Bruner asserts that it is vital, "Not…to detechnicalize sickness or health care, but you've got to rehumanize it as well—relate it to life" (7, p. 8).

NARRATIVE MEDICAL ETHICS

The repeated formulation of a narrative makes it possible for explanation and, hence, the understanding it can foster. This occurs regularly and is common practice in psychotherapy. Here the narration, as related to the analyst or therapist, makes it possible to elaborate a new perspective and come to new interpretations of the patient's history. By reconstructing the past in narration, this generates conditions that make possible an opening up to the future. Ricoeur reminds us that, in a story and in narration, the dynamics associated with initiative, imagination, and intersubjectivity come to the fore, building bridges between experience and memory of the past, initiative pertaining to the present, and choice, which allows access to the future. At the intersection of such dynamics, freedom from the constriction of pain may find expression (8).*

The narrative approach has gained significant attention in both ethics and medicine. Such interest has become increasingly important in the bioethical debate, which, in its hermeneutical and phenomenological reflections, underscores the moral dynamics that are at stake during the clinical encounter. These bioethical reflections have also emphasized the importance of bodily experience in illness and in the dynamics of care, with significant implications for meanings of pain and its care on an anthropological level (2).†

Rita Charon pointed out that the world as we know it undergoes a transformation when we are stricken with pain, because this has consequences not only "in the corporeal aspects of everyday life—now with pain…but in the deepest wells of meaning now with limits, regrets, forced separations, final plans" (14). Pain and its manifest illness are not just abstract concepts, says Charon, but entail material issues, isolation, loneliness, and sometimes hopelessness.

In such situations, forms of distancing and separation often evolve between patients and doctors, and also between patients and nursing staff. Another kind of separation is marked by the divide between *healthy* and *ill*, based on a profound norm, which when considering pain, is strong and often irrevocable. Susan Sontag stated: "Illness is the nightside of life, a more onerous citizenship" (15).

---

* See Ricoeur, P., 1986, *Time and Narrative*, Chicago: Chicago University Press (8); Ricoeur, P., 1986, *From Text to Action*, Evanston, IL: Northwestern University Press (9).

† See Brody, H., 1987, *Stories of Sickness*, New Haven, CT: Yale University Press (10); Zaner, R., 2004, *Conversations on the Edge*, Washington, DC: Georgetown University Press (11). It must be said that the narrative approach, which draws upon the narration of case histories as well as personal narratives, has also met with criticism within the North American bioethical debate. See, in particular, Lindemann Nelson, H., 1997, *Stories and Their Limits*, New York, Routledge (12); Chambers, T., 1999, *The Fiction of Bioethics*, New York, Routledge (13).

In medical contexts, Kathryn Montgomery Hunter suggested that the narrative perspective should be adopted starting from the reconstruction of the case history gathered during the clinical interview (16). A narrative vision of medicine could help to sustain physicians' understanding of the medical care process and would allow appreciation of the intersection between their life and the life of their patients. Such life history must bear careful evaluation when deciding upon what treatment to prescribe: to be sure, the ways the person in pain will integrate the physician's narrative within his or her life story may play a relevant role with regard to what, why, and the extent to which a particular therapeutic process will be undertaken.

As Pia Bülow observed, personal narratives draw on different genres, comprising life histories, anecdotes, case histories, and myths, and through each it is possible for individual and collective experiences to merge into the therapeutic process (17). Sociologists have referred to this as "experiential knowledge," where experiences and information are shared, and a sense of community grows out of such sharing.

Rereading ethical narrative from a psychological and cognitive perspective, Jerome Bruner pointed out how narration is tied both to telling a story and to knowing, and that narratives depend on language as much as on social and cultural dynamics.* A narrative is often constructed by using or referring to cases, but these are placed within a precise temporal framework. It is perhaps thanks to these rather well-defined and recurrent structures of a narrative that prototypic metaphors for the human condition may have arisen (19).

In narrative, Bruner observed, the presence of an audience or listener is always assumed, so if on the one hand a story defines an identity and ratifies its moral autonomy, on the other hand its relation to others, along with the commitment this implies, is just as essential to the act of narrating. The search for balance between autonomy and engaged action, difficult as it may be to achieve, is read by Bruner as the defining trait of the human condition. In many ways, pain impacts this balance. The construction of an identity, and hence of a moral identity in particular, would therefore seem to depend on our ability to narrate ourselves, a skill that draws upon the various cultural forms within which we have been brought up and in which we are immersed. The dialectical dimensions of the culture may at times be found directly reflected in the histories, narratives, and personal identities that are part of it. Thus, pain too, is interpreted as strongly bound to an individual's perception, to the influence of society, and the models society provides for dealing with it.

How then is the issue of pain and the illness it evokes to find collocation within a narrative approach? Pain is generally experienced as an alienating presence: "the patient often experiences pain as an intrusive foreign agent" (20, p. 5). Anthropologists remind us that the description of pain provided by medicine is based on recurrent reference to dichotomies: the juxtaposition of physiology and pathology, body and mind, objective and subjective, real and unreal, and natural and artificial. The experience of chronic pain cannot be reduced to the experience of an individual as a separate, isolated being, but can only be understood within an interpersonal dynamic; because in almost every case, pain and suffering concern and involve the family and the relational context of the individual.

* See also Stern, L. and L.J. Kirmayer, 2004, *Transcultural Psychiatry* 41(1): 130–142 (18).

Pain is expressed through both verbal and bodily language, and the caregiver should aim at understanding both forms of communication. The language of pain appears to have been modified by the ever-increasing presence of logic and technological rationality that are now typical of medical discourse; in such a way, however, the personal and relational components of pain risk are being silenced. In clinical practice, and in pain medicine in particular, one encounters histories and situations where the person undergoes a crisis that affects different areas of the body and embodiment, subjectivity and objectivity, time and the sense of time, and narrativity and the relational aspect that characterizes it. In such a scenario, a hermeneutical and narrative ethical approach may exercise an important function. Let us therefore try to read the experience of pain-as-illness attempting a kind of phenomenology.

A PHENOMENOLOGY OF PAIN-AS-ILLNESS

Illness can be characterized as a time of crisis for the person, in which the person assumed the subjective and objective mantle of *being* a patient. In other words, it is the point where the objective events of the body manifest subjective experience in the embodied person. At this crucial moment, medicine is often called upon to intervene and render care of the patient as a *suffering person*. Pain-as-illness can be defined on at least three different but complementary levels: a crisis of the body, a crisis of the sense of time, and a crisis of the subject. In this condition of compromised balance, the care process takes place and form.

EMBODIMENT

The experience of pain-as-illness heavily affects the body and, hence, may radically call into question our self-perception. Also, when the pain is persistent, the person goes through the repeated crisis of questioning many aspects and conditions of life that were previously taken for granted.

The body is our primary means of being in the world and of entering relationships: it is through our body that from birth—in fact even earlier—we relate to the world, and according to its motions, we develop and structure different relationships within it. The messages of society and culture are also mediated by our bodies. Psychologists consider body language to be one of the most important forms of nonverbal communication, to the point where the body is often regarded as a theater, a sort of stage upon which our lived experience—more or less explicit, in greater or lesser depth and detail—finds representation.

In every bodily experience, we can witness an ambiguity: the German language has two different words for the body: *Körper*, the body in its physical and organic aspects, and *Leib*, which refers to the body as living and lived, perceiving and perceived. Husserl draws upon this distinction, stating that our body is the only one that is not a mere physical body (*Körper*) but a proper living body (*Leib*), and thus underscoring the distinction between the body objectified by science, available for anatomical and physiological analysis and examination, and one's own body as concretely experienced and lived in real life.

From Husserl's phenomenological considerations, Gabriel Marcel stated that we are a *Leib* and have a *Körper*. The copresence of these two dimensions becomes even more relevant in some moments and experiences fundamental to human life, such as love, or pain. The body, which mediates our being in the world and our relating to it, always conveys a meaning; the different forms of alienation and expropriation of the body thus reflect a loss of meaning, and it is for this reason that the experience of pain-as-illness is difficult, wearisome, and often dramatic.

The body in pain is confronted with its own impotence: losing its habitual capability to function and interact with others and with the outside world, it risks no longer being able to function as a site and subject of definite actions or may even impose itself as an obstacle against the subject's will. The body is suddenly center stage, a different body than that which was familiar, it forcefully claims its own right to space, time, attention, and care.

The sense of estrangement that comes with this experience casts the body simultaneously as *Körper* and *Leib*, and inevitably calls into question one's sense of self. In pain, the body, which voices and represents the subject, becomes a simulacrum, no longer inhabited primarily by the self, as much as by the pain and its illness which enters the stage, marking the body with its organic presence. And yet it is precisely in, and coming from, such a radical modification of perception and lived experience, that each person then attempts to rewrite his or her identity as transformed by the illness experience of pain.

The intervention of the physician or nurse to take care of the pain patient's body places the physician or nurse into contact with the suffering person. The body represents the mediating element of this encounter, through which the care process acquires form.*

Temporality

Pain can also bring on a profound crisis in relation to the temporal dimension, for it is often accompanied by an alteration of time perception.† Thus, change can be perceived along three dimensions: with reference to one's own personal time and thus to one's own life history, with respect to the awareness of time, and in relation to the time of society and social relations.

The temporality of one's own life history changes as this personal history is interrupted and needs to find a new definition after the caesura produced by the illness of pain; the perception of time also undergoes a transformation: with the onset of pain, time is suddenly acutely perceived in its significance, its periodicity, and its finitude. During pain, time is also modified in that often one can no longer choose and manage it freely but has to take into account the claims that illness manifestations make upon it: there is a deep difference in the perception of temporality between someone who

* To call the body into question is to question the anthropological image. Hence, to approach the body requires the recognition of its concreteness and transcendence and its being the site of an intimate intentionality, to safeguard and affirm on many levels the dignity of human life. This requires particular attention in the present-day context, considering the characteristics that medical practice has acquired.

† Consider also the various forms and manifestations of psychic suffering.

lives at society's pace and with a sense of time shared by society, and someone who has to adapt to a new pace and perception of time, defined and measured by illness.*

This change in the perception of time influences the life and self-perception of the person in pain, affecting also the way one perceives and tells one's own story, as well as the kind of language that can emerge within this different timeframe. This change in temporality can also correspond to a change in meaning, a reconsideration of the meaning of one's own life.

Subjectivity

This is a double crisis: one of the body, which represents the locus of our being in the world, and of time, which mediates our individual and personal history within the broader context of community and universal history. Thus, subjectivity is in crisis. Through pain (and its manifest illness), the subject fully experiences his or her own subjectivity as rooted in experience, for here we experience our individuality and the limits of the body and the will. As well, we confront the radical crisis of meaning, hence entering the dynamics related to the quest and demand for reason. With respect to this, Ricoeur stated that "Suffering, along with joy, is the ultimate niche of singularity" (21).

According to Levinas, the initial experience of pain is one of absurdity, of lack of meaning, which is accompanied by the unexpected and unknown. This aspect of illness is precisely due to the experience of alteration it induces. But along with it comes the possibility of an opening up, a disclosure: the opportunity to find how our personal, subjective identity remains, even though the images or roles with which we identify may have changed.†

In such a crisis of subjectivity, encounters with the other and the opening up to a relationship such as the one involved in caring play a fundamental role. Although the objectification of pain as a phenomenon allows the physician to acquire a distance that enables him or her to intervene without being adversely affected by emotional impact, to limit one's perception of pain to its mere objectification risks separating its illness from the person who is experiencing it, and hindering effective communication with the patient, and in this way can impede the possibility of addressing and tailoring care to each patient as a particular, unique individual.

Communication and listening, both essential to good medical practice at the clinical-diagnostic level, thus also constitute ways of approaching the pain patient as a person. Attention to the patient's subjective history (along with good communication) becomes an essential part of the care process, and in this way, it becomes possible to look at pain, and the ill person, as well as medicine and its role from a different perspective, which configures an anthropology.‡

---

* The carer intervenes upon and relates to a modified temporality, which must be taken into account.

† According to Levinas, it is a matter of understanding how we retain our identity as subjects in the midst of our changing identifications.

‡ "Disease as embedded in life can only be represented through a creative conceptual response. Its 'thereness' in the body must be rendered 'there' in the life. And this process, even more than the referential or 'locutionary' process of biomedical representation, requires an aesthetic response, an active, synthetic process of constituting in an effort to grasp what is certainly there but is indeterminate in form" (3).

METAPHORS AND IMAGES IN THE EXPERIENCE OF PAIN-AS-ILLNESS

A patient reports: "My pain, it's like my legs were burning when I was going around," thus using the metaphor of fire to describe pain, and employing images when talking about the experience as lived. Some metaphors are significant, such as "The pain was like a monster, a giant thing." What are these people communicating and what do their images convey? Autobiography—the genre within which these narratives may most often be placed—is one that helps to give meaning to and make sense of our existence. The philosopher Alasdair McIntyre suggested that, in the end, we are merely coauthors of the stories of our lives, as it is only in the context of interpersonal dynamics that we can lend shape to our own story. In telling our story, we need someone to listen and accept our communication, and this recipient of narration can become an integral part of the story, after having constituted an essential prerequisite to its unfolding. Some authors suggest that identity is narrative, both in that we structure ourselves through narratives, and because we and our story are part of other stories that vary according to the narrator and point of view adopted. Hence, our own story merely represents the point of departure that may be integrated, modulated, or changed with respect to other voices.

Narrations could also have a reconstitutive function for the personality in those cases where identities have been damaged by stories that have bound them to stereotypes and therefore ask to be revisited.* Such stereotypes can be changed, and their marginalizing or penalizing aspects questioned or subverted. And so a personal representation can be retold through counternarratives that make room for new images and metaphors that can represent one's new positioning. This is particularly relevant to the extent that once in pain, a person's world often changes, not just physically, but also in existential and moral terms.

It has been emphasized that meaning and identity are also acquired by trying to produce a coherent picture of the experienced illness of pain; for this purpose, different narrative genres may be used, such as the narration of one's life history, the use of anecdotes, the narration of a case history, and myths or metaphors. Telling a story may also enable a mutual sharing of personal histories, furthering an encounter between individual and collective experiences, which can constitute a therapeutic process.

In some cases, particularly where the illness is uncertain, contested, or rare, such as so often with maldynic pain, narration and the sharing of experiences may be an important if not a preferential route toward healing. For patients, the pooling of experiences, information, and resources may reveal itself to be essential to relate to clinicians, given that maldynic pain remains difficult to explain and interpret (17).

Stories also help to cultivate moral sense and improve moral perception; they allow clarification and communication (both to ourselves and to others) of the reasons for our choices, and in sharing, can foster dialectic to reconfigure a different perspective. Why can narrative be significant with respect to the moral dimension and hence to ethics? There is a profound link between narrative, communicative

* This is the position sustained by Lindemann Nelson (22).

dynamics and the moral sphere: the descriptive moments of a story are rich in moral indicators, in the sense that they reveal a person's perspective on life and moral interpretations of it. Metaphor, for instance (i.e., "saying something in place of something else"), manages to articulate meanings that might otherwise remain unsaid or unexpressed. There is a deep connection between the experience a person undergoes and the images that person may choose to represent the phases of change and transition, and surely this is true for maldynic pain.

When asked, most patients are happy to tell their stories. Few ask to be dispensed from talking about a particularly negative situation from which they seek to gain distance, and even in this case, at times they use an image of distancing or separation.

The images and metaphors employed are many and varied; some convey peacefulness and a positive attitude, at times even hope, which could indicate a situation where pain is accepted or at least tolerated. Other images reflect the negativity that is being experienced, or its emotional or moral charge. Some of these metaphors are metaphors of movement, such as traveling, running, dancing, or spiritual journeys. The family is often present in these narrations in a positive sense, as resource and provider of company and support: relationships appear to play an essential role, with respect to the new condition one finds oneself in.

A significant difference is associated with the temporal dimension of pain: when pain is intense, attention is acutely focused on a very limited period of time, as if the present moment was absorbing the entire person. There appears to be no sense of personal history, and reference to events of an even immediate past or future is rare. In case of chronic conditions, however, people manifestly search for a way of living with pain-as-illness, and in their narratives the metaphor of the journey is more frequent. In some cases, metaphors convey social images of pain; in other cases, one finds personal representations of pain, as perceived on an existential level or in its organic manifestations; and other metaphors display the symbolic representations the caregivers provide of themselves and their function.

In both literature and poetry, there are many metaphors that talk about pain and illness: the Italian poet and philosopher David Maria Turoldo, at the end of his life, in his collection of poetry *O sensi miei*, referred to his disease, cancer, as "the dragon within me" (23), and the American poet Flannery O'Connor described her illness as "a mad dog." The importance of these images is not confined to mere literary value. Instead, the images can reveal a very particular instrument within the narration of a story or an experience of pain, illness, and suffering from a moral perspective.

To be sure, the experience of pain often brings about a change in the patient's sense of what is good and desirable. The patient's priorities may change, as well as his or her general perspective on life, and worldview. Ethical creativity, in the sense of an ability to reshape our own existence, is fostered and simultaneously at stake in the time and experience of pain and illness. Hence, while its role is prominent for the patient, it is also important for the clinician and caregiver. Images and metaphors make it possible to observe the moral changes that the patient undergoes as a moral agent, and in this way allows both access to the relationship and the ability to take communication to greater depth. The narrative process requires an intersubjective exchange of listening and attention, which constitutes a kind of practice. The experience of maldynic pain is one of facing great change and finding a new position in the world; narration can play an important role and provide useful tools for undertaking such passage.

REFERENCES

1. Todorov, T. 1996. *Facing the extreme: Moral life in concentration camps*. New York: Metropolitan Books.
2. Brody, H. 1987. *Stories of sickness*. New Haven: Yale University Press.
3. Good, B.J. 1994. *Medicine, rationality, and experience: An anthropological perspective*. Cambridge: Cambridge University Press.
4. Cattorini, P. 1994. *Malattia e alleanza. Considerazioni etiche sull'esperienza del soffrire e la domanda di cura*. Pontecorboli: Firenze.
5. Morris, D.B. 2002. Narrative, ethics and pain: Thinking with stories. In: *Stories matter. The role of narrative in medical ethics*, eds. R. Charon and M. Montello, 196–218. New York: Routledge.
6. Montello, M., and R. Charon. 2002. Memory and anticipation: The practice of narrative ethics. In: *Stories matter. The role of narrative in medical ethics*, eds. R. Charon and M. Montello, ix–xii. New York: Routledge.
7. Bruner, J. 2002. Narratives of human plight: A conversation with Jerome Bruner. In: *Stories matter. The role of narrative in medical ethics*, eds. R. Charon and M. Montello, 3–9. New York: Routledge.
8. Ricoeur, P. 1984–1988. *Time and narrative* (*Temps et récit*), 3 vols. [K. McLaughlin and D. Pellauer trans.]. Chicago: University of Chicago Press.
9. Ricoeur, P. 1986. *From text to action: Essays in hermeneutics II* [K. Blamey and J.B. Thompson trans.]. Evanston, IL: Northwestern University Press.
10. Brody, H. 1987. *Stories of sickness*. New Haven, CT: Yale University Press.
11. Zaner, R. 2004. *Conversations on the edge: Narratives of ethics and illness*. Washington DC: Georgetown University Press.
12. Lindemann Nelson, H., and J. Lindemann Nelson (eds.). 1997. *Stories and their limits. Narrative approaches to bioethics*. New York: Routledge.
13. Chambers, T. 1999. *The fiction of bioethics: Cases as literary texts*. New York: Routledge.
14. Charon, R. 2006. *Narrative medicine. Honoring the stories of illness*. New York: Oxford University Press.
15. Sontag, S. 1977. *Illness as metaphor*. New York: Farrar, Straus and Giroux.
16. Montgomery Hunter, K. 1991. *Doctors' stories: The narrative structure of medical knowledge*. Princeton, NJ: Princeton University Press.
17. Bülow, P.H. 2004. Sharing Experiences of Contested Illness by Storytelling. *Discourse and society* 15(1): 33–53.
18. Stern, L., and L.J. Kirmayer. 2004. Knowledge structures in illness narratives development and reliability of a coding scheme. *Transcultural psychiatry* 41(1): 130–142.
19. Bruner, J. 2002. *Making stories: Law, literature, life*. New York: Farrar, Straus and Giroux.
20. Kleinman, A., Brodwin, P.E., Good, B.J., and M.J. DelVecchio Good. 1992. Pain as human experience: An introduction. In: *Pain as human experience: An anthropological perspective*, eds. M.J. DelVecchio Good, P.E. Brodwin, B.J. Good, and A. Kleinman, 1–27. Berkeley: University of California Press.
21. Ricoeur, P. 1996. Les trois niveaux du jugement medical. *Esprit* 12: 21–33.
22. Lindemann Nelson, H. 2001. *Damaged identities, narrative repair*. Ithaca: Cornell University Press.
23. Turoldo, D.M. 1997. *O sensi miei*. Milano: Rizzoli.

12 Psychological Assessment of Maldynic Pain

The Need for a Phenomenological Approach

Michael E. Schatman

CONTENTS

INTRODUCTION

Chronic pain management, as it is currently practiced in its numerous medically sanctioned forms, is firmly anchored in *science*, with the *art* of the field becoming progressively more obscure. This, unfortunately, is likely true of all of modern medicine despite the perception that its practice as a whole has improved with time. Leder writes, "Tendencies toward over-specialization, over-reliance on technology, and loss of humanitarian concern have constituted the dark side of modern medicine's stunning achievements" (1, p. 19). Steen and Haugli (2) note that although the biomedical model may be an effective approach to disease states in which the cause is clearly defined, it is not appropriate for conditions such as chronic pain. Whether an approach to maldynic pain involves medication management, injections, implantable therapies, physical therapy, or even cognitive-behavioral psychology, practitioners consider themselves as professionals applying treatment based upon scientific principles and evidence in order to relieve pain. Sadly, chronic pain patients have come to embrace the same expectations regarding practitioners' approaches to their condition as those maintained by practitioners. This is not to suggest, however, that the maldynic pain patient *needs* or even *wants* to be treated from a purely scientific model, as such a model may be dehumanizing.

The phenomenological movement in psychiatry and psychology was a protest against such dehumanization in the fields of mental health. The phenomenological approaches offer "research and theory that faithfully reflects the distinctive characteristics of human behavior and first-person experience" (3, p. 167). One of these approaches, *phenomenological hermeneutics*, is a developing bioethical medical philosophical conceptualization that lends itself well to the treatment of maldynic pain. Based heavily upon the philosophy of Heidegger (4) and initially applied to medicine by Gadamer (5), "the hermeneutics of medicine is grounded in the *meeting* between doctor and patient—a meeting in which the two different horizons of medical knowledge and lived illness are brought together in an interpretative dialogue for the purpose of determining why the patient is ill and how he can be treated" (6, p. 419). Although this approach does not reject the scientific or technical components of medicine, the phenomenological hermeneutic approach to medicine considers the patient–healer dialogue to be a constitutive element of practice, without which health cannot be achieved. The clinical encounter is the basis for the development of a mutual understanding, whereby "Doctors (as well as representatives of other health care professions) are thus not first and foremost scientists who apply biological knowledge, but rather interpreters—hermeneuts of health and illness" (6, p. 416). The phenomenological hermeneutic approach has been applied to medical fields including psychiatry (7), cardiology (8), critical care (9–12), pediatrics (13), hospice care (14), rehabilitation (15), and surgery (16).

Despite a rich history of utilization of phenomenological approaches in psychotherapy by psychologists—for example, Combs (17), May (18), and Rogers (19)—interpretative phenomenology has been minimally used in the field of behavioral medicine. Smith writes, "There appears to be little work by psychologists exploring, through the detailed qualitative analysis of verbal reports, how particular individuals attempt to make sense of, or find meaning in, their illness and it seems somewhat ironic that in order to explore the phenomenology of illness one turns to sociological rather than psychological studies" (20, p. 266). Psychologists, particularly in medical settings, have become excessively technophilic, focusing on *evidence basis* as opposed to the *relationship* with the patient. Wampold and Bhati write, "Psychologist contributions to outcomes overwhelm treatment differences—the person of the psychologist is critical. The most researched common factor—the alliance between the psychologist and the patient—has been found to be a robust predictor of outcome, even when measured early in therapy" (21, p. 566). Accordingly, it is not surprising that nurses and medical sociologists, not psychologists, have been at the forefront of the phenomenological hermeneutic movement in medical practice.

Phenomenological hermeneutics should not be confused with Husserlian transcendental phenomenology, which suggests that any preconceptions should be put aside or *bracketed* (22). While philosophically robust and valuable as a framework for research, Husserl's phenomenology is perhaps too "pure" to apply to clinical models. Although it certainly does not emphasize preconception, phenomenological hermeneutics in medicine recognizes that it is not possible to completely bracket one's *being-in-the-world* in clinical practice (23). Paley also criticizes the feasibility of the Husserlian notion of bracketing assumptions as a means of disabling them. He states, "Presumably, 'setting aside' your assumptions is not like emptying your pockets, or taking your shoes and socks off. In fact, there is more than one problem here, because assumptions must be identified before they can be disabled" (24, p. 110). Recognition of a patient's maldynic illness does not need to detract from the development of an understanding of his or her subjective lived experience. Double-blind research has its benefits; however, double-blind assessment merely leaves both the patient and the clinician "blind."

In an essay on the hermeneutic role of consultation-liaison psychiatrists, Leder notes that an understanding of the biophysical model of medicine is important for the psychiatrist in medical settings (25). This is also true, however, for the psychologist working with patients with chronic pain. Such understanding of the biophysical model of chronic pain (even poorly understood maldynia) leaves chronic pain psychologists with preconceived notions regarding a patient's physical state, if nothing else. Although such preconception may preclude a Husserlian phenomenological approach, it does not prohibit the psychologist from consideration of the maldynic pain patient within his or her lived world.

The phenomenological pain psychologist is thought to be an *interpreter* who tries to understand data in the context of a specific situation, although Leder would perhaps argue that his or her role needs to surpass simple interpretation. Of consultation-liaison (C-L) psychiatrists, he writes, "the C-L psychiatrist is not simply an interpreter *per se*, but what I will term an inter-interpreter. An inter-interpreter, in my definition, is one whose hermeneutic work explicitly and centrally involves an interfacing between different conceptual worlds" (25, p. 372). Much like a C-L psychiatrist, the chronic pain psychologist is the liaison between biophysically directed physicians and patients striving not only for relief of their nociceptive discomfort, but for validation of their subjective experience of pain and relief of suffering on many levels. Although Leder emphasizes the importance of *interpretation* (25), an article by Engebretson, Monsivais, and Mahoney focuses on *narrative ethics* as central to pain patients' recoveries. The authors write, "The clinical ethic requires the clinician to enter into a meaningful dialogue with the patient to better understand not only the patient's pain and suffering but some of the contextual issues that are woven into the pain experience" (26, p. 25). An emphasis is placed on attending to cues the patient provides within his or her narrative as a tool for enhancing the patient's agency (26). Although phenomenological hermeneutic approaches may differ, to an extent, in their presuppositions and methods, their emphases share the notion of the importance of *understanding the spoken word* of the patient (27).

THE IMPORTANCE OF PHENOMENOLOGICALLY ASSESSING THE CHRONIC PAIN PATIENT

As practitioners of chronic pain management, it is important to draw a distinction between *chronic pain* and *persistent pain*, as chronicity is determined by the widespread dysfunction across a number of areas of the pain sufferer's life. These include not only the physical realm, but the emotional, behavioral, social, sexual, recreational, spiritual, vocational, financial, and legal realms as well (28). In providing psychological evaluation to the patient with maldynic pain, it is critical to examine all of these areas of likely dysfunction in order to understand the patient's *overall* experience, as it is always greater than the patient's pure nociceptive experience. Given the qualities of maldynia, temporary relief of patients' physical discomfort will not necessarily result in the restoration of the multiple areas of their lives in which they have experienced dramatic losses. Although the efficacy of *all* medical practice would likely improve through more holistic treatment, maldynic pain perhaps lends itself to the consideration of the patient as a *complex, multifaceted being* as opposed to a *disease state* more so than do many other types of pathophysiology, as it is a condition for which there is often no cure. It is an illness not of a *body part*, but of the *person*. Radford notes that *pain* is not likely to have a common meaning, and this premise certainly applies to pain's impact on one's life as well as to nociceptive experience (29). Therefore, it is imperative that the *person* is assessed as thoroughly as possible, and in a manner that provides the clinician with as much information as possible regarding that person's *subjective* pain experience. Leder (1) suggests that the *text* in clinical medicine is the *person, as ill*. In chronic pain management, however, the *text* is the *person, as suffering*, and accordingly, it is the suffering of the maldynic pain patient which merits assessment. A review of the literature yields hundreds of studies confirming the considerable variability in individuals' *responses* to and *correlates* of chronic pain, but it has proven difficult to assess the variability in the *experience* or the *meaning* of chronic pain to the unique, individual sufferer. Osborne and Smith note, "However, although in each case the researchers have identified important constructs and patient profiles that are characteristic of chronic pain, they have been unable to address how or why such behaviours and beliefs are formed or maintained" (30, p. 67).

Unfortunately, as mentioned above, many psychologists in chronic pain management have become technophilic in their assessment of patients, thereby precluding recognition of individual patient differences in their unique experience of pain and related circumstances. Typically, a battery of tests is administered, followed by a uniform clinical interview. Jacob and Kerns state:

> Regardless of theoretical perspective, clinicians involved in the assessment of the psychosocial context of the chronic pain experience are generally encouraged to adopt a hypothesis-testing approach to evaluation....Most commonly, this review is conducted in an interview format that is relatively standardized across individuals, regardless of the specifics of their pain complaints. (31, p. 364)

These statements represent a blatant disregard for the unique manner in which a patient experiences maldynic illness, yet the approach described has become the norm for psychologists working in the field of chronic pain management. Hypotheses are generated prior to the encounter, and psychologists, sadly, believe that their role is to make each patient fit into a standardized mold. Psychological testing is commonly utilized to classify maldynic pain sufferers into subgroups in order to determine the direction of treatment. Turk and Okifuji question whether there is actually any benefit to such classification, noting that efforts to evaluate the effects of matching treatments to the groups identified do not appear in the literature (32). This lack of empirical evidence supporting the classification of patients with chronic pain into subgroups provides additional support for the consideration of the patient with maldynic pain as a *unique individual* whose overall pain experience can be best understood through an exploration of his or her *subjective life-world*. Turk and Okifuji ultimately call for the identification of subgroups of chronic pain patients, but they do so with caution. The authors state, "Subgroups should be viewed as prototypes, with significant room for individual variability with the subgroup. Thus treatments designed to match subgroup characteristics will also need to consider and address the unique characteristics of individual patients" (32, p. 411).

In an analysis of individualized assessment in phenomenological psychology, Fischer examines the dilemma of making psychological assessment relevant to the particular patient while simultaneously safeguarding against the assessment process losing its systematic quality and validity (33). While focused on psychological assessment in general, Fischer's emphasis on the *lived experience* of patients in conducting psychological evaluation certainly applies to the assessment of maldynic pain. For example, she discusses the importance of *contextualizing the referral* by going directly to the referral source in order to understand "the decisions facing him/her, the concrete events that led to that dilemma and the particular meanings to him/her of the referral categories" (33, p. 117). As psychologists working with patients with maldynic pain, it is easy and common to make certain assumptions regarding the purpose of a referral and the referral source's difficulties with the patient. Even if we read reports and office notes from the referring physician prior to interacting with the patient, the strong possibility exists that the written record does not fully capture the physician's *purpose* for actually making the referral. The literature is replete with accounts of difficulties that primary care physicians, who provide the majority of the treatment to patients with chronic pain (34–36), experience in dealing with the emotional and behavioral sequelae of their patients' discomfort (37–39). A 2001 study indicated that only 15% of primary care physicians surveyed actually enjoyed treating chronic pain (40). Another study has indicated that physicians and other health care professionals attribute lower status to chronic musculoskeletal pain patients than to any other group to whom they provide care (41). Often, a physician will refer a patient with maldynic pain to a psychologist as a response to his or her own frustration with the relationship, which is typically expressed to the patient, either subtly or directly. By speaking directly to the referral source, the chronic pain psychologist can learn about the quality of the relationship the patient has with him or her, thereby gaining access to data critical to the patient–psychologist dynamic. In certain cases, direct discussion can help to understand that the evident *dysfunction* is actually that of the referral source rather than the patient, or at least that the pain syndrome has been exacerbated by conflict in the physician–patient relationship.

Individualized phenomenological psychological assessment also emphasizes the importance of making the patient an *informed participant* from the beginning of the encounter (33). In working with patients suffering from maldynic pain, their signatures of informed consent are clearly not sufficient to make them informed participants. It is common practice among pain psychologists to provide formal psychological assessment, often administered by a pyschometrician, prior to actually *meeting* a patient. The training and clinical skills of psychometricians are variable, and the risk of alienating the patient with chronic pain from the assessment and treatment process exists when the patient's initial contact with psychological services does not involve a clinician understanding of his or her unique circumstances. Perhaps even more deleterious is the practice of some pain treatment facilities of sending patients psychometric assessment measures by mail prior to their initial appointments. Such a strategy is likely to be beneficial to a patient only if the patient's postal carrier is a trained psychometrician and has sufficient time to guide the patient through the testing process. Obviously, this is rather sarcastically uncommon to say the least. Initiation of psychological services through immediate formal testing does not address maldynic pain patients' anxieties relating to whether they are heard, seen, validated, and healed. Standardized testing can exacerbate already heightened senses of anxiety, as patients are likely to experience angst associated with providing the "right" versus the "wrong" answer. Additionally, as patients' somatic focus increases during the test-taking process through exposure to *pain words* (e.g. the McGill Pain Questionnaire), symptoms may be exacerbated, even without the individual's awareness of the process (42). *Process*—that is, the evolution of a physically visible outcome (33)—is the critical element of individualized phenomenological psychological assessment and is a result of *collaboration* between the pain psychologist and the patient. By initiating the assessment process through immediate psychometric testing, patients may feel invalidated and might see psychological services as noncollaborative, and there is potential for the premature development of an adversarial relationship.

In discussing the illness experience in general, Zaner writes, "Patients are, moreover, invariably strangers to those professing to be able to help in situations that, on the other hand, frequently involve quite intimate actions regarding the patient's body, self, personal and family relations, and social life. The relationships with helpers are essentially asymmetrical, with power....on the side of the helpers" (43, p. 315). This asymmetry (described as *powerlessness*) was one of the prominent emergent themes identified through a rare example of qualitative research on the experiences of patients with chronic back pain (44). As they are frequently told that their pain is of myriad etiologies by the myriad of physicians they have seen, patients with chronic pain are understandably often confused, thereby contributing to their emotional distress. To maldynic pain sufferers, accurate information *is* power. It is also essential if a healing relationship is to be developed between the patient and the psychologist. Results of a study by Williams and Thorn (45) indicate that patients' lack of understanding of the nature of their pain is associated with poor treatment outcomes. Physicians often claim that they do not have sufficient time to educate patients regarding their pain conditions (46). Accordingly, Davidoff and Florance (47) call for the utilization of "informationists" (i.e., specially trained professionals who retrieve the information that physicians are too busy to retrieve for patients themselves). Psychologists should certainly not masquerade as physicians and provide medical diagnoses to their patients, but it is appropriate to share genuine and accurate general information regarding the nature of maldynic pain with them. For example, letting a patient know at the onset of evaluation that the psychologist can help the patient learn how to manage his or her pain and that healing is possible even if a cure is not, can change the patient's view of his or her situation, reducing feelings of hopelessness that are typically present. Doing so is consistent with Engebretson et al.'s position that "an awareness of medical resources and their divergent notions about health, illness, and treatment" (26, p. 25) is critical to the development and implementation of a narrative ethic. Stressing collaboration through a sharing of power as the theme of the assessment process with patients suffering from maldynic pain can help reduce their sense of impotence, thereby enhancing the likelihood that a healing bond will be developed through the assessment process.

Even though psychological testing of patients with maldynic pain should not be prematurely initiated, its inclusion in the evaluation process should not be totally dismissed, as numerous studies (48,49) have supported its benefit in the overall assessment of pain sufferers. Even if we recognize problems with technophilism in the treatment of chronic pain, clinicians cannot afford to simplistically regard the biomedical and phenomenological perspectives as an either/or dynamic (50). However, Fischer notes that life events, not test scores, are the primary data. She states, "Test scores, categories, and diagnoses are abstractions from particular actions, and these abstractions already are grounded in assumptions about the orderliness of human affairs" (33, p. 117). This is of great importance in assessing patients with maldynic pain, as too often they are lumped into a homogeneous group rather than seen as unique individuals whose pain experiences are understood neither through medical nor psychological diagnosis. Turk (51) emphasizes the notion that the *homogeneity myth* has the potential to contribute to negative outcomes in the treatment of chronic pain.

In a treatise on the psychological evaluation of posttraumatic stress, Briere (52) calls for the utilization of a phenomenological approach. Even if it is not the direct result of a specific trauma, maldynic pain *is* trauma. Much like the stressful events that precipitate posttraumatic stress disorder (PTSD), maldynia alters the sufferer's life-world in profound ways. Accordingly, some of Briere's recommendations regarding the psychological assessment of patients in posttraumatic states apply to the assessment of patients with maldynic pain. For example, Briere states, "The evaluator should take this reactive dimension of trauma assessment into account, so that the client is not unduly stressed by the interview... In some instances, this approach may mean that certain psychological tests (e.g., projective instruments) are not administered until the client is more stable and less distressed" (52, p. 84). Briere emphasizes the importance of providing an assessment environment for the patient which is manifestly safe and allows for the building of the strongest possible rapport. Such an approach is clearly fitting for a patient with maldynic pain, as well.

THE MINNESOTA MULTIPHASIC PERSONALITY INVENTORY (MMPI): A TARNISHED "GOLD STANDARD"

At the time that Fischer presented her thesis, the formalized assessment of chronic pain sufferers was in its infancy. For the most part, psychologists relied upon the Minnesota Multiphasic Personality Inventory (MMPI), with knowledge of interpretation of the measure based upon norms established by Hathaway and McKinley (53) over 60 years ago. The criterion groups for the development of the MMPI consisted of psychiatric inpatients at the University of Minnesota Hospital, and the test was criticized for many years with regard to its lack of a representative normative sample (54–56). After almost two decades of discussion and research, the revised version of the MMPI was released by Butcher and colleagues in 1989 (57). The MMPI-2 was normed on a considerably larger and more nationally representative sample than was the original MMPI, and is considered psychometrically superior to the original test, at least in terms of the types of populations for which the test can provide valid psychological assessment (58). For example, the original MMPI normative sample did not include members of minority groups, although a meta-analytic review (59) failed to identify any racial validity advantage of the MMPI-2 over the original measure. The authors suggest that the inclusion of ethnic minorities in a standardization sample does not necessarily result in an assessment tool's cultural sensitivity.

Although the original normative sample for the MMPI was not known to represent patients with chronic pain, a considerable body of research using the MMPI to assess patients with chronic pain exists, with a Medline search yielding approximately 200 such studies published between 1981 and 1991. There is also no information on the extent to which chronic pain sufferers were represented in the restandardization sample of the MMPI-2, yet it continues to be the most commonly used instrument for the assessment of the psychological status of patients with chronic pain (60). Even though the data are more than 20 years old, a study by Hickling Sison, and Holtz (61) indicated that 77.7% of psychologists in multidisciplinary pain clinics considered the MMPI to be one of the five most valuable assessment tools, while only 33.3% reported that the clinical interview was one of the five most valuable tools. Clinicians' continued reliance upon this test is curious in light of research indicating a high risk of misinterpretation and inadequate predictive validity (summarized by Bradley and McKendree-Smith [60]), as it has never been normed for chronic pain patients. What exactly is it that the MMPI and the MMPI-2 tell us about the person who suffers from chronic pain? Originally, the MMPI was designed to identify and classify psychopathology, a function for which it continues to be heavily utilized today. Pain psychologists, however, often rely upon the MMPI in order to understand the *inner workings* or *flavor* of the patient. Yet this lengthy (567 items) and potentially intrusive (48) test tells us very little about the maldynia sufferer's experience of his or her disease. Numerous investigators (62–64) have attempted to identify *pain personality profiles* through empirical studies. However, as Novy notes, "most of these have been inconclusive, unreplicated, or strongly criticized" (65, p. 282).

There are many critics of the use of the MMPI in the assessment of patients with chronic pain, with a review (66) suggesting that its use with this population is perhaps limited to that for which the test was designed (i.e., identification of psychopathology). Given the nature of most of the items of the MMPI, patients are likely to recognize that the test is measuring psychopathology as opposed to an element of maldynic illness, thereby serving to reinforce the Cartesian mind–body dualism that is such an inappropriate framework for this type of condition. Kugelmann writes, "The problematic of responsibility, legal, moral and even existential, arises because pain defies western categories, challenging and simultaneously affirming the dualities that define western medicine" (67, p. 1665). From a phenomenological perspective, an assessment measure that reinforces these defining dualities, such as the MMPI, ought to be considered with caution.

UNIDIMENSIONAL ASSESSMENT TOOLS

Numerous psychological measures that assess a single dimension of the chronic pain experience are routinely used by psychologists who treat maldynia. Despite proclamations to the contrary, there exists no *gold standard* of a test battery which can, as a group of measures, capture the essence of what the patient with chronic pain experiences. In discussing the use of assessment tools with chronic pain psychologists, it appears that each has his or her own unique battery, the elements of which are based upon that clinician's training history, understanding of the tests, their availability and convenience of administration, scoring and interpretation, and the supportive literature. However, a group composed almost exclusively of academic, pharmaceutical industry, and government agency chronic pain researchers, referred to as the Initiative on Methods, Measurement, and Pain Assessment in Clinical Trials (IMMPACT) group (68) in 2005 released consensus recommendations for outcome domains and specific measurements that should be used in chronic pain outcome trials. The six core outcome domains that the group determined should be measured were pain; physical functioning; emotional functioning; participant ratings of improvement; and satisfaction with treatment, symptoms, and adverse events; and participant disposition. In evaluating potential core outcome measurements, criteria utilized included reliability, validity, responsiveness, and interpretability. Specific core outcome measures recommended included a 0 to 10 numerical pain rating scale, the Beck Depression Inventory (69), the Multidimensional Pain Inventory (70) Interference Scale, the Brief Pain Inventory (71) pain interference items, the Profile of Mood States (72), the Patient Global Assessment of Pain Scale (73), and the use of rescue analgesics. A review of these commonly used assessment tools indicates that the IMMPACT group chose measures that met their criteria psychometrically, but there is no mention in their article of assessing patients' lived experiences. Their recommended measures quantify nociceptive experience, depression, anxiety, and functional interference, yet fail to address suffering. To their credit, the IMMPACT group specifically state that their selected outcome measures "should be considered when designing clinical trials of chronic pain treatments" as opposed to utilizing them in actual clinical practice (68). However, given the academic distinction and reputations of members of the group, the risk that clinicians will interpret their consensus as the *gold standard* for assessment in clinical practice should not be underestimated. These tools are already popular among chronic pain psychologists, and the IMMPACT group's endorsement of their use in clinical trials may be seen as an imprimatur for truncating or even foregoing a patient interview. In the current climate of evidence-based medicine, clinicians treating maldynic pain are likely to become overly focused on psychometrically sound tests, even if doing so limits or precludes important dimensions of the healing encounter with the patient.

MULTISCALE MULTIDIMENSIONAL ASSESSMENT TOOLS

It has been suggested that multiscale multidimensional assessment tools are superior to unidimensional tools in chronic pain assessment due to their potential provision of an array of topics for clinician–patient communication (74), as well as to the obvious multidimensionality of the chronic pain experience (75,76). Unfortunately, literature on the utility (i.e., validity) of multiscale multidimensional tools used in the assessment of patients with chronic pain has reported mixed results, at best.

One multiscale, multidimensional assessment tool that has been considered a substitute for the MMPI in evaluating patients with chronic pain is the Millon Behavioral Health Inventory (MBHI) (77). As the MBHI was developed specifically for use with medical patients, it has been considered more relevant in the chronic pain management setting than the MMPI (48). Additionally, it is considerably shorter than the MMPI (150 items) and may be seen as less intrusive due to its perceived relevance to patients with chronic pain, as its items relate directly to patients' medical status. However, like the MMPI, the MBHI was developed to assess the psychological functioning of medical patients, not their pain experience. Additionally, research on the MBHI has not supported its predictive validity for a number of chronic pain syndromes (78–81). Thus, given its inability to assess the maldynic pain patient's experience of pain in his or her lived world in conjunction with problems with predictive validity, the MBHI should not be considered a particularly valuable tool for the assessment of the chronic pain sufferer.

Another multiscale, multidimensional measure of psychological functioning frequently used in the assessment of patients with chronic pain is the Symptom Checklist-90 Revised (SCL-90R) (82). Like the MBHI, this test measures only psychological disturbances and distress, although it is even less adequate as an assessment tool with patients with maldynic pain, as it was standardized with psychiatric patients and a nonpatient sample as opposed to medical patients. Its use in chronic pain populations is likely based on its purported measurement of somatization, among other forms of psychopathology. Although relatively brief (90 items), studies by Buckelew et al. (83) and Shutty et al. (84) have suggested that its standardization with psychiatric patients results in limited validity within a chronic pain population.

The Pain Patient Profile (P-3) (85) was developed specifically for the identification of pain patients who could potentially benefit from the inclusion of psychological services as an aspect of their overall treatment program, and was normed on pain patients, as well as a community sample. It is composed of 44 items, and assesses depression, anxiety, and somatization. The P-3 is meant to serve as a screen, with its ultimate goal being facilitating appropriate referral to a psychologist. Although certainly not a measure of the maldynic pain patient's lived pain experience, the goals of the measure are certainly consistent with a more phenomenological approach to assessment. Unfortunately, reviews of the psychological and medical literature yield a paucity of studies confirming the validity of the P-3, thereby potentially limiting its clinical utility in the assessment of patients with maldynic pain.

The Behavioral Assessment of Pain Questionnaire (BAP) (86) was developed to measure disability associated with persistent pain. Among the numerous factors it measures are activity level, sleep disturbance, mood disturbance, narcotic usage, and suffering. In its measurement of suffering, the BAP is unique among multiscale, multidimensional measures of pain experience, and is accordingly more sensitive to the spirit of phenomenology than the other assessment tools that have been reviewed. The measure's normative sample was over 1,000 subacute and chronic pain patients, which supports its use with a maldynic pain population. Campbell and Schatman (87) refer to BAP as "the most comprehensive measure of emotional and behavioral responses to chronic pain." Unfortunately, the BAP's considerable length (390 items) results in some of the same problems with administration and potential alienation of patients with chronic pain as were mentioned in regard to the MMPI. Additionally, as a review of the psychological and medical literature indicates a paucity of independent empirical support of the test's validity and reliability, its utilization for the assessment of patients with maldynic pain remains questionable.

The final multiscale, multidimensional measure commonly used in the assessment of chronic pain patients that will be reviewed in this chapter is the West Haven–Yale Multidimensional Pain Inventory (WHYMPI) (70). As a 52-item self-report measure whose items pertain directly to pain experience, the WHYMPI is not as likely to alienate patients with chronic pain as the MMPI or the BAP. In a review of the instrument, Jacob and Kerns write, "it places an emphasis on patients' idiosyncratic beliefs or appraisals of their pain problems, the impact of pain on their lives, and the responses of others" (31, p. 366), suggesting that the WHYMPI is phenomenologically based to a greater extent than are measures that were previously discussed. Numerous studies (88,89) have supported the test's validity and reliability, and its factor structure when translated into foreign languages has been supported in studies with patients with chronic pain in a number of other countries (90–92). Detracting from the phenomenological viability of the WHYMPI is its empirically derived classification of pain patients (93), with one of the supposedly homogeneous subgroups identified as *dysfunctional*. Such a classification fails to recognize the patient with maldynic pain as possessing a distinct realm of experience and knowledge, and may be insulting and alienating to the patient. Nevertheless, given the limitations of the other multiscale, multidimensional assessment measures discussed, it appears that the WHYMPI is the most phenomenologically robust.

THE STRENGTH AND MEANING OF THE CLINICAL INTERVIEW

The clinical interview is the most valuable assessment tool that the psychologist can share with patients with maldynic pain. Unlike standardized tests, the clinical interview focuses on the encounter with the patient, as well as providing an opportunity to gather the individual's specific history, which is not accessible through standardized testing. Of taking a history from a patient, Leder writes, "The very telling of this story may have not only diagnostic but therapeutic significance. For it counteracts two primary features of illness that give rise to suffering: senselessness and isolation" (1, p. 13). Information such as history of severe childhood illness and hospitalizations (94), and exposure to a family environment of chronic pain and illness (95,96), both linked to somatic pain in adult life, are not available through formal assessment tools typically utilized in the assessment of patients with maldynic pain. Steen and Haugli (50) note that self-awareness and self-discovery are necessities if patients are to identify the aspects of their lives that impact pain, and it is certainly more likely that these requisites will be achieved through the encounter with the clinician than through formal, standardized testing. Results of a study by Street et al. (74) suggest that patients prefer medical professionals who express an interest in the patient's perspective on physical health and its impact on daily living. A patient with maldynic pain can tell the clinician how his or her life has been impacted by pain, provided that the clinician is willing to listen. In situations in which patients with chronic pain have become so depressed that they are withdrawn and do not spontaneously offer information regarding pain-related changes in their lives, gentle questioning by the clinician may be necessary. Inherently prejudiced structured clinical interviews should be avoided at all costs, as they do not encourage dialogue, and accordingly preclude the psychologist from intuiting the patient's conduct and experience of health and illness.

Nevertheless, it is naïve to suggest that a pain psychologist can enter a clinical interview without any preconceptions whatsoever, as the encounter is not likely to take place if the patient is not suffering, with such suffering preconceived by the psychologist. A complete lack of preconceptions prior to encountering the patient with maldynic pain would be consistent with the Husserlian phenomenological approach criticized as being incongruous with clinical reality earlier in this chapter. In discussing phenomenological research, Lowes and Prowse suggest that "since the products of phenomenological interviewing are co-created by both interviewer and participant, the demonstration of rigour and trustworthiness depends upon researchers fully explicating their preconceptions and their contribution to the interview process" (97, p. 472). This suggestion holds true of the clinical interview as well as the research interview, although all phenomenological interviews represent *research*, as they serve the purpose of data collection. The explication of the pain psychologist's preconceptions can be important to the development of dialogue, which, as discussed above, can be an essential aspect of the collaborative healing process that is central to the phenomenological hermeneutic approach.

Other research (98–100) that supports the primacy of the clinical interview over psychometric testing in assessing maldynic pain sufferers has indicated that patient satisfaction is predicted by medical professionals' nonverbal behaviors (e.g., body language, ability to express emotions through facial expressions and tone of voice, nodding, etc.). Physical distance and level of eye contact have been found to predict the extent to which psychotherapy clients believe that their therapists like and accept them (101,102). Obviously, the benefit of nonverbal behavior to the therapeutic alliance is not likely to be experienced if a psychologist initiates the assessment of the maldynic pain sufferer through psychometric testing, particularly if the psychologist is not the person who administers the measures.

Numerous studies (14,103,104,105) suggest that patients with chronic pain experience a sense of isolation, and this is often detrimental to their emotional status. Kugelmann (67) surmised that such a sense of isolation results in feelings of disengagement from life, causing a feeling of being *marooned*, even when in the physical presence of others. Given psychologists' training and experience in emotionally connecting with others, they should possess the skills to approach the patient with maldynic pain in his or her life space, thereby reducing the patient's sense of isolation and aloneness. Isolation serves a defensive function among patients with chronic pain, allowing them to avoid the risk of alienating others through their irritability and general misery (30). By emphasizing the clinical interview in psychological assessment, psychologists have the potential to reduce maldynic pain patients' uncomfortable feelings of alienation from the world around them. The use of psychological testing as the initial phase of evaluation, on the other hand, can serve to increase the patient's self-absorption and is certainly not likely to reduce feelings of isolation.

Smith (20) notes that feelings of loss of control frequently occur in medical patients. Herein lies another problem associated with reliance upon the technophilic mode of assessment that emphasizes standardized testing as opposed to the clinical interview as its primary means of data collection. As numerous quantitative (106–112) as well as qualitative (113,114) studies of patients with chronic pain of nonmalignant origin have suggested, there is a strong relationship between chronic pain and perceived loss of control over one's life; thus, a strategy for assessment that enhances rather than detracts from the maldynic pain patient's sense of control is needed. Standardized testing is *done to* patients, as opposed to the clinical interview, which is *done with* patients. When the loss of control inherent to the standardized testing situation is considered in conjunction with clinician–patient asymmetry, the argument for emphasizing the encounter of the clinical interview as opposed to impersonal standardized testing is further strengthened. Empirical support for the primacy of the clinical interview as opposed to paper-and-pencil testing is provided through a study (115) that found that primary care patient attitudes toward questionnaire screening were less positive than those attitudes regarding physician interview.

The meaning of the clinical interview and its importance in the psychological assessment of patients with maldynic pain appears to be greater among male patients than among females. A study of female partners of fibromyalgia sufferers (116) identified a prominent theme of frustration experienced as a result of men's reluctance to communicate with them. The authors found that patients with fibromyalgia tended to become socially withdrawn, which is unfortunate given the established relationship between social withdrawal and decreased physical well-being (117–120). Additionally, they found that partners' demonstrations of compassion were effective in emotionally comforting the fibromyalgia sufferers. The assessment of the maldynic pain patient's lived experience through the encounter of the clinical interview can frequently provide comfort. A review of the literature fails to yield any studies suggesting emotional comforting through standardized psychometric testing. Additional support for an emphasis on the phenomenological hermeneutic approach among male patients with maldynic pain can be found in a discussion of the lived experiences of males with fibromyalgia by Paulson, Danielson, and Soderberg (121). These authors identify a theme of "not being really understood" among their male sample, while simultaneously noting a strong tendency for male patients to be guarded and fear being considered *whiners*. Males with chronic pain are likely to set up *walls* between themselves and all others who are not in pain (122,145). Given studies (123–126) suggesting that males tend to be more guarded and less open to experience than females, the phenomenological hermeneutic approach (as opposed to relying primarily upon psychometric testing) is likely to be essential if we are to understand the lived pain experiences and ultimately help male patients with maldynia.

A "PATIENT-CENTERED" APPROACH TO ASSESSMENT OF PATIENTS WITH MALDYNIC PAIN

Carl Rogers, the originator of patient-centered psychotherapy, was considered by some to be an *existential psychologist*, as his approach emphasized allowing the patient to become the primary locus for the evaluation of his or her experience

(127–129). Rogers states, "Experience is, for me, the highest authority. The touchstone of validity is my own experience. No other person's ideas, and none of my own ideas, are as authoritative as my experience.... Neither the Bible nor the prophets—neither Freud nor research—neither the revelations of God nor man—can take precedence over my own direct experience" (130, p. 23). Rogers' three conditions for relationship health, *congruence, unconditional positive regard,* and *empathic understanding* (130), should serve as the basis for the psychological assessment of patients suffering from maldynic pain. In examining these conditions, it becomes evident that the standard psychological assessment of patients with maldynic pain violates each of them profusely.

Congruence refers to the therapist being real and genuine, open, integrated, and authentic while interacting with the patient. Rogers (130) considers this to be the most important of his conditions, suggesting that without it a patient cannot progress toward health. In assessing patients with chronic pain, how congruent are psychologists? Clair and Prendergast (131) discuss the importance of being direct and open with patients in therapy as a means of more quickly developing the integrity of the relationship. As much of the work that psychologists do with patients with maldynic pain is in pain clinics, treatment is often considered time-limited. Accordingly, congruence must be demonstrated from the initial stage of the relationship (i.e., the assessment process). Pain psychologists can be congruent in their assessments of patients with chronic pain, regardless of whether formal psychometric measures are employed. For example, initiating psychological testing prior to actually meeting the patient is not good clinical practice. By doing so, the psychologist risks being seen as "behind the curtain," a distant observer, as opposed to the accessible individual who is striving for collaboration with the patient with maldynic pain. Fischer (33) suggests that the person being assessed is the coevaluator of his or her state and person. Patients with chronic pain should be invited to discuss their understanding of the purpose of the psychological referral, and what they would like to gain through their involvement in the process. Although the standardized means of providing psychological testing should not be disregarded, being open and honest with patients with maldynic pain facilitates the collection of interpersonal and existential information, supplementing information gained through psychometric tests.

Unconditional positive regard refers to the clinician's experience of positive attitudes of warmth, caring, liking, interest, and respect (130) toward a patient. The psychological assessment process represents a critical juncture at which to provide the maldynic pain sufferer with positive regard. As mentioned earlier, it is likely that the patient has been referred by a physician who has not necessarily liked him or her, with the physician's inability to *fix* the patient potentially resulting in negative countertransference and, at times, hostile or passive-aggressive acting-out behavior by the health care provider. Rogers (130) acknowledges that unconditional positive regard does not necessarily come easily or quickly. Certainly, a psychologist cannot be expected to experience genuine warmth for a patient at first encounter, and feigning such feelings would contradict the (perhaps primary) condition of *congruence.* Doing so could have catastrophic consequences for the therapeutic alliance and, accordingly, the outcome of future treatment. However, trusting the pain patient throughout the process of psychological assessment is within the realm of possibility.

Frequently, the epistemic authority between subjectivity and objectivity of patients' reports of pain is questioned by medical professionals (132). Regarding exaggeration or even feigning of chronic pain symptoms, patients should "be considered innocent until proven guilty." A 1999 review (133) of 12 studies of patients with chronic pain found evidence of malingering to be weak, despite physician perceptions of malingering ranging from 1% to 75%. Certainly, false-positive determination of a patient feigning pain can be more deleterious than a false negative.

Another way in which regard can be demonstrated is through timely feedback regarding the outcomes of assessment. Berg (134) considered the feedback process in psychotherapy as an interpersonal transaction that can strengthen a therapeutic relationship. Chronic pain psychologists owe their patients the respect of providing timely and accurate feedback regarding testing results, with such feedback mandated by the profession's ethical standards (Section 9.10 of the American Psychological Association [APA] *Ethical Principles of Psychologists and Code of Conduct*) (135). Chronic pain sufferers often complain about the lack of information that they receive pertaining to their illness (136,137), with limited patient understanding of their conditions potentially contributing to negative mood (138). However, rather than simply providing interpretation of test results, the psychologist should make more complete and meaningful feedback an integral part of the assessment process (131). Asking patients with maldynic pain about their experience of the test-taking process can help them feel like valued human beings as opposed to objects. The pain psychologist should emphasize strengths that are identified through test results, with any identified psychopathology and maladaptive emotional or behavioral responses to maldynia presented as areas for future exploration in the therapeutic relationship. Regardless of psychometric tests utilized, it is beneficial to ask patients which assessments they find best capture the experience of pain within their subjective life-world. Doing so can provide the psychologist with information regarding which tests may be most appropriate for use in discharge and follow-up evaluations.

Fischer suggests that making a formal psychiatric diagnosis in phenomenological individual assessment is permissible, "provided that such diagnosis does not substitute for individualized, collaboratively interventional assessment" (33, p. 120). Diagnosing patients is consistent with the phenomenological hermeneutic approach to medicine espoused by Leder (1), who considers it (along with prognosis and proper treatment) to be a part of the full interpretation of a patient. However, diagnosing psychopathology in patients with maldynic pain can be stigmatizing, and the possibility of future repercussions (e.g., access to health care insurance) exists. Accordingly, it is both respectful and ethical to provide an APA *Diagnostic and Statistical Manual of Mental Disorders* (*DSM*)-based diagnosis of Pain Disorder Associated with Both Psychological Factors and a General Medical Condition to all patients suffering from maldynia, as their illness is always experienced psychologically as well as physically. Of course, the actual codification of pain disorder will be modified to some extent upon the release of *DSM-V*. Still, any such diagnosis is a relatively benign one and is not likely to result in third-party payers' attributions of preexisting psychopathology as responsible for the patient's chronic pain condition. Overpsychopathologizing is likely to be alienating to maldynia patients, as medical professionals who have treated them prior to psychological evaluation may have suggested that the problem is *in their head.* Patients are likely to feel respected when the psychologist moves away from a stance of Cartesian dualism and takes time to explain the interacting physical and emotional factors that create, and *are* maldynic pain. Frischenschlager and Pucher note, "Patients, if they feel that they and their chronic pain are taken seriously without having to fear becoming psychologically pathologized, are prepared for a self-exploring co-operation" (139, p. 419).

Finally, the pain psychologist's evaluation of the maldynic patient should be one in which *empathic understanding* is demonstrated. Given that the patient–healer relationship is asymmetrical, with power typically in the hands of the healer, Svenaeus writes, "This [asymmetry] necessitates empathy on the part of the doctor. He must try to understand the patient, not exclusively from his own point of view, but through trying to put himself in the patient's situation… It is only through empathy that the doctor can reach an independent understanding that is truly productive, in the sense of shared *and* independent" (6, p. 416). It is important to share with the maldynic patient that more than anything else, the clinician recognizes that the patient is suffering, and that the goal is to help relieve that suffering, even if the pain cannot completely be ameliorated. Discussing the distinction between pain and suffering is crucial, as suffering is at the core of misery. Addressing the role of suffering in clinical ethics, Loewy states, "Suffering is something experienced by individuals. Its existential quality is subjective and peculiar to the one suffering" (140, p. 84). A patient's suffering will never be understood without holistic assessment, as a pain level of "a 10 out of 10" on a numerical rating scale tells little about the dysfunction that pain causes in the patient's life. It is far easier for a psychologist to take the *clinical* approach, maintaining distance from the patient's pain and suffering. Loewy notes, "Humans are social beings, and all humans…when confronted with the misery of another, feel a sense of unease, a sense that there is something very wrong, and that they too are, in a sense, suffering" (140, p. 84). This is consistent with Leder's assertion that the virtue of a shared humanity allows for the clinician's ability to empathetically participate in the patient's experience (1).

The psychologist can never fully experience a patient's pain, but he or she can certainly respect it. Patients with chronic pain suffer during the psychological assessment process, as it involves attending not only to their physical pain, but to their dysfunction and emotional misery as well. Being in the patient's world throughout the psychological assessment process is a demonstration of respect and is as close to unconditional positive regard as one can be in the assessment process. Standardized tests do not bring the psychologist into the patient's world; rather, they provide quantitative data, which in itself tell little about what a patient is experiencing. Svenaeus notes that focusing on dialogue with patients is critical, as only through verbal interaction can the doctor understand the patient's own interpretation of his or her illness situation. *Authentic interpretation* (6) is emphasized as crucial to the practice of clinical medicine. Computerized interpretation of tests such as the MMPI, upon which psychologists are becoming more reliant (141,142), reduces the likelihood that the psychologist will be-in-the-world of the pain patient. The most commonly used comprehensive interpretive report, the Minnesota Report (143), is based upon a database of 40,000 psychiatric cases in a variety of mental health settings. There is no evidence, however, that it captures the essence of the maldynia sufferer's subjective life-world, and thus it tends to perpetuate the technophilic manner in which patients with chronic pain are treated. Such practice can be seen to further remove the psychologist from the patient's phenomenological experience of his or her maldynic illness.

In a qualitative study by Miller et al. (144), all participating patients reported that physicians did not listen to them when they made an effort to describe their pain and its overall impact on their lives. *Listening* had a very different meaning, however, to the physicians who were interviewed in the study, as they reported *hearing words* as a means of obtaining diagnostic cues. Doing so is likely less emotionally taxing to physicians. Walker and colleagues state, "Medicalisation of a condition such as pain allows health professionals to reproduce knowledge which supports their social position as educated experts and justifies the treatment they give to patients" (44, p. 623), regardless of the treatment's proven efficacy and the patient's response to it. To individuals suffering from maldynic pain, the refusal or inability of physicians to acknowledge narrative to communicate the patient's context of their life-world is likely to be a cause of the mistrust of physicians which has been identified in several qualitative studies of chronic pain patients' attitudes (44,145). Irrespective of physicians' listening styles, pain psychologists are obligated to listen to the patient in a manner in which the *patient* sees as respectful. Engebretson et al. take the argument for listening to patients even further, suggesting that "A biocultural model allow patients' voices to be heard, and defines listening to these stories as a moral act. Ineffective listening skills should not only be considered disrespectful, but also unethical" (26, p. 22).

CROSS-CULTURAL SENSITIVITY

As a final note, it is essential that pain psychologists recognize the growing proportion of ethnic and racial minorities in the U.S. population (146), and that every possible effort is made to assure that assessment, including psychometric tests, is culturally sensitive. Strengthening this point is literature suggesting that racial and ethnic minorities in the United States who suffer from chronic pain are likely to be underserved (147–151), that they trust physicians less (152–155), that they find mental health services to be less useful (156), that they have more negative expectations of mental health services (157), and that they are less likely to receive a referral for mental health services, even when controlling for severity of symptoms (158) than are nonminorities. Although Portenoy, Ugarte, Fuller, and Haas (159) and Schatman (160) have suggested that socioeconomic status rather than racial and ethnic minority status, per se, is the cause of the undertreatment of pain, a study by Tait et al. (161) suggest that sociocultural biases may be the cause of inadequate treatment of minorities with chronic pain. As the Tait et al. (161) study was conducted in Missouri, regional differences in sociocultural biases still need to be examined. Regardless, racial and cultural sensitivity in the psychological assessment of patients with maldynic pain is a crucial aspect of the phenomenological approach.

CONCLUSION

In response to the historically technophilic approach to the assessment of patients with chronic pain, an argument has been made for a change to a more phenomenological approach, emphasizing the clinical interview. This essay does not necessarily advocate the discontinuation of all standardized tests in the assessment of the emotional and behavioral features of maldynia, although only the provision of an environment in which the patient can truly be heard will allow the psychologist to recognize the meaning of pain within the sufferer's life-world. Such an understanding is crucial if the psychologist intends to recognize the meaning of the patient's pain, and an inability to do so will likely result in a failure to facilitate relief. Psychologists should not underestimate the potential that their assessment of the pain patient may have in her overall recovery. Steihaug (162) suggests that the physiotherapist has the potential to physically assess pain patients more accurately than does the typical physician, as the physiotherapist is more attuned to changes in breathing, posture, and muscular tension, while physicians tend to be focused on identifying pathophysiology. Likewise, phenomenologically oriented psychologists are in a position to share in a patient's lived experience of pain, provided that they avoid the pitfall of reducing the patient to diagnosable psychopathology.

Teaching patients to manage maldynic pain is more of an art form than a science, and it is essential that psychologists (as well as other treating professionals) recognize this. The assessment process provides the psychologist not only with an opportunity to understand the meaning of pain to the specific patient, but affords the first, and perhaps only, occasion in which to build the foundations of a working therapeutic bond. As the assessment process is ideally the first encounter in a longer helping relationship, the psychologist may be given only one invitation into the pain patient's life-world, and failure to recognize this window of opportunity can and often does preclude an opportunity to share in the patient's process of healing.

Intensive reliance upon standardized psychological tests in the assessment of maldynic pain patients may be driven by the marketplace economy, to a certain extent. As discussed, psychologists do not have to be involved in the actual testing process, with less costly surrogates often assuming the psychometrician role. Although a review of the literature fails to yield information on the length of the actual psychologist–patient clinical interview, economics would suggest that the allotted period is likely to decrease with time. To understand the phenomenological meaning of a chronic pain patient's discomfort, considerable face-to-face interaction is needed. Ideally, the clinical interview becomes multiple clinical interviews, as 30 minutes or 1 hour is generally insufficient to understand what chronic pain means to a particular patient. While running the risk of engaging in social constructionism, marketplace realities cannot be ignored. In discussing the importance of effective listening and reflection in treating patients with pain, Engebretson et al. clearly state, "the health-care system must support this ethical practice with adequate time and reimbursement provided for those who engage in it. Without such system support and use of a biocultural model as the standard, the meaning behind the words of the pain sufferer will remain hidden" (26, p. 26). However, irrespective of marketplace realities, it would behoove the pain psychologist to dedicate less effort to being a technician in his or her assessment of patients with maldynic pain, as the patient is more likely to seek—and need—a meaningful connection with an empathetic professional who remains, first and foremost, a person.

REFERENCES

1. Leder, D. 1990. Clinical interpretation: The hermeneutics of medicine. *Theoretical Medicine* 11: 9–24.
2. Steen, E., and L. Haugli. 2000. Generalised chronic musculoskeletal pain as a rational reaction to a life situation? *Theoretical Medicine* 21: 581–599.
3. Wertz, F.J. 2005. Phenomenological research methods for counseling psychology. *Journal of Counseling Psychology* 52: 167–177.
4. Heidegger, M. 1996. *Being and Time: A Translation of Sein and Zeit* (J. Stambaugh, trans.). Albany: State University of New York Press.
5. Gadamer, H.G. 1993. *Truth and Method,* 2nd Revised Edition (J.W. Weinsheimer and D.G. Marshall, trans.). New York: Continuum.
6. Svenaeus, F. 2003. Hermeneutics of medicine in the wake of Gadamer: The issue of phronesis. *Theoretical Medicine and Bioethics* 24: 407–431.
7. Ingleby, D. 1981. Understanding "mental illness." In: *The Politics of Mental Health,* ed. D. Ingleby, 23–71. Harmondsworth, UK: Penguin.
8. Dickerson, S.S. 2002. Redefining life while forestalling death: Living with an implantable cardioverter defibrillator after a sudden cardiac death experience. *Qualitative Health Research* 12: 360–367.
9. Alasad, J. 2002. Managing technology in the intensive care unit: The nurses' experience. *International Journal of Nursing Studies* 39: 407–413.
10. Andrew, C.M. 1998. Optimizing the human experience: Nursing the families of people who die in intensive care. *Intensive & Critical Care Nursing* 14: 59–65.
11. Walters, A.J. 1994. An interpretative study of the clinical practice of critical care nurses. *Contemporary Nurse* 3: 21–25.
12. Thornton, J., and A. White. 1999. A Heideggerian investigation into the lived experience of humour by nurses in an intensive care unit. *Intensive & Critical Care Nursing* 15: 266–278.
13. Totka, J.P. 1996. Exploring the boundaries of pediatric practice: Nurse stories related to relationships. *Pediatric Nursing* 22: 191–196.
14. Duke, S. 1998. An exploration of anticipatory grief: The lived experience of people during their spouses' terminal illness and in bereavement. *Journal of Advanced Nursing* 28: 829–839.
15. Dickerson, S.S., Stone, V.I., Panchura, C. et al. 2002. The meaning of communication: Experiences with augmentative communication devices. *Rehabilitation Nursing* 27: 215–220.
16. Torjuul, K., Nordam, A., and V. Sørlie. 2005. Action ethical dilemmas in surgery: An interview study of practicing surgeons. *BMC Medical Ethics* 6: 7.
17. Combs, A.V. 1948. Phenomenological concepts in nondirective therapy. *Journal of Consulting Psychology* 12: 197–208.
18. May, R. 1964. On the phenomenological bases of psychotherapy. *Review of Existential Psychology & Psychiatry* 4: 22–36.
19. Rogers, C. 1957. The necessary and sufficient conditions of therapeutic personality change. *Journal of Consulting Psychology* 21: 95–106.
20. Smith, J.A. 1996. Beyond the divide between cognition and discourse: Using interpretative phenomenological analysis in health psychology. *Psychology and Health* 11: 261–271.
21. Wampold, B.E., and K.S. Bhati. 2004. Attending to the omissions: A historical examination of evidence-based practice movements. *Professional Psychology: Research and Practice* 35: 563–570.

22. Husserl, E. 1954. *The crisis of European sciences and transcendental phenomenology* (D. Carr trans.). Evanston, IL: Northwestern University Press.
23. Walters, A.J. 1995. A Heideggerian hermeneutic study of the practice of critical care nurses. *Journal of Advanced Nursing* 17: 317–327.
24. Paley, J. 2005. Phenomenology as rhetoric. *Nursing Inquiry* 12(2): 106–116.
25. Leder, D. 1988. The hermeneutic role of the consultation–liaison psychiatrist. *The Journal of Medicine and Philosophy* 13: 367–378.
26. Engebretson, J., Monsivais, D., and J.S. Mahoney. 2006. The meaning behind the words: Ethical considerations in pain management. *American Journal of Pain Management* 16: 21–27.
27. Grondin, J. 1990. Hermeneutics and relativism. In: *Festivals of Interpretation: Essays on Hans-Georg Gadamer's Work,* ed. K. Wright, 42–62. Albany: State University of New York Press.
28. Schatman, M.E. 2003. *Utilizing outcome data to help chronic pain patients recognize their successes.* Paper presented at the 14th Annual Clinical Meeting of the American Academy of Pain Management, Denver, CO.
29. Radford, C. 1972. Pain and pain behaviour. *Philosophy: The Journal of the Royal Institute of Philosophy* 47: 189–205.
30. Osborne, M., and J.A. Smith. 1998. The personal experience of chronic benign lower back pain: An interpretative phenomenological analysis. *British Journal of Health Psychology* 3: 65–83.
31. Jacob, M.C., and R.D. Kerns. 2001. Assessment of the psychosocial context of the experience of chronic pain. In: *Handbook of Pain Assessment,* eds. D.C. Turk and R. Melzack, 362–384. New York: Guilford Press.
32. Turk, D.C., and A. Okifuji. 2001. Matching treatment to assessment of patients with chronic pain. In *Handbook of Pain Assessment,* eds. D.C. Turk and R. Melzack, 400–414. New York: Guilford Press.
33. Fischer, C.T. 1979. Individualized assessment and phenomenological psychology. *Journal of Personality Assessment* 43: 115–122.
34. Crombie, I.K., and H.T. Davies. 1998. Selection bias in pain research. *Pain* 74: 1–3.
35. Marketdata Enterprises. 1995. *Chronic pain management programs: A market analysis.* Valley Stream, NY: Marketdata Enterprises.
36. Roper Starch Worldwide Inc. 1999. *Chronic pain in America: Roadblocks to relief.* Retrieved August 6, 2010, from http://www.ampainsoc.org/links/roadblocks/.
37. Deyo, R.A., and W.R. Phillips. Low back pain: A primary care challenge. *Spine* 21: 2826–2832.
38. Smith, B.H. 2001. Chronic pain: A challenge for primary care. *British Journal of General Practice* 51: 524–526.
39. Sullivan, M.D., Turner, J.A., and J. Romano. 1991. Chronic pain in primary care: Identification and management of psychosocial factors. *Journal of Family Practice* 32: 193–199.
40. Potter, M., Schafer, S., Gonzalez-Mendez, E. et al. 2001. Opioids for chronic nonmalignant pain. Attitudes and practices of primary care physicians in the UCSF/Stanford Collaborative Research Network, University of California, San Francisco. *Journal of Family Practice* 50: 145–151.
41. Album, D. 1991. The prestige of diseases and medical specialties. *Tiddsskr Nor Laegeforen* 111: 2127–2173.
42. Pennebaker, J., and J.W. Skelton. 1981. Selective monitoring of physical sensations. *Journal of Personality & Social Psychology* 41: 213–223.
43. Zaner, R.M. 1990. Medicine and dialogue. *The Journal of Medicine and Philosophy* 15: 303–325.

44. Walker, J., Holloway, I., and B. Sofaer. 1999. In the system: The lived experience of chronic back pain from the perspectives of those seeking help from pain clinics. *Pain* 80: 621–628.
45. Williams, D.A., and B.E. Thorn. 1989. An empirical assessment of pain beliefs. *Pain* 36: 351–358.
46. Bredart, A., Bouleuc, C., and S. Dolbeault. 2005. Doctor-patient communication and satisfaction with care in oncology. *Current Opinion in Oncology* 17: 351–354.
47. Davidoff, F., and V. Florance. 2000. The informationist: A new health profession? *Annals of Internal Medicine* 132: 996–998.
48. Murphy, J.K., Sperr, E.V., and S.H. Sperr. 1986. Chronic pain: An investigation of assessment instruments. *Journal of Psychosomatic Research* 30: 289–296.
49. Wilcoxson, M.A., Zook, A., and J.J. Zarski. 1988. Predicting behavioral outcomes with two psychological assessment methods in an outpatient pain management program. *Psychology & Health* 2: 319–333.
50. Steen, E., and L. Haugli. 2000. The body has a history: An educational intervention programme for people with generalised chronic musculoskeletal pain. *Patient Education and Counseling* 41: 181–195.
51. Turk, D.C. 2005. The potential of treatment matching for subgroups of patients with chronic pain: Lumping versus splitting. *Clinical Journal of Pain* 21: 44–55.
52. Briere, J. 2004. *Psychological assessment of adult posttraumatic states: Phenomenology, diagnosis, and measurement* (2nd ed.). Washington, DC: American Psychological Association.
53. Hathaway, S.R., and J.C. McKinley. 1943. *The Minnesota Multiphasic Personality Schedule* (Revised). Minneapolis, MN: University of Minnesota Press.
54. Dahlstrom, W.G. 1972. Whither the MMPI? In: *Objective personality assessment: Changing perspectives,* ed. J.N. Butcher, 85–115. New York: Academic Press.
55. Loevinger, J. 1972. Some limitations of objective personality tests. In: *Objective personality assessment: Changing perspectives,* ed. J.N. Butcher, 45–58. New York: Academic Press.
56. Pancoast, D.L., and R.P. Archer. 1989. Original adult MMPI norms in normal samples: A review with implications for future developments. *Journal of Personality Assessment* 53: 376–395.
57. Butcher, J.N., Dahlstrom, W.G., Graham, J.R. et al. 1989. *Manual for the Restandardized Minnesota Multiphasic Personality Inventory: MMPI-2.* Minneapolis: University of Minnesota Press.
58. Ben-Porath, Y.S., and J.R. Graham. 1991. Resolutions to interpretive dilemmas created by the Minnesota Multiphasic Personality Inventory 2 (MMPI-2): A reply to Strassberg. *Journal of Psychopathology and Behavioral Assessment* 13: 173–179.
59. Hall, G.C.N., Bansal, A., and I.R. Lopez. 1999. Ethnicity and psychopathology: A meta-analytic review of 31 years of comparative MMPI/MMPI-2 research. *Psychological Assessment* 11: 186–197.
60. Bradley, L.A., and N.L. McKendree-Smith. 2001. Assessment of psychological status using interviews and self-report instruments. In: *Handbook of Pain Assessment,* eds. D.C. Turk and R. Melzack, 292–319. New York: Guilford Press.
61. Hickling, E.J., Sison, G.F., and J.L. Holtz. 1985. Role of psychologists in multidisciplinary pain clinics: A national survey. *Professional Psychology: Research and Practice* 16: 868–880.
62. Leavitt, F., and D.C. Garron. 1982. Rorschach and pain characteristics of patients with low back pain and "conversion V" MMPI profiles. *Journal of Personality Assessment* 46: 18–25.
63. Chapman, S.L., and J.S. Pemberton. 1994. Prediction of treatment outcome from clinically derived MMPI clusters in rehabilitation for chronic low back pain. *Clinical Journal of Pain* 10: 267–276.

64. Nelson, D.V., and D.M. Novy. 1996. The psychological characteristics of reflex sympathetic dystrophy versus myofascial pain syndromes. *Regional Anesthesiology* 21: 202–208.
65. Novy, D.M. 2004. Psychological approaches for managing chronic pain. *Journal of Psychopathology and Behavioral Assessment* 26: 279–288.
66. Vendrig, A.A. 2000. The Minnesota Multiphasic Personality Inventory and chronic pain: A conceptual analysis of a long-standing but complicated relationship. *Clinical Psychology Review* 20: 533–559.
67. Kugelmann, R. 1999. Complaining about chronic pain. *Social Science and Medicine* 49: 1663–1676.
68. Dworkin, R.H., Turk, D.C., Farrar, J.T. et al. 2005. Core outcome measures for chronic pain clinical trials: IMMPACT recommendations. *Pain* 113: 9–19.
69. Beck, A.T., Ward, C.H., Mendelsohn, et al. 1961. An inventory for measuring depression. *Archives of General Psychiatry* 4: 561–571.
70. Kerns, R.D., Turk, D.C., and T.E. Rudy. 1985. The West Haven–Yale Multidimensional Pain Inventory (WHYMPI). *Pain* 23: 345–356.
71. Cleeland, C.S., and K.M. Ryan. 1994. Pain assessment: Global use of the Brief Pain Inventory. *Annals of the Academy of Medicine* 23: 129–138.
72. McNair, D.M., Lorr, M., and L.F. Droppleman. 1992. *POMS Manual: Profile of mood states*. San Diego, CA: Industrial Testing Service.
73. Guy, W. 1976. *ECDEU Assessment Manual for Psychopharmacology*. (DHEW Publication No. ADM 76–338). Washington, DC: U.S. Government Printing Office.
74. Street, R.L., Gold, W.R., and T. McDowell. 1994. Using health status surveys in medical consultations. *Medical Care* 32: 732–744.
75. Caudill, M., Schnable, R., Zuttermeister, P. et al. 1991. Decreased clinic use by chronic pain patients: Response to behavioral medicine intervention. *Clinical Journal of Pain* 7: 305–310.
76. Mease, P. 2005. Fibromyalgia syndrome: Review of clinical presentation, pathogenesis, outcome measures, and treatment. *Journal of Rheumatology* (suppl.) 75: 6–21.
77. Millon, T., Green, C.J, and R.B. Meagher. 1979. The MBHI: A new inventory for the psychodiagnostician in medical settings. *Professional Psychology* 10: 529–539.
78. Gatchel, R.J., Deckel, A.W., Weinberg, N. et al. 1985. The utility of the Millon Behavioral Health Inventory in the study of chronic headaches. *Headache: The Journal of Head and Face Pain* 25: 49–54.
79. Gatchel, R.J., Mayer, T.G., Capra, P. et al. 1986. Millon Behavioral Health Inventory: Its utility in predicting physical function in patients with low back pain. *Archives of Physical Medicine and Rehabilitation* 67: 878–882.
80. Herron, L., Turner, J.A., Ersek, M. et al. 1992. Does the Millon Behavioral Health Inventory (MBHI) predict lumbar laminectomy outcome? A comparison with the Minnesota Multiphasic Personality Inventory (MMPI). *Journal of Spinal Disorders* 5: 188–192.
81. Sweet, J.J., Breuer, S.R., Hazlewood, L.A. et al. 1985. The Millon Behavioral Health Inventory: Concurrent and predictive validity in a pain treatment center. *Journal of Behavioral Medicine* 8: 215–226.
82. Derogatis, L.R. 1983. *SCL-90R Administration scoring and procedures manual* (2nd ed.). Towson, MD: Clinical Psychometric Research.
83. Buckelew, S.P., DeGood, D.E., Schwartz, D.P. et al. 1986. Cognitive and somatic item response pattern of pain patients, psychiatric patients, and hospital employees. *Journal of Clinical Psychology* 42: 852–856.
84. Shutty, M.S. Jr., DeGood, D.E., and D.P. Schwartz. 1986. Psychological dimensions of distress in chronic pain patients: A factor analytic study of symptom checklist-90 responses. *Journal of Consulting and Clinical Psychology* 54: 836–842.

85. Tollison, D.C., and J.C. Langley. 1995. *Pain patient profile manual.* Minneapolis: National Computer Services.
86. Tearnan, B.H., and M.J. Lewandowski. 1992. The Behavioral Assessment of Pain Questionnaire: The development and validation of a comprehensive self-report instrument. *American Journal of Pain Management* 2: 181–191.
87. Campbell, A., and M.E. Schatman. 2005. *Development and validation of the Pain Outcomes Profile: A brief, cost effective clinical outcomes tool.* Paper presented at the 16th Annual Clinical Meeting of the American Academy of Pain Management, San Diego, CA.
88. Mikail, S.F., DuBreuil, S.C., and J.L. D'Eon. 1993. A comparative analysis of measures used in the assessment of chronic pain patients. *Psychological Assessment* 5: 117–120.
89. Riley, J.L., Zawacki, T., Robinson, M.E., et al. 1999. Empirical test of the factor structure of the West Haven–Yale Multidimensional Pain Inventory. *The Clinical Journal of Pain* 15: 24–30.
90. Bergstrom, K.G., Jensen, I.B., Bodin, L. et al. 1998. Reliability and factor structure of the Multidimensional Pain Inventory—Swedish language version (MPI-S). *Pain* 75: 101–110.
91. Flor, H., Rudy, T., Birbaumer, N., and M. Schugers. 1990. Zur anwend bar keit des West Haven–Yale Multidimensional Pain Inventory in Deutschen sprachraum. *Der Schumerz* 4: 82–87.
92. Lousberg, R., Van Breukelen, G.J., Groenman, N.H. et al. 1999. Psychometric properties of the Multidimensional Pain Inventory, Dutch language version (MPI-DLV). *Behavior Research and Therapy* 37: 167–182.
93. Turk, D.C., and T.E. Rudy. 1988. Toward an empirically derived taxonomy of chronic pain patients: Integration of psychological assessment data. *Journal of Consulting and Clinical Psychology* 56: 233–238.
94. McBeth, J., Macfarlane, G.J., Benjamin, S. et al. 1999. The association between tender points, psychological distress, and adverse childhood experiences: A community-based study. *Arthritis & Rheumatism* 42: 1397–1404.
95. Craig, T., Boardman, A., Mills, K. et al. 1993. The South London somatisation study I: Longitudinal course and the influence of early life experiences. *British Journal of Psychiatry* 163: 579–588.
96. Schanberg, L.E., Keefe, F.J., Lefebvre, J.C. et al. 1998. Social context of pain in children with juvenile fibromyalgia syndrome: Parental pain history and family environment. *Clinical Journal of Pain* 14: 107–115.
97. Lowes, L., and M.A. Prowse. 2001. Standing outside the interview process? The illusion of objectivity in phenomenological data generation. *International Journal of Nursing Studies* 38: 471–480.
98. DiMatteo, M.R., Taranta, A., Friedman, H.S. et al. 1980. Predicting patient satisfaction from physicians' nonverbal communications skills. *Medical Care* 18: 376–387.
99. DiMatteo, M.R., Hays, R.D., and L.M. Prince. 1986. Relationship of physicians' nonverbal communication skills to patient satisfaction, appointment noncompliance, and physician workload. *Health Psychology* 5: 581–594.
100. Bensing, J. 1991. Doctor-patient communication and the quality of care. *Social Science and Medicine* 32: 1301–1310.
101. Bensing, J., Kerssens, J.J., and M.A.A. van der Pasch. 1995. Patient-directed gaze as a tool for discovering and handling psychosocial problems in general practice. *Journal of Nonverbal Behavior* 19: 223–242.
102. Robinson, J.D. 1998. Getting down to business: Talk, gaze and body orientation during openings of doctor–patient consultations. *Health Communication* 25: 97–123.
103. Peek, L., and J.P. Sawyer. 1988. Utilization of the Family Drawing Depression Scale with pain patients. *Arts in Psychotherapy* 15: 207–210.

104. Bowman, J.M. 1994. Reactions to chronic low back pain. *Issues in Mental Health Nursing* 15: 445–453.
105. Passchier, J., de Boo, M., Quaak, H.Z.A. et al. 1996. A health-related quality of life of chronic headache patients is predicted by the emotional component of their pain. *Headache: The Journal of Head and Face Pain* 36: 556–560.
106. Dalton, J.A., Feuerstein, M., Carlson, J., and K. Roghman. 1994. Biobehavioral pain profile: Development and psychometric properties. *Pain* 57: 95–107.
107. Lenhart, R.S., and J.S. Ashby. 1996. Cognitive coping strategies and coping modes in relation to chronic pain disability. *Journal of Applied Rehabilitation Counseling* 27: 15–18.
108. Jensen, I.B., and L. Bodin. 1998. Multimodal cognitive-behavioural treatment for workers with chronic spinal pain: A matched cohort study with an 18-month follow-up. *Pain* 76: 35–44.
109. Nicassio, P.M., Schuman, C., Radojevic, V. et al. 1999. Helplessness as a mediator of health status in fibromyalgia. *Cognitive Therapy and Research* 23: 181–196.
110. Gibson, S.J., and R.D. Helme. 2000. Cognitive factors and the experience of pain and suffering in older persons. *Pain* 85: 375–383.
111. Grant, L.D., Long, B.C., and J.D. Willms. 2002. Women's adaptation to chronic back pain: Daily appraisals and coping strategies, personal characteristics and perceived spousal responses. *Journal of Health Psychology* 7: 545–564.
112. Tan, G., Jensen, M.P., Robinson-Whelen, S. et al. 2002. Measuring control appraisals in chronic pain. *Journal of Pain* 3: 385–393.
113. Gullacksen, A.C., and J. Lidbeck. 2004. The life adjustment process in chronic pain: Psychosocial assessment and clinical implications. *Pain Research & Management* 9: 145–154.
114. Sofaer, B., Moore, A.P., Holloway, I. et al. 2005. Chronic pain as perceived by older people: A qualitative study. *Age and Ageing* 34: 462–466.
115. Lish, J.D., Kuzma, M.A., Lush, D.T. et al. 1997. Psychiatric screening in primary care: What do patients really want? *Journal of Psychosomatic Research* 42: 167–175.
116. Paulson, M.A., Norberg, A., and S. Soderberg. 2003. Living in the shadow of fibromyalgia pain: The meaning of female partners' experiences. *Journal of Clinical Nursing* 12: 235–243.
117. Orth-Gomer, K., Unden, A.L., and M.E. Edwards. 1988. Social isolation and mortality in ischemic heart disease: A 10-year follow-up study of 150 middle-aged men. *Acta Medica Scandinavica* 224: 205–215.
118. Fees, B.S., Martin, P., and L.W. Poon. 1999. A model of loneliness in older adults. *Journal of Gerontology: Series B: Psychological Sciences and Social Sciences* 54: 231–239.
119. Levenstein, S., Smith, M.W., and G.A. Kaplan. 2001. Psychosocial predictors of hypertension in men and women. *Archives of Internal Medicine* 161: 1341–1346.
120. Eng, P.M., Rimm, E.B., Fitzmaurice, G. et al. 2002. Social ties and change in social ties in relation to subsequent total and cause-specific mortality and coronary heart disease incidence in men. *American Journal of Epidemiology* 155: 700–709.
121. Paulson, M., Danielson, E., and S. Soderberg. 2002. Struggling for a tolerable existence: The meaning of men's lived experiences of living with pain of the fibromyalgia type. *Qualitative Health Research* 12: 238–249.
122. Madjar, I. 1997. The body in health, illness, and pain. In: *The body in nursing,* ed. J. Lawler, 53–73. Melbourne, Australia: Churchill, Livingstone.
123. Ragsdale, J.D. 1996. Gender, satisfaction level and the use of relational maintenance strategies in marriage. *Communication Monographs* 63: 354–369.

124. Komiya, N., Good, G.E., and N.B. Sherrod. 2000. Emotional openness as a predictor of college students' attitudes toward seeking psychological help. *Journal of Counseling Psychology* 47: 138–143.
125. Costa, P., Jr., Terracciano, A., and R.R. McCrae. 2001. Gender differences in personality traits across cultures: Robust and surprising findings. *Journal of Personality and Social Psychology* 81: 322–331.
126. Misra, I. 2003. Openness to experience: Gender differences and its correlates. *Journal of Personality and Clinical Studies* 19: 141–151.
127. Snyder, D.M. 1982. Perspective and engagement in counseling and psychotherapy: The phenomenological approach. *International Journal for the Advancement of Counseling* 5: 95–107.
128. Griffin, E. 1991. *A first look at communication theory*. New York: McGraw-Hill.
129. Worsley, R. 2002. *Process work in person-centred therapy: Phenomenological and existential perspectives*. London: Palgrave.
130. Rogers, C. 1961. *On becoming a person*. Boston: Houghton Mifflin.
131. Clair, D., and D. Prendergast. 1994. Brief psychotherapy and psychological assessments: Entering a relationship, establishing a focus, and providing feedback. *Professional Psychology: Research and Practice* 25: 46–49.
132. Aydede, M., and G. Guzeldere. 2002. Some foundational problems in the scientific study of pain. *Philosophy of Science* 69 (Suppl. for the proceedings of PSA 2000): 265–283.
133. Fishbain, D.A., Cutler, R., Rosomoff, H.L. et al. 1999. Chronic pain disability exaggeration/malingering and submaximal effort research. *Clinical Journal of Pain* 15: 244–274.
134. Berg, M. 1985. The feedback process in diagnostic psychological testing. *Bulletin of the Menninger Clinic* 49: 52–69.
135. American Psychological Association. 2002. Ethical principles of psychologists and code of conduct. *American Psychologist* 57: 1060–1073.
136. Glenton, C. 2002. Developing patient-centred information for back pain patients. *Health Expectations* 5: 319–329.
137. Lipton, R.B., and W.F. Stewart. 1999. Acute migraine therapy: Do doctors understand what patients with migraine want from therapy? *Headache* 39 (suppl. 2): S20–S26.
138. Sofaer, B., and J. Walker. 1994. Mood assessment in chronic pain patients. *Disability and Rehabilitation* 16: 35–38.
139. Frischenschlager, O., and I. Pucher. 2002. Psychological management of pain. *Disability and Rehabilitation* 24: 416–422.
140. Loewy, E.H. 1991. The role of suffering and community in clinical ethics. *Journal of Clinical Ethics* 2: 83–89.
141. Butcher, J.N., Perry, J.N., and M.M. Atlis. 2000. Validity and utility of computer-based test interpretation. *Psychological Assessment* 12: 6–18.
142. Butcher, J.N., Perry, J., and J. Hahn. 2004. Computers in clinical assessment: Historical developments, present status, and future challenges. *Journal of Clinical Psychology* 60: 331–345.
143. Butcher, J.N. 2002. *User's guide for the MMPI-2 Minnesota Report,* 4th ed. Minneapolis, MN: Pearson Assessments.
144. Miller, W.L., Yanoshik, M.K., Crabtree, B.F. et al. 1994. Patients, family physicians, and pain: Visions from interview narratives. *Family Medicine* 26 :179–184.
145. Thomas, S.P. 2000. A phenomenological study of chronic pain. *Western Journal of Nursing Research* 22: 683–705.
146. U.S. Bureau of the Census. 2001. *Statistical Abstract of the United States: 2000.* Washington, DC: U.S. Government Printing Office.
147. Cleeland, C.S., Gonin, R., Hatfield, A.K. et al. 1994. Pain and its treatment in outpatients with metastatic cancer. *New England Journal of Medicine* 330: 592–596.

148. Cleeland, C.S., Gonin, R., Baez, L. et al. 1997. Pain and pain treatment in minority outpatients with metastatic cancer. *Annals of Internal Medicine* 127: 813–816.
149. Bernabei, R., Gambassi, G., Lapane, K. et al. 1998. Management of pain in elderly patients with cancer. *Journal of the American Medical Association* 279: 1877–1882.
150. Kramer, B.L., Harker, J.O., and A.L. Wong. 2002. Description of joint pain by American Indians: Comparison of inflammatory and noninflammatory arthritis. *Arthritis and Rheumatism* 47: 149–154.
151. Kramer, B.J., Harker, J.O., and A.L. Wong. 2002. Arthritis beliefs and self-care in an urban American Indian population. *Arthritis and Rheumatism* 47: 588–594.
152. Lipton, J.A., and J.J. Marbach. 1984. Ethnicity and pain experience. *Social Science and Medicine* 19: 1279–1298.
153. Doescher, M.P., Saver, B.G., Franks, P. et al. 2000. Racial and ethnic disparities in perceptions of physician style and trust. *Archives of Family Medicine* 9: 1156–1163.
154. Corbie-Smith, G., Thomas, S.B., and D.M. St. George. 2002. Distrust, race, and research. *Archives of Internal Medicine* 162: 2458–2463.
155. Boulware, L.E., Cooper, L.A., Ratner, L.E. et al. 2003. Race and trust in the health care system. *Public Health Reports* 118: 358–365.
156. Snell, C.L., and J.S. Thomas. 1998. Young African American males: Promoting psychological and social well-being. *Journal of Human Behavior in the Social Environment* 1: 125–136.
157. Richardson, L.A. 2001. Seeking and obtaining mental health services: What do parents expect? *Archives of Psychiatric Nursing* 15: 223–231.
158. Borowsky, S.J., Rubenstein, L.V., Meredith, L.S., Camp, P., Jackson-Triche, M., and K.B. Wells. 2000. Who is at risk of nondetection of mental health problems in primary care? *Journal of General Internal Medicine* 15: 381–388.
159. Portenoy, R.K., Ugarte, C., Fuller, I., and G. Haas. 2004. Population-based survey of pain in the United States: Differences among white, African American, and Hispanic subjects. *Journal of Pain* 5: 317–328.
160. Schatman, M.E. 2006. Racial and ethnic issues in chronic pain management: Challenges and perspectives. In: *Weiner's Pain Management: A Practical Guide for Clinicians* (7th ed.) eds. M.V. Boswell and B.E. Cole, 83–98. Boca Raton, FL: CRC Press.
161. Tait, R.C., Chibnall, J.T., Andresen, E.M. et al. 2004. Management of occupational back injuries: Differences among African Americans and Caucasians. *Pain* 112: 389–396.
162. Steihaug, S. 2005. Can chronic muscular pain be understood? *Scandinavian Journal of Public Health* 33 (suppl. 66): 36–40.

13 Painism—A New Ethics
Richard Ryder's Moral Theory and Its Limitations

Hans Werner Ingensiep

CONTENTS

INTRODUCTION

The British psychologist Richard D. Ryder (born 1940) is famous for his role as one of the pioneers of the modern animal liberation and animal rights movement. Ryder is well known for the introduction of new ethical ideas in the past 30 years and especially for coining of the term *speciesism*, which he first used in a privately printed leaflet in Oxford 1970 and then in many subsequent publications (1,2). In Cavalieri and Singer's "Great Ape Project" and other papers, Ryder argued for *Sentientism* (2) against the leading moral concepts emphasizing that only the capacity to feel pain is morally relevant. In 1990 Ryder coined the term *Painism* to describe "A Modern Morality" (3–5). For Ryder, pain is always pain of an individual and cannot be aggregated. Painism argues not only against the traditional *anthropocentric* personism based on qualia of persons like consciousness, intelligence, or rationality (e.g., Kant) but also against modern utilitarianism (6,7) and against an animal rights position based on an inherent value of each individual (8). Ryder, Singer, and Regan try to integrate pain into their ethical concepts in different ways. They all criticize the speciesism of traditional ethics and try to extend their approach to general ethics and politics.

"CAN THEY SUFFER?"

These contemporary approaches to animal ethics and bioethics are the recent products of a long historical and cultural debate in Western societies about the role of sentience, pain, and suffering. Concerning animals, for instance, Plutarch or Montaigne inspired the discussion about Descartes' concept of beasts as machines and led, in the Enlightenment, to extensive philosophical debates on animal pain that became an important ethical bridge between animals and humans (9). Pain, in general, played an important role in culture (10) and has also been recognized as a factor in the "history of the modern sensibility" (11) (e.g., concerning the problem of vivisection). While perhaps most famous is Jeremy Bentham's crucial question in a footnote: "Can they suffer?" (12, p. 412), less known is the dictum of Humphry Primatt, one of the pioneers in animal ethics in the eighteenth century: "Pain is pain...whether it be inflicted on man or beasts" (13, p. 21). But, on the other hand, according to John Stuart Mill, it "is better to be a human being dissatisfied than a pig satisfied" (14, p. 18), indicating that there seems to be a different quality concerning pleasure and pain in animals and humans. These statements illustrating the ethical and cultural bridge between humans and animals are an important foundation for current efforts to integrate all sentient beings into general ethics.

This chapter focuses on ethical aspects and problems in Ryder's *Painism*. There are many problems with the definitions and distinct boundaries for nonhuman organisms having a moral status within bioethics (15,16), and these give rise to conflicts with traditional concepts of ethics (e.g., Kant's deontology concerning animals and plants) (17,18). Kantian ethics seems to be a prototype for a deontologic personism that attracts reproaches of *anthropocentrism* and *speciesism*. Conversely, the new ethical model of an *expanding circle* (19) of entities having values or rights leads to boundary problems, reductions, and fallacies in argumentation, and unicriterial approaches to what and why beings have *moral status* (16). In one respect, the new concept of painism seems to be such a reductionistic and unicriterial approach to ethics. Yet, pain and suffering are important factors in modern bioethics (20,21), both in theory and practice (22,23). "The patient with pain becomes the person defined by pain," as James Giordano points out the complex situation of pain as a disease process versus pain as an illness phenomenon (23, p. 407). In general, it is not easy to bridge the dualism of pain as a disease with objective manifestation, and pain as a subjective illness phenomenon. Far more difficult is the attempt to integrate all kinds of problems in human and nonhuman ethics (abortion, euthanasia, war, biotechnology, environmental protection, etc.) into a unifying concept of painism, as Ryder does. In order to connect these domains, Ryder uses the term *pain* in a very broad meaning. Nevertheless, he is able to knowledge the differences and difficulties of (the term) *pain*. Thus, he interprets "pain broadly to include all negative experiences, that is to say, all forms of suffering, mental as well as 'physical.' So, the words 'pain' and 'suffering'... are interchangeable" (5, p. 26). Only under these conditions is it possible to include moral problems related to sentient animals and humans within an expanding circle of beings having a *moral status*.

I believe that it is important to consider some general philosophical aspects and questions regarding Ryder's proposal of painism. First, a crucial systematic question for ethics: is pain necessary or sufficient to establish moral status? I compare the three approaches of Singer, Regan, and Ryder to provide some insight to questions concerning the role of pain and speciesism in modern ethics. This question is not crucial for a pure biomedical ethics, where pain often is subordinate to the principle of nonmaleficience (24) and to the question "Why do no harm?" (25). But it is equally important to note that pain is not an important factor in nonhuman ethics of biocentrism or ecocentrism. As Warren thinks, the human and animal privilege of self-consciousness, sentience, or pain, biocentric ethicists like Paul Taylor (26) could ask: "why should the non-sentience of organisms automatically mean that a moral agent cannot have moral obligations towards them" (16, p. 433)? From an anthropocentric, biocentric, or ecocentric point of view, ethics sentience and pain are not necessary for a moral status.

In light of this, I will discuss examples for the application of Ryder's painism in applied ethics, human ethics, biotechnology, and environmental ethics. In particular, I look at borderline cases in medicine that illuminate an old thinking pattern about humans in a *vegetative* state as being beyond pain and sensibility. My question is: what kind of role does pain play in ethical judgements about persons sometimes called *human vegetables* (42)?

In short, my thesis is that Ryder's frame of painism is too narrow for a general theory of ethics. To build his argument, I compare Ryder's painism with the positions of Singer and Regan, and then deliver examples, exegesis, and criticism in more detail. The general problem is that an ethics reliant solely upon pain may be insufficient, or perhaps at the very least, unsatisfying. The most difficult question for psychologists and biologists seems to be: what is pain? And for painists: how far does pain extend within the expanding circle of sentient beings? In sum, I ask if painism, as an unicriterial approach, is a useful starting point for ethics.

RYDER'S PAINISM: A MIDDLE WAY BETWEEN THE SCYLLA OF UTILITARIANISM AND THE CHARYBDIS OF RIGHTS THEORY?

A crucial question for ethicists seems to be: is pain necessary or sufficient to establish moral status? In answering the question, Ryder steers between the Scylla of Singer's utilitarianism and the Charybdis of Regan's rights theory. In common with Singer, Ryder's painism holds that pain is a basic criterion. Ryder's painism also shares Regan's emphasis on the individual and rejection of utilitarian aggregation. Ryder tries to find a position in between and to establish painism as new morality for guiding the treatment of both humans and animals. Thus, he rejects the utilitarian justification of pain by Singer, "if the aggregated benefits of all those affected by the action outweigh the pain" (5, p. 48). According to Ryder's pain-based fundamentalism, "pain is the only evil" and the "moral objective is to reduce the pain of others" (5, p. 27). An aggregation of pains across individuals does not make sense. For each individual, only that individual's pain is real. For painists, pain within the same individual is paramount; thus, "the quantity of pain suffered by the maximum sufferer matters far more than the quantity of individuals effected" (5, p. 28). For Ryder, this is an implication from his rule: "speciesism is always wrong" and "the aggregation of pains and pleasures across individuals is meaningless" (5, p. 29). Ryder sets an example. "Killing 100 people painlessly becomes less wrong than torturing one of them to death." For Ryder, the key question is: "how much pain was experienced by the individual who suffered most" (5, p. 28)?

Against Regan's rights theory, Ryder sees no clear justification of an *inherent value* and a fundamental right "to be treated with respect." Moreover, he criticizes the problem of *precedence* in conflicts when Regan says that "the only thing that can override one right is another right." For Ryder, there is a problem with the "hierarchy of rights": "Without such rules, rights theory has a hollow centre. When faced with conflicting rights, which right do we choose" (5, p. 48)? But the main failure in Regan's rights theory, Ryder points out, is its foundation in "mysterious references to *telos* (purpose) or to intrinsic values" (5, p. 50).

IS PAIN NECESSARY OR SUFFICIENT FOR MORAL STATUS?

As mentioned, in a utilitarian approach to pain and pleasure, Bentham's question "Can they suffer?" is crucial. Pain and suffering are the main antagonists of pleasure, and both are connected to the capacity for sentience. According to Warren's description of Singer's concept, this capacity is both necessary and sufficient for having the same moral status applied to all sentient beings (16, p. 442). The argument is that only sentient beings can be cognitively aware of what happens to them, so that "Although the ability to experience pain is valuable to mobile organisms, serious pain is itself a harm. Pain is an inherent evil to the being that experiences it, and pleasure an inherent good." Yet Warren objects to both strong sentience theory and a weaker version, because although sentience may be sufficient for moral status of some organisms, it may not be necessary to instantiate moral regard for plants and nonsentient animals (as in biocentric or ecocentric approaches to ethics) (16, p. 443).

In his "Practical Ethics," Singer's preference utilitarianism describes three classes of living beings: self-conscious beings like most adult humans, great apes, and perhaps other higher animals like whales, dogs, or cats; non-self-conscious animals like most birds or fish; and beings without any sentience like mollusks, microorganisms, or plants. Singer's concept guarantees primary moral status to persons with self-consciousness; that is, they are guaranteed full protection against being killed because of their future options, interests, and preferences. This indicates that it is not necessary to have pain for moral status. In the second class, "the case

TABLE 13.1
Is Pain Necessary or Sufficient for Moral Status?

Position	Is Pain Necessary?	Is Pain Sufficient?
Utilitarianism (Singer)	No	Yes
Painism (Ryder)	Yes	Yes
Rights Theory (Regan)	No	No

against killing is weaker" (7, p. 132) because of the possibility that these animals could be killed painlessly such that "the killing of non-self-conscious animals may not be wrong," if they "can be killed painlessly" (7, p. 132). But this is only an argu ment at the level of theoretical ethics, not of practice in animal farming or animal experimentation. The painless-replaceability argument indicates that moral status for Singer allows painless killing if all interests are weighted and if the ethical calculation of interests leads to more pleasure in the world. Thus, sentience and pain are necessary to take these beings into account, but neither are sufficient to protect them against being harmed or killed. The third class (beings without capacity for sentience and pain) of beings are, for Singer, morally irrelevant, and no necessity exists to take them into account.

In summary, from Singer's point of view, either sentience or self-consciousness is sufficient to establish moral status. Thus, the capacity for pain is not necessary. A survey allows comparison of different positions and answers to the crucial question (see Table 13.1). Further embellishment of Ryder's account will be provided later in this chapter.

PAINISM AND SENTIENTISM

Ryder claims that "Painism supersedes sentientism" and explains why he rejects the earlier preferred term *sentientism* as a descriptor for his moral position (2,5, p. 34). The advantages of the new term are very different, in both principle, and on a pragmatic level. First is "that sentientism might be deemed to refer to *any* sort of feeling or sensation—or even to sentient beings incapable of experiencing pain at all" (5, p. 34). He speculates that, for instance, life-forms on other planets may have evolved systems in order to avoid danger without the subjective experience of pain. For Ryder, the subjective experience of pain is necessary and sufficient "regardless as to who or what that experiences it"; if one day painient machines may be produced, they, too, will become morally considerable (5, p. 34). Thus, for Ryder, the term *individual* includes all existing and future systems that are capable of feeling pain. We must conclude that existing biological things, for instance, plants, or even future iterations of machines or organic-synthetic chimeras—if possessing the capacity for pain—can be integrated in Ryder's painism. Obviously, Ryder wants to avoid any kind of speciesism and a narrow concept of a moral circle, and thus prefers the term *painism*.

The terms *painism* and *painient* seem to be more popular and easier to understand than *sentientism*, and pragmatically, "these words usefully fill some significant gaps in the English language" (5, p. 34). This is why Ryder rejects the term *pathocentrism* from the Belgian Philosopher Johan Braeckman; its wider meaning inclusive of disease and emotion without any attendant pain becomes confusing. For Ryder, the capacity for pain is not only necessary but also sufficient for conferring moral status.

HURTING AND HARMING

Tom Regan attempts to elucidate the difference between *hurting and harming* so as to allow focus on the role of pain in his rights theory (27, p. 202). Regan's assumption is that every "subject of life" has "dignity" or an "inherent value" as an individual. Regan tells a fictional story about a scientist testing a new drug on chimpanzees, and argues against such research when anesthetics or other palliatives are not used to eliminate or reduce suffering: "to cause an animal to suffer is to harm that animal—that is to diminish that individual animal's welfare." But there may be an important difference between harming and suffering. "An individual's welfare can be diminished independently of causing him or her to suffer, as when, for example, a young woman is reduced to a 'vegetable' by painlessly administering a debilitating drug to her while she sleeps." According to Regan, harm has been done to her, although she did not "suffer." "Not all harms hurt, in other words, just as not all hurts harm" (27, p. 202). From his point of rights theory, this is a deprivation of individual freedom and life, and it does not matter that this "subject of life" has been killed "humanely," without pain. This fiction illustrates that pain is neither necessary nor sufficient for the moral status of individuals in Regan's theory (see Table 13.1), rather, on the contrary, grounding ethics in "pain (i.e. suffering)…as the only evil" (5, p. 27) would be impossible for Regan.

PAIN—THE ONLY EVIL?

This brings us to address possible philosophical problems within Ryder's concept of painism, and its connection to naturalism, Darwinism, neurocentrism, and indeterminism. Ryder prefers both a very broad concept of pain to include all kinds of suffering and a very general concept of morality, which "is only about the right and wrong treatment of *others*" (5, p. 4). This seems to be a morality without a free moral agent who has responsibilities and duties, for instance, the duty to act morally. Here, he defaults to a reductive materialist stance in his claim that we are "entirely the slaves of our brains" and our "free will" may be a product of unpredictable events at the quantum level" (5, p. 51). Yet, he argues for a natural source of ethics: "ethical standards are thus inspired by compassion," an "innate compassion" exists, and the "history of ethics can be seen as an ever-expanding circle of compassion" (5, pp. 15, 17, 62).

For Ryder, to answer the question, why is pain bad? (5, p. 35), he states that pain is bad because it is bad—seemingly a circular answer. From this, he derives that "a bad thing is that which causes pain," and so, "morality derives from this fact" (5, p. 36). If the moral premise of painism is based on facts or causes, it seems to be not far from a naturalistic fallacy starting with natural or psychological facts (pain is bad) and then deriving value judgments or duties: "for painists the primary duty is to alleviate pain, specifically the pain of the greatest sufferers" (5, p. 37).

There is a difference between pain (at the physiological or psychological level) as a natural evil and pain as a moral evil. Toothache can be a physiological and psychological evil, but it is only a moral evil if the pain is induced by a moral agent, for instance, a torturer. Then we need another ethical criterion and a value judgement, for instance, that a person should be treated as an end in oneself (Kant). Ryder grounds his *moral revolution* in analogies between the "crucial similarity" of humans with other animals: "from this flows the conclusion that whatever is right or wrong in our treatment of other human beings is also right or wrong in our treatment of painient individuals of other species, however small they may be" (5, p. 118). Despite the ambiguity—or perhaps fallacy—of Ryder's argument in structure, it is important to assess its content as relates to the value of pain in the moral consideration of both human and nonhuman others.

PAIN IN NONHUMAN OTHERS: SPECIESISM—NORMATIVE AND DESCRIPTIVE DIMENSIONS

The term *speciesism* was introduced in the 1970s as a normative term in ethics and politics. For Ryder, this term is still very important to his concept of painism. "A difference in species is as morally irrelevant as are differences in race, age or gender" (5, p. 118). When introduced, the term *speciesism* also had a political function in the animal rights movement, to eliminate the unfair conditions of animal mass farming and animal experiments in biological or medical research. Thus, it was created and mediated to prevent discrimination and to demand equal moral treatment and rights to humans and nonhumans. But the term has a theoretical background and a descriptive dimension as often emphasized by Ryder, Singer, and others. Antispeciesism and painism belong to the "moral implications of Darwinism" and indicate "that animals and humans are in the same moral category" (5, p. 44). Darwinian evolution implies "that the human species is but one of many species" and that "pain is the common enemy of all animal species" (5, p. 45). In this context, *species* is used as a biological term with descriptive dimensions, and still we must consider that the term has been used to convey very different meanings in history and contemporary biological contexts. For example, in pre-Darwinian times, *species* had sometimes been understood as an essentialistic logical, or ontological term. After Darwin's *Origin of Species*, it became more of a classificatory unit for biological entities with special morphological, physiological, or genetic properties. There is a complex discussion in philosophy and evolutionary biology as to whether *species* refers to an individual or a set of organisms (28). Whatever *species* is, it is a theoretical concept in biology and not a normative term to engender judgement as often used in applied ethics. As a descriptive term, it characterizes all kinds of organisms—humans, animals, plants, and microorganisms. Thus, the biological term *species* is not restricted in its use.

Does the reproach *speciesism* in modern ethics mean more than *anthropocentrism*? It obviously does not, if speciesism is defined as in the *Oxford English*

Dictionary of 1985, as "discrimination against or exploition of certain animal species by human beings, based on an assumption of mankind's superiority" (5, p. 44). But from the theoretical point of view, *speciesism* is not limited to *homo sapiens sapiens* but must be extended to all species of organisms that try to dominate others, for instance, when "higher" sentient animal species discriminate or exploit "lower" nonsentient animals or plants. In other words, if the principle "regardless of species" is introduced into ethics, it concerns all organisms, not only sentient animals as Ryder and others posit. From an evolutionary point of view, there is no reason to discriminate against any species—the continuity of the *species* does not allow us to draw discontinuous borders in organic nature. From this point of view, the border between painient and nonpainient individuals could be seen as arbitrary.

SPECIES, INDIVIDUALS, AND PAIN

Ryder asserts that "Painism focuses upon individuals" and not on the conservation of *species* as a biological entity. He criticizes the anthropocentric and ethnocentric views of some environmentalists as "irrational" if they favor the extermination of *species* of *foreign* plants, which are not endemic in the ecosystem (5, p. 116). From his point of view, this seems to be a kind of speciesism and anthropocentric discrimination. Thus, this critique does not follow from his construct of painism, because painism does not include "non-painient plant life" (5, p. 46). This makes clear the tension between *speciesism* and *painism* in Ryder's concept. As well, the concept produces other difficulties. Ryder maintains that "moral standards should apply equally to all painient individuals regardless of species" (5, p. 119). If "regardless of species" means only regardless of the human species and not of all species (including sentient animals), then we must recognize an inconsistent use of the biological term *species*. If "regardless of species" refers only to the human species, it is used as a normative political term and not as a biological term. In this case, it does not afford any deeper meaning than the notion of *anthropocentrism* in animal ethics. Why should we only consider painient individuals of some species? Is that not another version of *speciesism* being asked by a collective of higher painient animals with nervous systems? Ryder's painism may render the same question as raised by Heta and Matti Häyri—who is like us? (30). We should classify this position as a form of neurocentric speciesism. Yet, if we accept that pain is the only evil, and thus is foundation of ethics, why do we not have a duty to protect all victims of natural predators that kill painient animals for food? If pain is the *only* quality upon which to establish moral status, we should prevent these evils in any circumstance. I know, this is not the intention of painism. But such considerations demonstrate that painism as a modern ethics must also acknowledge some form of natural order if it is to be meaningful.

If pain is a natural phenomenon and an arbitrary product of natural selection that confers advantages for some species of higher animals, then painism can be seen as a new kind of zoocentric speciesism, if not a new naturalism. But the problem for ethicists is not pain as a natural phenomenon; the crucial question is whether pain, as a unicriterial approach, is a necessary and sufficient starting point for ethics.

PAINISM—SOME APPLICATIONS AND QUESTIONS

PAINISM AND BIOTECHNOLOGY

Fred Gifford begins an address on biotechnology with a fictitious dialogue about "Football Birds" or "Egg Machines" created by agricultural science. These entities lack any ability to move and have no sensory input to the brain (31). Gifford stated: "The main issue might *seem* to be the way the 'animals' are treated but in fact no harm comes to the animals for they have no conscious experience of pain or stress... So just what could be wrong about it" (31, p. 197)? How would Ryder's painism decide this question?

For Ryder, it is "surely, morally desirable to alter nature in certain ways." There is "no moral objection per se" against genetic engineering and cloning of designer organisms under the following conditions: that the process of production entails "no suffering," that the produced individuals "do not suffer increased pain or distress due to their altered natures," and that such organisms do not cause harm to others (5, p. 114). Rather, Ryder explicitly welcomes the potential of genetic engineering to "*reduce the capacity for pain*, if, for example, it becomes possible to produce people and animals with reduced painience. What a revolution this will be! Ultimately there is the prospect of a painless future!" (5, p. 114). In this case, Ryder expresses his anthropological utopia, a vision of a "painless future," and he speculates about the "trick" of eliminating pain on the one hand, and maintaining the feeling of "joy and compassion" on the other hand, so as to enable, if not fortify, moral concern for others. But, as Giordano (32) noted, while an emphasis upon the capacity to feel pain may be essential to any moral regard or ethic, it is also important to recognize that the elimination of pain, in part or in whole, raises additional philosophical, ethical, and practical issues and concerns. These include the extent of pain that should be eliminated; in whom; by what criteria; and how the lack of painience might affect or alter notions of existential gain and loss, nonharm, and, ultimately, the regard and treatment of others.

PAINISM AND ENVIRONMENTAL ETHICS

Ryder's work indicates a distance and difference between modern environmental ethics and painism. Modern environmentalism began earlier than animal ethics, but as Ryder points out, its philosophical basis is often thin and anthropocentric, driven often by assuming a self-interest human or "mystical" quality concerning life in general, including nonpainient plant life (5, p. 46).

Painism is grounded in the pain of individuals; the classical approaches of animal liberation, humanism, and environmentalism are "subordinate to painism" in that they give rise to compassion (5, p. 46). In this case, Ryder's painism is similar to Singer's preference utilitarianism, where "the environment" is only relevant if sentient beings are concerned: as Singer asks, "is there value beyond sentient beings? Once we abandon the interests of sentient creatures as our source of value, where do we find value" (7, p. 277)? For Ryder, it is not the utilitarian aggregation and calculation, but only the existence of a maximum sufferer, whether the sufferer is an animal or a human being. For both Ryder and Singer, plants have no value, no interest, and no right to moral regard because they are not capable of sentience or feeling pain.

Thus, Ryder and Singer would oppose those environmental ethicist like J. Baird Callicott, Paul Taylor, or Christopher Stone, who argue for rights or respect for plants. For instance, Christopher Stone (33) argues for guardianship, legal rights, moral worth, and dignity of nonsentient beings like plants. Is this tenable within the context of painism?

Painism and Plants

Mary Anne Warren summarizes the skeptics' position concerning plants in modern ethics: As far as we can tell these organisms have neither sense organs nor nervous systems; and their behaviour rarely seems indicative of a capacity for pleasure or pain. Thus, while it is impossible to be absolutely certain that these organisms are not sentient, it is a good bet that they are not (15). Here, Warren adressed Tompkins and Bird's (34) "Secret Life of Plants"—indeed a dubious concept from a modern scientific point of view. But serious scientists organized the first symposium on Plant Neurobiology in Florence in 2005, and the acknowledged plant physiologist Trewavas tried to defend "plant intelligence" (35,36) in a discussion about cell-to-cell communication by means of phytohormones as neurotransmitter-like substances, the role of electric membrane potentials, and ion channels, functioning like "brain proteins" to form "...plant synapses: actin-based domains for cell-to-cell communication" (37,38). I mention these theories and facts in order to point out the hypothetical "boundary of sentience" of Ryder's and Singer's ethical position. If science would someday find something like *sentience* or *pain*, it could initiate problems with plant experiments or vegetarianism. An ethical approach like Ryder's cannot accommodate environmental value and protection, but Singer's can if it serves utility but is still neurocentric. I do not expect that plants can feel pain, but the hypothetical question clearly indicates that ethical theories like Ryder's or Singer's are founded in zoocentrism and neurocentrism. In short, no nervous system, no sentience, no pain, no interests, no moral value, no rights.

However, this kind of neurocentric approach may exclude not only plants from any *moral status,* but also humans beyond the *boundary of sentience* (e.g., patients in a *persistent vegetative state* sometimes called *human vegetables*) (see below). It is curious that the approach to environmental ethics advocated by Christopher Stone was just the other way around. If it is possible that *human vegetables* have rights, Stone concludes analogously that it is also possible for plants to have some type of moral status. He considers the case that legal guardianship is possible for both human vegetables and real vegetables (33). This consideration is not provided in Ryder's painism; therefore, it becomes important to address painism in the context of euthanasia and concepts of negative states and human vegetables.

Painism and Euthanasia

Ryder's treatment of euthanasia illuminates new aspects of painism and the role of pain in ethics. Ryder provides cases of brain-dead persons being used as organ donors, in which organs were removed without the use of anesthesia. This is dubious because even a brain-dead person could have intact neural pathways that are activated by extreme stimulation. Such cases illustrate the call for "the precautionary principle that, wherever painience is uncertain in a living animal it should be assumed to exist" (5, p. 70).

Yet, painism justifies euthanasia, for nonhumans and those human cases where unconsciousness seems to be irreversible in accordance with some of the prior work of Callahan, such that "Wherever death is perceived to be the greater good or lesser evil for the patient (except where there is evidence of specific opposition to euthanasia by that individual patient) then it should be allowed to proceed." Concerning "passive euthanasia," Ryder maintains that "If withholding anything causes pain or distress then it ought not to be done." The argument that suffering does not matter because this patient is dying anyway is "entirely irrational and deeply immoral" (5). For Ryder, anesthesia, analgesia, and euthanasia should be developed, and discrimination of age is "irrational and wrong" (5, p. 72).

Brain, Pain, Persistent Vegetative State, and Extreme Cases

This leads to other aspects and problems of painism as relates to extreme cases in human ethics. For example, what about persons near or beyond the boundary of sentience, such as in a persistent vegetative state (PVS)? Does pain play a role in this ethical issue? According to Ryder, we must expect that even rudimentary pain guarantees a moral status, but the question remains: what is the consequence? From the classical frame of the four principles (24), there often is no clear answer possible because of the complexity of the situation. Sometimes it seems that principles of autonomy and justice dominate beneficence and nonmaleficence. With respect to extreme cases (like brain death or PVS), ethicists must consider whether "burdens of treatment outweigh benefits" and that "the principle of nonmaleficience does not imply the maintenance of biological life, nor does it require the initiation or continuation of treatment without regard to the patient's pain, suffering, and discomfort" (24, p. 135).

In general, it is accepted that PVS is "a form of deep unconsciousness" caused by failed operations in the cortex responsible for activities "that we recognize as specifically human" (39, p. 197). Nevertheless, the brain stem is still functioning and controlling involuntary functions. The crucial question in this context is whether PVS patients have pain. It is not easy to address, because it depends on the way in which we view pain (40). If we distinguish pain as a physical sensation, the response of the PVS patient to noxious stimuli would indicate that there is a physical response to pain or discomfort but "that the affective level of human suffering is not present. Experience with such patients shows no behavioural indication of such suffering" (39, p. 205). Yet, Caplan admits that "we have no way of knowing what is going on in the mind of the unconscious person. If we could indeed establish that there is pain, and that there is, in fact, considerable pain, then our answers might be quite different. That question, however, remains to be answered, although present consensus argues against the existence of such pain, mental or physical" (39, p. 206). Even the issue of physical pain sometimes seems to be presented inconsistently.

In the PVS case of Nancy Cruzan, the State Supreme Court described her medical condition as "oblivious to her environment except for reflex responses to sound and perhaps painful stimuli," suggesting that "her highest cognitive brain function is exhibited by her grimacing perhaps in recognition of ordinarily painful stimuli, indicating the experience of pain and apparent response to sound" (39, p. 52). In this case, the Supreme Court considered "best interest" standards and declared that the burden of a prolonged life measured by the experience of pain and suffering markedly outweighed its satisfactions; therefore, treatment could be terminated under a "limited-objective standard" (39, p. 48). This indicates the complexity of the situation but clearly stresses "the burden of a prolonged life from the experience of pain and suffering" as an important factor in balancing different principles in ethical judgements about PVS.

From the viewpoint of Ryder's painism, two statements are relevant. The first concerns the approach to euthanasia and brain death. Wherever painience is uncertain (as in the case of "brain death"), the precautionary principle confirms a duty "not only to administer enough analgetics to remove all pain but also to administer all appropriate psychoactive drugs…not only to alleviate pain but to induce euphoria" (5, p. 51). Is it ethically allowable to actively induce the dying process of a PVS patient whose life could be extended several decades? This question leads back to the roots of Ryder's painism and his engagement for the treatment of animals. As Ryder emphasizes, nonhuman euthanasia is accepted and justified in the West. In human cases like severe dementia, severe head injury, "and whenever unconsciousness is expertly deemed to be irreversible, the moral case is similar" (5). But what about the cases of dementia, PVS, brain death, and so forth? If the patient, significant others, and kin would "suffer greater distress if euthanasia occurs than the distress they are *already* suffering," then euthanasia is unethical, but if not, "euthanasia should proceed." This statement sounds like a utilitarian weighting of interests on the basis of pain. However, if the patient has issued informed directives regarding termination of life support, then euthanasia is permissible. Ryder asked, "Where is the ethical difficulty" (5, p. 71)? I hold that one difficulty is the confusion of heterogeneous principles. In these cases, Ryder introduces the criterion "respect for autonomy" to solve ethical problems, but neither a utilitarian weighting of pain nor the principle of autonomy is justifiable in a pure painism that is grounded solely on the presence of pain in human and nonhuman beings.

WHO IS THE "MAXIMUM SUFFERER?"

With respect to the important role of individual pain, we have to remember Ryder's other basic statement: "the moral objective is to reduce pains of others." But who are these "others?" Ryder rejects any aggregation of pains and pleasures across individuals as is possible in utilitarianism (5, p. 27), claiming that "The moral priority is to try to reduce the pain of the maximum sufferer in each case" and not to maximize the pleasure of all affected individuals. The number of individuals is irrelevant, but the aggregation of pain and pleasure in the same individual is meaningful. According to this rule, "killing 100 people painlessly becomes less wrong than torturing one of them to death." This sentence indicates that there is no "right to life," "protection of life," or "sanctity of life" in painism, per se. In this case, the question seems to be only: "how much pain was experienced by the individual who suffered most" (5, p. 28)? This is a narrow monocriterial approach and leads to the question, who is the maximum sufferer? For instance, from the viewpoint of painism, if we know that a particular PVS patient (e.g., with no pain) has a dog who loves her, and who would pine if the PVS patient died, then it would be justified to keep the PVS patient alive to protect this animal as the potential *maximum sufferer*, given Ryder's rule that we "try to act, therefore, as though human interests count for no more than nonhuman interests" (5, p. 29). But here we see an apparent imbalance that touches moral sentiments. Obviously, some clarification is needed if we are to take painism as a valid ethical construct.

Concerning the "Vegetative Language" about Human Beings beyond Sentience and Pain

The fictitious case shows that animals count in painism, but nonpainient beings like plants do not. In this context, I would like to add an observation about the different language used to define people in these borderline cases. Obviously, we use a special language to characterize natural beings that lack the capability for pain. This language is of bioethical relevance. For instance, the life of humans in a persistent vegetative state has been compared with a vegetable's life, and in cases of brain death or dementia, often we find the statement that they "vegetate." This kind of language includes descriptive and normative judgements. As mentioned, for Singer or Ryder, there exists no inherent or intrinsic value beyond the border of sentience (e.g., as in the case of plants). The moral status of plants seems to be similar to the situation of the PVS patients: "If they have no experiences at all, and can never have any again, their lives have no intrinsic value. Their life's journey has come to an end. They are biologically alive, but not biographically" (7, p. 191). In ordinary language, these persons sometimes are called *human vegetables*, and even the scientific term *vegetative state* contains an allusion to the plant metaphor. In Internet dialogues about the Schiavo case, the following sentences appeared (March 20, 2005): (A) "I would think the most offensive comparison would be comparing Terri to a piece of broccoli…" (B) "A vegetable, which she is, and courts and doctors have determined she is." I call this kind of language about human beings beyond sentience or pain *vegetative language* and here feel it is important to add a few remarks from historical and ethical points of view.

A common assumption in this kind of vegetative language is that it refers to a state beyond sentience and pain. Since the end of the 1960s, discussions about brain death included the term *human vegetable*, and this has been criticized (41). The origins of this vegetative language derive from the Aristotelian and scholastic term *anima vegetativa*, the *vegetative soul*, which is responsible for nutrition, growing, and propagation of organisms (42). According to Aristotle, all human beings possess a vegetative soul, but their *telos* is to live a rational life. The *telos* of animals is to realize the capacities of a sensitive soul, including pleasure and pain. This tradition in metaphysics influenced the introduction of the term *vegetative nervous system* in the early nineteenth century by the German physician Reil, which was then replaced in the English language by the term *autonomous nervous system*. Nevertheless,

Jennet and Plum still favored the new term *persistent vegetative state* in their 1972 paper, which has led to several misunderstandings. They wrote: "the word *vegetative* itself is not obscure: *vegetate* is defined in the *Oxford English Dictionary* (29, p. 76) as: 'to live a merely physical life.'" But Jennett and Plum emphasized the point of being "devoid of intellectual activity or social intercourse." As further explanation, they added: "It suggests even to the layman a limited and primitive responsiveness to external stimuli; to the doctor it is also a reminder that there is relative preservation of autonomic regulation of internal milieu" (43, p. 736). This is a version of special language about a status beyond pain, but it suggests to both the layman and the physician that there is no pain to consider.

From a philosophical point of view, these vegetative terms and statements can allow one to recognize an ontological paradigm for the moral status of real vegetables and human vegetables. In this context, only the ethical role of pain is important. Whatever is possible or impossible concerning the state of pain and suffering of PVS patients, it seems to be incorrect to describe the state of these patients by vegetative terminology. They are not plants, they do not live the life of plants; from a biological point of view, plants are completely different organisms than animals. Further, it remains an open question whether PVS patients have a biography, because it is possible that some of these patients live in a border region with rudimentary cognition and pain (44).

Thus, experts should be cautious when using this vegetative language in statements about people in borderline situations (e.g., PVS, brain death, dementia, or within the dying process), because this language can easily lead to a total misunderstanding of the complex medical and ethical situation. The animal ethicist Tom Regan used this vegetative terminology merely as a fictitious analogy and metaphor to illustrate his view about hurting and harming a person who is killed in his or her sleep without pain. The environmentalist Christopher Stone used this vegetative language as a strategic analogy to defend rights for trees, corresponding to the rights for human vegetables as represented by their guardians. But for Peter Singer, it is more than a metaphor or strategy if he compares plants with humans who are beyond the border of pain. Here it is an essential ontological analogy, indicating a substantial parallelism of the moral status of plants and those patients in situations beyond sentience and pain. This is a kind of neurocentrism and zoocentrism, which now follows the traditional speciesism in anthropology (45).

CONCLUSION

Richard Ryder's *painism* is one of the more interesting concepts in modern bioethics, and indeed Ryder was an important thinker in the last 30 years. His term *speciesism* was inspiring for animal ethics. For Ryder, pain is necessary and sufficient to establish moral status. This monocriterial basis of ethics leads to several tensions with Ryder's speciesism and allows only a narrow framework for the solution of the difficult problems in modern environmental and biomedical ethics. Ryder's painism provides no starting point for the principle of autonomy and for the protection of life. Ryder emphasizes the aggregation of pain only within each individual, but rejects any aggregation of pain between individuals as utilitarianism does. This position leads to the question: who is the maximum sufferer? When considering biotechnology, euthanasia, or PVS, principles other than pain are likely to be necessary to understand the complex situations and engage descriptive and normative judgments.

ACKNOWLEDGMENTS

I would like to thank Virginia A. Sharpe, Heike Baranzke, and James Giordano for their helpful comments on earlier versions of this chapter.

REFERENCES

1. Ryder, R. 1972. Experiments on animals. In: *Animals, men and morals. An enquiry into maltreatment of non-humans,* eds. S. Godlovitch, R. Godlovitch, and J. Harris, 41–82. New York: Taplinger.
2. Ryder, R.D. 1993. Sentientism. In: *The Great Ape Project. Equality beyond humanity,* eds. P. Cavalieri and P. Singer, 220–222. London: Fourth Estate.
3. Ryder, R.D. 1998. Painism. In: *Encyclopedia of applied ethics, Vol. 3 (J–R),* ed. R. Chadwick, 415–418. San Diego, CA: Academic Press.
4. Ryder, R.D. 1999. Painism: Some moral rules for the civilized experimentator. *Cambridge Quarterly of Healthcare Ethics* 8: 35–42.
5. Ryder, R.D. 2001. *Painism. A modern morality*. London: Centaur Press.
6. Singer, P. 1975/1990. *Animal liberation* (2nd ed.). New York: Review/Random House.
7. Singer, P. 2005. *Practical ethics* (2nd ed.). Cambridge: Cambridge University Press.
8. Regan, T. 1984. *The case for animal rights*. Berkeley: University of California Press.
9. Ingensiep, H.W. 1996. Tierseele und tierethische Argumentationen in der deutschen philosophischen Literatur des 18. Jahrhunderts (The animal soul and arguments in animal ethics in the German philosophical literature of the 18th century). *International Journal of History and Ethics of Natural Sciences, Technology and Medicine.* N.S. 4(2): 103–118.
10. Morris, D.B. 1994. *Geschichte des Schmerzes. Aus dem Amerikanischen von Ursula Gräfe*. Frankfurt am Main: Insel Verlag.
11. Thomas, K. 1983. *Man and the natural world. A history of the modern sensibility*. New York: Pantheon Books.
12. Bentham, J. 1948. *A fragment on government with and introduction to the principles of morals and legislation (1789/1823)*, ed. W. Harrison. Oxford: Basil Blackwell.
13. Primatt, H. 1992. *The duty of mercy (1776/1834),* ed. R.D. Ryder. Fontwell, West Sussex: The Kinship Library, Centaur Press.
14. Mill, J.S. 1863/1895. *The works of John Stuart Mill. III Utilitarianism* (12th ed.). London: The New Universal Library, Routledge.
15. Warren, M.A. 1997. *Moral status: Obligations to persons and other living things.* Oxford: Oxford University Press.
16. Warren, M.A. 2003. Moral status. In: *A companion to applied ethics*, eds. R.G. Frey and C.H. Wellman, 439–450. Malden, MA: Blackwell.
17. Ingensiep, H.W. 1997. Personalismus, Sentientismus, Biozentrism—Grenzprobleme der nicht-menschlichen Bioethik. (Personism, sentientism, biocentrism—boundary problems in non-human bioethics). *Theory in Bioscience* 116: 169–191.
18. Baranzke, H. 2004. Does beast suffering count for Kant: A contextual examination of Section 17 of *The Doctrine of Virtue. Essays in Philosophy. A Biannual Journal.* 5(2): 1–16.

19. Singer, P. 1981. *The expanding circle. Ethics and sociobiology*. Oxford: Clarendon Press.
20. Cassel, E.J. 2004. Pain and suffering. In: *Encyclopedia of Bioethics, Vol. 4* (3rd ed.). ed. S.G. Post, 1961–1969. New York: MacMillan Reference, Thomason Gale.
21. McCloskey, H.J. 2001. Pain and suffering. In: *Encyclopedia of Ethics* (2nd ed.). eds. L.C. Becker and C.B. Becker, 1269–1271. New York: Routledge.
22. Giordano, J. 2006. Moral agency in pain medicine: Philosophy, practice and virtue. *Pain Physician* 9: 41–46.
23. Giordano, J. 2004. Pain research: Can paradigmatic expansion bridge the demands of medicine, scientific philosophy and ethics? *Pain Physician* 7: 407–410.
24. Beauchamp, T.L., and J.F. Childress. 2001. *Principles of biomedical ethics* (5th ed.). Oxford: Oxford University Press.
25. Sharpe, V.A. 1997. Why "do no harm?" *Theoretical Medicine* 18: 197–215.
26. Taylor, P.W. 1986. *Respect for nature. A theory of environmental ethics*. New Jersey: Princeton University Press.
27. Regan, T. 1993, Ill-gotten gains. In: *The Great Ape Project. Equality beyond humanity,* eds. P. Cavalieri and P. Singer, 194–205. London: Fourth Estate.
28. Wilkins, J.S. 2009. *Species. A history of the idea*. Berkeley: University of California Press.
29. *Oxford English Dictionary, the Compact Edition, Vol II P–Z.* 1971. Oxford: Oxford University Press.
30. Häyri, H., and M. Häyri. 1993. Who's like us? In: *The Great Ape Project. Equality beyond humanity*, eds. P. Cavalieri and P. Singer, 173–182. London: Fourth Estate.
31. Gifford, F. 2002. Biotechnology. In: *Life science ethics,* ed. G. Comstock, 191–224. Ames, IA: Iowa State Press.
32. Giordano, J. 2010. From a neurophilosophy of pain to a neuroethics of pain care. In: *Scientific and philosophical perspectives in neuroethics*, eds. J. Giordano and B. Gordijn, 172–189. Cambridge: Cambridge University Press.
33. Stone, C. 1985. Should trees have standing? Revisited: How far will law and morals reach? A pluralist perspective. *Southern California Law Review* 1–154.
34. Tompkins, P., and C. Bird. 1974. *The secret life of plants*. London: Allen Lane.
35. Trewavas, A. 2003. Aspects of plant intelligence. *Ann Bot* 92: 1–20.
36. Trewavas, A. 2004. Aspects of plant intelligence: An answer to Firn. *Ann Bot* 93: 353–357.
37. Baluska, F., Volkmann, D., and D. Menzel. 2005. Plant synapses: Actin-based domains for cell-to-cell communication. *Trend Plant Sci* 10: 106–111.
38. Baluska, F., Mancuso, S., and D. Volkmann. 2006. *Communication in plants: Neuronal aspects of plant life*. Berlin: Springer-Verlag.
39. Caplan, A.L., McCartney, J. J., and D.S. Sisti. 2006. *The case of Terri Schiavo*. Amherst, NY: Prometheus Books.
40. Giordano, J. 2009. *Pain: Mind, meaning and medicine.* Glen Falls, PA: PPM Press.
41. Jonas, H. 1974. Against the stream: Comments on the definition and redefinition of death. In: *Philosophical essays,* ed. H. Jonas, 132–140. New Jersey: Prentice Hall.
42. Ingensiep, H.W. 2006. Leben zwischen "Vegetativ" und "Vegetieren." Zur historischen und ethischen Bedeutung der vegetativen Terminologie in der Wissenschafts- und Alltagssprache. (Life between "vegetative" and "vegetate." Concerning the historical and ethical meaning of the vegetative terminology in scientific and ordinary language). *International Journal of History and Ethics of Natural Sciences, Technology and Medicine* 14(2): 65–76.
43. Jennett, B., and F. Plum. 1972. Persistent vegetative state after brain damages. *The Lancet* (April 1): 734–737.

44. Gupta, A., and J. Giordano. 2007. On the nature, assessment, and treatment of fetal pain: Neurobiological bases, pragmatic issues and ethical concerns. *Pain Physician* 10: 525–532.
45. Ingensiep, H.W. 2009. Speciesismos. In: *Handbuch antrhopologie der menschzwischen natur, kultur und tecknik*, eds. E. Bohlken and C. Thies, 418–422. Stuttgart: Metzler.

14 Maldynia

*Chronic Pain, Complexity, and Complementarity**

James Giordano and Mark V. Boswell

CONTENTS

INTRODUCTION: A COMPLEX PROBLEM

An expanding epistemology has generated enhanced understanding of the mechanistic basis and existential impact of chronic pain as a complexity-based systems event. The basic and clinical sciences, humanities, and experiential narratives of patients all contribute lenses through which we can examine and demystify the enigma of persistent pain. It is only through the combination of distinct domains of knowledge that we can both comprehend pain as a dysfunction of the dynamical, nonlinear adaptability of the nervous system, and at the same time apprehend the manifestations of these changes within the networked hierarchy of interacting systems that is the patient. These are concepts that are inherent to and derived from complexity theory, and the use of a complexity-based model of pain may be important to fully reconcile notions of disease and illness, and fit these within a more encompassing framework of diagnostics and therapeutics. Thus, the study of pain conjoins neuroscience to a burgeoning discourse addressing concepts of brain–mind, disease-illness, and the ethical dimensions of care.

* This chapter is adapted, with permission from Boswell, M.V., and J. Giordano, 2009, Reflection, Analysis And Change: The Decade of Pain Control and Research and Its Lessons for the Future of Pain Management, *Pain Physician* 12: 923–928.

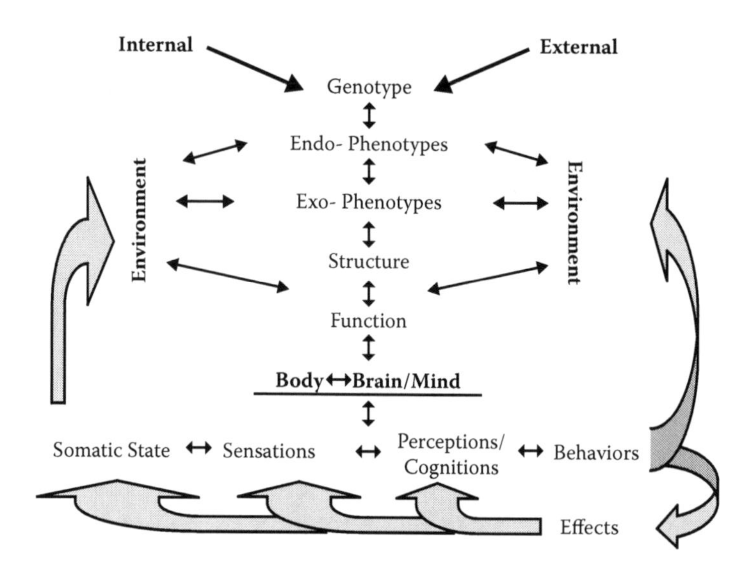

FIGURE 14.1 Conceptualization of interacting genotypic, phenotypic, and environmental factors. Note dynamical and reciprocal (bidirectional) interactions of factors, causes, and effects. (Image © Loveless/Giordano, 2010. Printed with permission.)

CHRONIC PAIN—A SPECTRUM DISORDER?

We believe that chronic pain can be seen as (a component of) a spectrum disorder that is characterized by common genetic and environmental predispositions that induce disruption and loss of adaptive nonlinear properties within and between various elements of the system (1–5). It is probable that no one single factor is universally causal, but instead, multifarious changes occur in molecular,* chemical,† and structural‡ components of various peripheral (PNS) and lower central neuraxes subtending pain (i.e., "bottom-up" effects). Modification of central nervous system (CNS) functions and (micro/macro) structure can affect systems' properties of supratentorial neural networks to manifest "top-down" effect(s) in which the altered functions of brain networks evoke changes in the neural systems that mediate physiologic effects, cognition, emotions, and behaviors that are manifested from biologic to social levels (6–9).

This conceptualization allows for cogent intertheoretical reduction, and depicts strongly biopsychosocial dynamics that are operative in pathology and are, therefore, important to, if not necessary for treatment (10) (see Figure 14.1).

It is important to note that spectrum mechanisms may not necessarily cause each other so much as coactivate each other, thereby allowing for multiple presentations

* For example, multiple gene expressions; transcription/translational factor activities and expression, sodium, calcium, and perhaps other channelopathies, and receptor sensitization/desensitization and up-/downregulation.

† Including opioid, glutamate, substance-P, and other peptidergic, monoaminergic (i.e., serotonergic, noradrenergic, dopaminergic), and perhaps hormonal systems.

‡ Such as neuronal membrane and synaptic remodeling, axonal and dendritic rearborization, and differentiation of glial structure and glial-neuronal interfaces, as well as more macroscale restructuring.

along the spectrum to be (mechanistically) associated, while possibly having little resemblance to one another (3,4). So, for example, pain and certain psychopathologies (e.g., depression, anxiety, somatoform disorders, substance abuse/addiction) may have common activating factors as a result of shared anatomical pathways, overlapping pathophysiologic processes, and the coinvolvement of specific neurochemical systems (e.g., serotonin, norepinephrine, glutamate, etc.). Such aberrant changes do not occur uniformly and may reflect dispositions to affect and alter the nervous system in ways that cause differing expressions of pain, and pain syndromes. Moreover, the distinct presentations of particular spectrum effects sustain pain as unique in each individual, both in the specificity of biological mechanisms, and the "feeling" and effects that these biological events incur and evoke (11). Given this uniquity of pain, its impact upon the lived body and life-world of the being who is the patient (i.e., its manifestation as illness), will also be unique in many ways (1,12,13).

A CALL FOR COMPLEMENTARITY IN PAIN CARE

For the pain practitioner, this knowledge serves a multifold purpose in the clinical encounter. First, it affords new insights to the pathophysiologic mechanisms of pain, and in this way provides information to better depict and predict the limits, delimitations, and viability of particular therapeutics. Second, it asserts the importance of viewing the patient as the medium or nexus for the expression of biological pathology as both disease and illness; in other words, it contextualizes Sydenham's nosological method to the life of the patient.* Thus, the goals are first to determine how the combination of genotype and environmental factors establish, affect, and enable the type and extent of pathologic phenotypes in certain individuals and populations, and thereupon, to identify those physiologic and environmental variables that are viable targets for therapeutic interventions against the initiation and progression of chronic pain.

Clearly, then, this entails a complementary viewpoint that appreciates genetics *and* environment, biological *and* psychosocial factors, objectivity *and* subjectivity, and the recognition of disease processes as occurring and evoking illness in the living being who becomes a patient (14). This ultimate regrounding to the patient prompts and upholds the need for keen objective evaluation *and* intersubjectivity and employment of technology *and* humanitarian values in pain care. This is important to all fields of medicine, and the balanced engagement of these elements is critical to pain medicine, given the frank subjectivity, individuality, and oftentimes objective elusiveness of pain. Simply put, the knowledge that no two pains—or their effects—are

* After Thomas Sydenham (1624–1689), English physician renowned for the development and use of the nosological method, a doctrinal approach to characterizing the natural features of disease. The method strives to align conformity of type (of presentation) with uniformity of cause (of the disorder), and in this way, match characteristic signs and symptoms to category and type of disease. The nosological method sought to relate intellectual observation, analyses, and comparison to the moral responsibilities of diagnosis and treatment. Often absent from discussions of Sydenham's doctrine is the (stated) explicit requirement to recontextualize the natural features of the disease to their expression in a particular patient.

identical mandates that the clinician must therefore know *both* the disorder *and* the patient to genuinely comprehend and treat the patient's pain.

PRECIPITATING CHANGE

The Congressionally declared *Decade of Pain Control and Research* (2000–2010) did much to instigate new directions and advances in basic and clinical studies of pain, not only in the United States, but internationally as the infusion of U.S. federal subsidies were aligned and paired with multinational research programs. Yet, it remains to be seen whether this research agenda has truly enabled, or will lead to, translational efforts in pain care. We believe that progress was made in pain research—fueled in part by prior efforts of the Human Genome Project, and the Decade of the Brain (1990–2000). Yet, as noted by Bruno Latour, scientific answers only serve to generate new and deepen remaining questions (15). Despite the sophistication of research approaches, questions remain about the nature of pain; its effect upon—and the effects of—the brain, mind, consciousness, and self of the pain patient; and what these contexts portend for the development and translation of diagnostic and therapeutic techniques, technologies, and approaches (16). In short, we may need to reexamine *meanings* and *values* inherent to the experience and expression of pain, and equally examine such values in relation to the definition and conduct of evidence-based pain medicine (17).

Without doubt, technology provides important tools for diagnoses and treatments. But, technology alone does not *provide* the diagnosis, heal the patient, or sustain the profession and practice of pain medicine (18–20). This remains within the humanitarian domain of *care*, in the literal sense, here construed as concern, worry, and regard. In this way, we argue that any regard for the pain patient—as the subject of the clinicians' moral responsibility—must address pain as a physical event and a phenomenal experience, its meanings, and the needs and vulnerabilities it incurs in each individual, and from this determine what and how existing and new technologies and techniques may be employed to serve the patient's best interests (21). In other words, the good of science in pain medicine is inextricably woven into its prudent use for the good of the pain patient.

NAVIGATING THE TECHNOLOGIC TREND

It is not so much that the practice of pain medicine has become overly dependent upon technology, but rather that other, extramedical forces may sway (i.e., pull) the use, if not misuse, of technology away from the core imperative to use such devices and approaches as wholly for the patients' best medical interests. Inarguably, one of these pulling forces has been misaligned economic incentives that have commodifed translational applications of science and technology, and instantiated conditions of "medicine-for-profit"; more simply put, the misappropriation of scientific and technologic resources in accordance with a business or market ethos, rather than one of patient-centered beneficence (22,23).

As a result, technology is increasingly viewed as an end, rather than a means by which to enable or enact the ends of what Pellegrino refers to as a "right and good healing" (24). Perhaps a bit of clarification is required here; first, it is important to note that good entails right. The good of pain care mandates that technology be used in ways that are methodologically "right" (i.e., appropriate and maximally effective) based upon the knowledge (regarding mechanism, action and effect) at hand. However, technical rectitude alone does not constitute "good"; to sustain beneficent use, any and all techniques and technologies must be applied in those ways that maximize benefits *to the patient*. To do otherwise would be to bastardize or refute the knowledge and capability conferred by science and technology in ways that are inconsistent with the core philosophical premises of medicine.

A number of factors complicate the use and utility of technology in pain medicine. Although relief of suffering appears as a central tenet of medicine, in practice it receives little attention, and the illness and suffering of pain—particularly chronic pain—remain elusive variables to evaluate (25,26). Illness and suffering are not readily quantified; thus, measurement remains problematic. Based upon this objective vagary, the treatment of chronic pain remains equally difficult, even given the most advanced technologic means. Although Stanley Reiser has lauded the invention of the stethoscope as a turning point in the technologization of medicine (27), we should not forget that there is no stethoscope with which to "auscultate pain." Instead, the "sounds of pain" are conferred through the subjective reports and narratives of the patient. Pain by its nature occupies the realm of first-person experience, and while certainly a symptom, it is often *felt* and *known* as illness rather than as disease. This is axiomatic to maldynic pain, and it is this subjective dimension that has rendered it so difficult to both assess and treat.

REALIZING COMPLEMENTARITY IN PRACTICE

For many reasons, the concept of pain as disease falls flat, and as such the disease-based, curative models of medicine can be ineffectual—if not at times wholly unsuitable—to address and "care" for the chronic pain patient. Pain is a biopsychosocial event, experience, and expression. We posit that in light of this, pain medicine, as a specialty, must be equally biopsychosocial in its orientation and approach to chronic pain-as-illness. A unidimensional model will not work; rather, an approach that is foundationally complementary in its curative and healing capacity appears to be best aligned with current concepts of pain, and the predicament of pain patients (28). This *Asclepian-Hygieian* model would be built upon a maxim of "cure when possible, heal as capable, and care always," and would be inclusive, multi- (if not trans-) disciplinary, and integrative (29).

This is not to imply that technology cannot or should not be used in such a complementary approach. Quite the opposite—it may well be that technology offers assets that can be employed to better define the substrates and effects of pain-as-illness, and may be instrumental to enhancing the delivery of care. But if this is to be the case, it is not sufficient that this technology exists. Instead, it must be made available to the clinician as an implement of practice when and where appropriate (to the needs of a given patient), and must be accessible and affordable to those patients who require its benefits (1).

On one hand, this establishes the requirements for health care reforms that empower physicians to utilize state-of-the-art diagnostics and therapeutics as necessary (reiteratively, to fulfill the best interests of the patient). We have described how certain economic systems of resources allocation and distribution might best accommodate these goals (30). On the other hand, the use of any technique or technology (regardless of whether new or old) requires that the pain physician act as steward of knowledge both to inform and guide patient insight, and shepherd the nature and direction of care. We argue that these constructs must act as a complementarity, or else any system of technological advancement and infrastructural health care reform will likely remain unsuccessful.

Admittedly, the stewardship and use of techniques and technologies in pain care in ways that are both right and good often becomes problematic. The admonition to do no harm to a patient who manifests a disorder that is difficult (if not impossible) to objectively measure poses profound challenges for the pain physician (see Galvagni, Schatman, and Venuti, this volume). This is particularly true for the physician who relies solely upon technology and has no grounding in the philosophy and ethics of medicine, or interest in pain medicine as a social good. This challenge is compounded by a lack of medical school education about pain and training in pain management (and this is especially true for chronic pain). The current norm is that instruction in pain care consists of little more than a postscript to the basic sciences. So, while mechanisms of pain and analgesia are taught during basic neuroscience courses, there is no direct link to how the complexities of these systems are relevant to the illness of chronic pain and challenges of chronic pain management.

THE CURRENT CONDITION OF PAIN CARE

So, where are we now? We may be beyond the halcyon days of pain management, the 1980s and 1990s. In many ways, pain care represented a frontier specialty, and like the Old West, certain claims were made upon territories and practices. Some of the apparent advances were developed with a cowboy mentality. Most assuredly, the treatments were well intended and seemed like good ideas: spinal cord stimulators and intrathecal drug infusions were deemed major advances; epidural steroid injections and vertebroplasty became popular, despite minimal evidence for effectiveness beyond merely short-term pain relief (31); and seemingly limitless protocols for opioid use in chronic nonmalignant pain became the norm (32). Today, however, newfound knowledge compels us to retreat from naïve élan, as the potential harm of such treatments becomes apparent (33).

What can the pain physician now offer the patient? Most state-of-the-art technical treatments for pain are still somewhat contentious (if not unproven), and long-term effects of many of the more innovative technological approaches remain as yet unknown. Some argue that interventional approaches provide only short-term effect, unclear benefit, and a defined risk of injury (31). This leaves the physician with a limited armamentarium of valid treatment options. Pharmaceuticals can be effective but have proven problematic: nonsteroidal anti-inflammatory drugs (NSAIDs) may cause adverse gastrointestinal, cardiovascular, and CNS effects (34); and opioids can lead to tolerance and dependence, if not frank addiction (35). Moreover, the prolonged or escalative use of opioids is wrought with litigious issues, and the clinical relationship can become frankly adversarial under such circumstances (12).

We previously identified three primary ethical problems that arise from these issues: undertreatment of pain; overutilization of technologies and pharmacologic treatments; and conflict between physician and patient, and physician and regulatory and payment authorities (1,36,37). In each of these scenarios, pain remains inappropriately treated. When new treatment options are made available, they often are highly technological and expensive. Moreover, economic factors frequently restrain, if not preclude, the direct provision of clinical care to the individual patient. Technological interventions may obviate the need for dangerous, addictive drugs, and to some extent, the relative popularity of new technological innovations reflects a growing consensus that pharmaceuticals are largely ineffective. That is not to say that technology denies or refutes the benefit of drugs. In fact, cutting-edge bioscience (e.g., geno- and nanotechnologies) offers great potential to produce more target-specific, selective, nonabusable, and, hence, increasingly effective pharmacological agents (38). But the average patient cannot afford such treatments.

New technologies and techniques may be free of the adverse effects of older pharmacologic therapies but are not without risk, in part because of their relative novelty, and the contingent nature of our understanding of the brain–mind, and the capacity of pain to affect the CNS. Moreover, as we have noted, even the best technique or technology is of little benefit if it is not accessible and affordable to those who need it. So, as it stands, pain clinicians must generally rely upon those therapeutics that are subsidized by insurance plans or that are not exorbitantly priced. These are largely pharmacologic and incur the aforementioned risks for the physician and patient alike. The perceived high risks of legal sanctions and malpractice suits inherent to pain care are particularly threatening to primary care physicians, given that practice scenarios frequently preclude the time and effort necessary to diligently monitor pain patients. Thus, primary care physicians become reluctant to treat pain and may refrain from treating pain patients in general. This has resulted in an increasing tendency to refer pain patients to pain specialists for management.

But herein lies another problem, as the profession of *pain management* is not specific to a particular discipline and, in this way, may be inchoate. The *pain specialist* may be an anesthesiologist, neurologist, physiatrist, psychiatrist, allied health provider, or even complementary/alternative medicine practitioner (e.g., chiropractor, acupuncturist). The cynic may opine that pain management is, in some respects, a faux specialty. It lacks homogeneity and, in this heterogeneous composition, may lack the uniform scientific rigor necessary to justify its independent existence, and it does not possess the breadth or depth of knowledge required of other specialties. In many respects, the cobbling together of pain medicine from parts of other disciplines may obviate its capacity to achieve independent clinical autonomy or life (and therefore a detractor might be tempted to label the specialty after Mary Shelley's Prometheus).

PROPOSING A FUTURE

Given the deficiencies of the specialty of pain medicine, the question, then, is how to provide appropriate care for patients in chronic pain? The current paradigm does not appear to be working, and, in any case, given the present means of funding medical care, physicians may experience drastic cuts that will threaten the way pain currently is treated (at least in the United States). However, this may not be an altogether bad thing. The crisis in pain care may lead to new treatment approaches. In light of the strong growth of consumerism affecting the practice of medicine today, it appears unlikely that the specialty of pain medicine will disappear entirely.

Pain medicine may ultimately achieve independent specialty status, but the end result will appear different than what we see now. Competent pain management requires time and commitment to the patient, and a solid physician–patient relationship. But this is not the current paradigm; the incentive model of reimbursement has forced physicians to work piecemeal; each pain patient becomes a billable procedure.

A unified approach to addressing the crisis in pain medicine has been proposed that attempts to reinstate the role and importance of the primary care physician (39,40). To some extent, this already is happening in palliative care (as championed by the specialties of family and internal medicine), and we believe that it can take place in pain medicine as well. Either the pain management specialist must become a primary care physician, or the pain specialist must engage the primary care physician to become a part of the solution, rather than merely abdicating responsibility to the specialist (who is ill-equipped to properly, effectively, and singularly treat the pain patient on a long-term basis).

In many ways, this approach is a plea to return to the basics of medical practice, "consistent with, and adherent to the epistemic, anthropologic, and ethical domains of the core philosophy of medicine" (41). Simply, we must learn to integrate the biopsychosocial model into the paradigm of pain care.

AN ETHICAL STANCE

Our previously proposed plan for reforming and integrating pain care assumes a meta-ethical stance that calls for affirmations and rules of the profession, and demands moral agency in practice (1,42). What ethical systems can, and perhaps should, be used by clinical agents in the daily articulations of the clinical encounter can vary based upon individual moral compass, circumstances, and needs (see Figure 14.2). In Chapter 15 of this volume, Edmund Pellegrino describes a set of core precepts that are necessary, although admittedly not sufficient, for pain care. Yet, these core constructs serve as both a starting point for the use of any system of ethics in pain medicine, and integral premises that are to be woven into the fabric of any such ethics.

Can or will such a paradigm be adopted? The current crisis of confidence in medicine in general, and pain medicine in particular, ensures that change must occur. If nothing else, funding for pain care specialists will decline or evaporate. Of financial necessity, pain medicine may be subsumed back into primary care. As well, it may be that pain specialists are up to the challenge of instantiating a new field. Currently, there is ongoing tension among the different disciplines classified under the rubric of pain management about which group will achieve superiority and assume the mantle of authority. Unfortunately, the current argument is not primarily about what is best for the patient, but more about hegemony in the medical market.

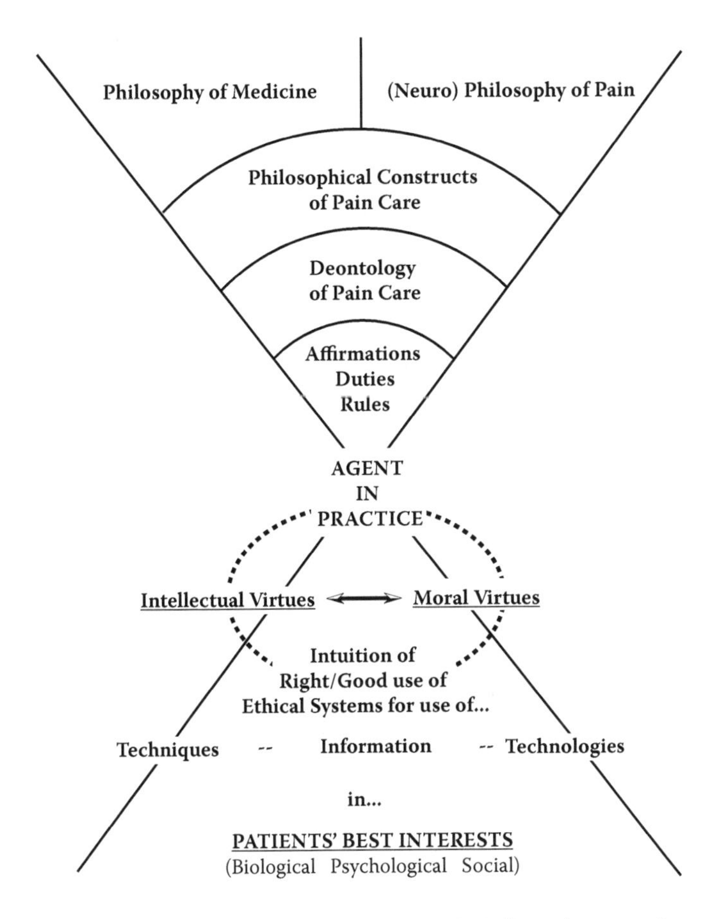

FIGURE 14.2 Representation of philosophical and ethical dimensions contributory to therapeutic and moral agency of the pain clinician-in-practice. As a therapeutic and moral agent, the clinician engages intellectual and moral virtues to appropriate various ethical systems and approaches to engage right and good use of information, techniques, and technologies to implement care in pain patients' best interests. (Image © Loveless/Giordano, 2010. Printed with permission.)

We hope that wiser members of the field will prevail in the end. Some practitioners of pain medicine recognize the error of conflict and appreciate the damage it will do to the cause and articulation of pain care as a profession. Some recognize that pain medicine, whether pharmacologically based, or focused on interventional techniques, is actually a part of a (larger, more inclusive) palliative care paradigm. Palliative care differs from pain medicine in that it has not jettisoned patient centeredness in exchange for technology. The patient remains the focus of palliative care, and the primary goal is symptomatic relief, whether accomplished by high- or low-tech means.

A key question that remains is whether the philosophy expounded here will survive to inform and ultimately become policy. Although pain medicine *prima facie* seems to serve a public good, pain control is not a central tenet of organized medicine. Despite lobbying by proponents of pain medicine to the contrary, patients do not have a right to pain relief (43). If pain relief is not a right, then funding for pain therapeutics will remain uncertain. Unless such interventions and practices appear to be affordable and economical, it may not be seen as viable and, thus, will not be provided by generalists or specialists; pain relief must be "a good deal." Certainly, at present, pain care appears neither affordable nor economical.

CONCLUSION: A WAY FORWARD

We are fond of paraphrasing the contemporary virtue ethicist Alasdair MacIntyre's definition of practice as an exchange of goods between, and relevant to agents in a relationship (44). That said, we hold that the primary *good* of the clinical relationship in pain medicine is the *care* of pain, as this reflects its essential *raison d'être*. What course of action will be required to instantiate competent pain care in practice? As stated in the Hippocratic corpus, clinicians "must be able to tell the antecedents, know the present, and foretell the future" (45). We posit that pain physicians should return to their roots, with the explicit goal of helping the patient (and, at very least, do no harm in both profession and practice). To this end, we advocate that clinicians play a greater role in guiding the course of medical practice. This entails working together with scientists to advance appropriate research and development of effective, affordable treatments, participating in guideline development, working to inform and formulate policy, and in this latter regard, partner with lawmakers to ensure continued patient access to proper pain care. And as befitting a classical but nonetheless perdurable definition of *care*, the primary goal must always be an unwavering regard for the good of the patient.

REFERENCES

1. Giordano, J. 2009. *Pain: Mind, meaning, and medicine.* Glen Falls, PA: PPM Press.
2. Giordano, J. 2009. The neuroscience of pain and the neuroethics of pain care. *Neuroethics* 3(1): 89–94.
3. Giordano, J., and R.Wurzman. 2008. Neurological disease and depression: The possibility and plausibility of putative neuropsychiatric spectrum disorders. *Depression: Mind and Body* 4(1): 2–5.
4. Giordano, J., and R. Wurzman. 2008. Chronic pain and substance abuse: Spectrum disorder and implications for ethical care. *Practical Pain Management* 8(6): 53–58.
5. Giordano, J. 2005. Neurobiology of nociceptive and anti-nociceptive systems. *Pain Physician* 8(3): 277–291.
6. Giordano, J. 2005. The neuroscience of pain and analgesia. In: *Weiner's pain management: A guide for clinicians,* eds. B.E. Cole and M.V. Boswell. Boca Raton, FL: CRC Press.
7. Giordano, J., and M.V. Boswell. 2007. Neurobiology of nociception and anti-nociception. In: *Handbook of chronic spinal pain management,* ed. L. Manchikanti. Paducah, KY: ASIPP Press.
8. Giordano, J. 2010. From a neurophilosophy of pain to a neuroethics of pain care. In: *Scientific and philosophical perspectives in neuroethics*, eds. J. Giordano and B. Gordijn, 172–189. Cambridge: Cambridge University Press.

9. Chapman, C.R. 2005. Psychological aspects of pain: A conscious studies perspective. In: *The Neurological Basis of Pain,* ed. M. Pappagallo, 157–170. New York: McGraw-Hill.
10. Patil, T., and J. Giordano. 2010. On the ontological assumptions of the medical model of psychiatry. *Philosophy, Ethics, and Humanities in Medicine* 5(3).
11. Damasio, A. 1999. *The feeling of what happens. Body and emotions in the making of consciousness.* New York: Harcourt.
12. Giordano, J. 2008. Pain, the patient and the physician: Philosophy and virtue ethics in pain medicine. In: *Ethical issues in chronic pain management*, ed. M.E. Schatman, 1–18. New York: Informa.
13. Leder, D. 1990. *The absent body*. Chicago: University of Chicago Press.
14. Kelso, J.A.S., and D.A. Engstrøm. 2009. *The complementary nature*. Cambridge, MA: MIT Press.
15. Latour, B. 1987. *Science in action: How to follow scientists and engineers through society*. Cambridge, MA: Harvard University Press.
16. Giordano, J. 2004. Pain research: Can paradigmatic revision bridge the needs of medicine, scientific philosophy and ethics? *Pain Physician* 7: 459–463.
17. Boswell, M.V., and J. Giordano. 2009. Evidence-based or evidence-biased: On the need to consider stakeholder values in research assessment. *Pain Physician* 12: 9–12.
18. Eaton, M.L., and D. Kennedy. 2007. *Innovation in medical technology: Ethical issues and challenges.* Baltimore MD: Johns Hopkins University Press.
19. Giordano, J. 2008. Technology in pain medicine: Research, practice, and the influence of the market. *Practical Pain Management* 8(3): 56–59.
20. Giordano, J. 2007. Techniques, technology and tekne: On the ethical use of guidelines in the practice of interventional pain medicine. *Pain Physician* 10: 1–5.
21. Herzlinger, R. 1997. *Market driven healthcare: Who wins, who loses in the transformation of America's largest service industry.* Cambridge, MA: Perseus Books.
22. Giordano, J. 2006. Cassandra's curse: Interventional pain management, policy and preserving meaning against a market mentality. *Pain Physician* 9: 167–170.
23. Brock, D. 2004. Ethical issues in the use of cost effectiveness analysis for the prioritization of health resources. In: *Handbook of bioethics: Taking stock of the field from a philosophical perspective,* ed. G. Khushf, 353–380. Dordrecht: Kluwer.
24. Pellegrino, E.D. 1987. Toward a reconstruction of medical morality. *Journal of Medical Humanities* 8(1): 7–18.
25. Kaspar, J., M.V. Boswell, and J. Giordano. 2009. Assessing chronic pain: Facilitating objective access to the subjectivity of pain. *Practical Pain Management* 9(3): 55–59.
26. Giordano, J., and D.K. McBride. 2009. Anticipating biotechnological trends in pain care. *Practical Pain Management* 9(5): 74–78.
27. Reiser, S.J. 1988. *Medicine and the reign of technology*. Cambridge: Cambridge University Press.
28. Giordano, J. 2008. Complementarity, brain-mind, and pain. *Forsch. Komplementarmed.* 15: 2–6.
29. Giordano, J., and W.B. Jonas. 2007. Asclepius and Hygieia in dialectic: Philosophical, ethical and pragmatic bases of an integrative medicine. *Integrative Medical Insights* 2(3): 89–101.
30. Giordano, J., R. Benedikter, and M.E. Schatman. 2009. Toward pain care for a global community: From philosophy to economic considerations. In: *Pain medicine: Philosophy, ethics and policy,* eds. J. Giordano and M.V. Boswell, 39–51. Oxon, UK: Linton Atlantic Books.
31. Boswell, M.V., A.M. Trescot, S. Datta et al. 2007. Interventional techniques: Evidence-based practice guidelines in the management of chronic spinal pain. *Pain Physician* 10(1): 7–111.

32. Manchikanti, L., S. Atluri, A. Trescot, and J. Giordano. 2008. Monitoring opioid adherence in chronic pain patients: Tools, techniques and utility. *Pain Physician* 11: 1–26.
33. Cole, B.E. 2009. Opioid prescriptions and proscriptions: Twenty-one questions. In: *Pain medicine: Philosophy, ethics and policy*, eds. J. Giordano and M.V. Boswell, 177–199. Oxon, UK: Linton Atlantic Books.
34. MacDonald, T.M. 1995. Side-effects of non-steroidal anti-inflammatory drugs: Studies from the tayside medicines monitoring unit. *Inflammopharmacology* 3(4): 321–326.
35. Ballantyne, J.C., and K.S. LaForge. 2007. Opioid dependence and addiction during opioid treatment of chronic pain. *Pain* 129: 235–255.
36. Giordano, J., and M.E. Schatman. 2008. A crisis in chronic pain care—An ethical analysis. Part one: Facts, issues, and problems. *Pain Physician* 11: 54–62.
37. Giordano, J., and M.E. Schatman. 2009. A crisis in chronic pain care—An ethical analysis. Part two: Proposed structure and function of an ethics of pain medicine. *Pain Physician* 11: 589–595.
38. Basbaum, A. 2006. *Emerging strategies for the treatment of neuropathic pain*. Seattle, WA: IASP Press.
39. Giordano, J., P. LeRoy, and U. Uthaman. 2006. On the role of primary care within a system of integrative multi-disciplinary pain management. *Practical Pain Management* 6(8): 66–69.
40. Giordano, J., and M.E. Schatman. 2008. A crisis in chronic pain care—An ethical analysis: Part three: Toward an integrative, multidisciplinary pain medicine built around the needs of the patient. *Pain Physician* 11: 64–73.
41. Giordano, J. 2006. Moral agency in pain medicine: Philosophy, practice and virtue. *Pain Physician* 9: 41–46.
42. Maricich, Y., and J. Giordano. 2009. Chronic pain, subjectivity, and the ethics of pain medicine: A deontic structure and the importance of moral agency. In: *Pain medicine: Philosophy, ethics, and policy,* ed. J. Giordano and M.V. Boswell, 85–94. Oxon, UK: Linton Atlantic Books.
43. Hall, J., and M.V. Boswell. 2009. Ethics, law and pain management as a patient right. In: *Pain medicine: Philosophy, ethics and policy,* ed. J. Giordano and M.V. Boswell, 125–136. Oxon, UK: Linton Atlantic Books.
44. MacIntyre, A.C. 2006. *The Tasks of Philosophy: Selected Essays*. Cambridge: Cambridge University Press.
45. Hippocrates. 2004. *Epidemics* Book I, XI–XIII (W.H.S. Jones trans.). Cambridge, MA: Harvard University Press, p. 165.

15 A Clinical Ethics of Chronic Pain Management

Basis, Reason, and Responsibilities

Edmund D. Pellegrino

CONTENTS

ON PAIN, BRIEFLY

Pain, illness, and suffering are complex biopsychosocial phenomena. The way they are manifest in individuals is the result of an interplay between age, social, cultural, ethnic, psychological, physiological, and even genetic dispositions. These vary qualitatively for each person and are almost never fully quantifiable. Therefore, while we can measure pain thresholds, we cannot measure pain, per se, or predict the limits of pain tolerance. Likewise, we can empathize with those who are in pain, but we can never wholly penetrate their experience. Pain is as Leriche said, "the resultant of a conflict between a stimulus and the whole individual" (1). It is becoming evermore clear that each person's pain—as an experience of their lived body and event of their consciousness—is unique (2,3). Therefore, regardless of its nature or magnitude, each person's pain is both real and personal in its experience. To speak of "imaginary" pain is to run the risk of moral irresponsibility, even though we readily acknowledge that there can be overreactors and malingerers.

The complexity of these relationships is multiplied in chronic pain—either *in* illness, or *as* illness—as pain becomes inextricably entangled with anguish, anxiety, and fear. It is in this complex relationship with chronic pain that some of the more difficult ethical questions in pain care arise (4). Because the treatments for pain can differ widely, the physician is under obligation to dissect them as completely as possible. This necessity shapes the attitudes and responses of physicians, and other care providers, as well as patients, and gives particular importance to the ethical choices they must make.

THE MORAL MANAGEMENT OF PAIN-AS-ILLNESS

Ethics is based in right and good relationships between humans, and takes into account what each person owes the other by virtue of the fact that they share the experience of being human. It is beyond the scope of this chapter, and indeed this book, to derive the fundamental principles of moral transactions between humans (such as beneficence, truth-telling, promise-keeping, respect for autonomy, and justice); I addressed these issues elsewhere (5,6). We must assume these principles to be the starting points from which we can examine their special applicability in the relief of pain. Detailed discussion of the relative merit and limitations of various ethical approaches to chronic pain care have been admirably reviewed by James Giordano and colleagues (7–10).

When considering the effects of chronic pain-as-illness, we must bear in mind that illness of any serious kind wounds our capacity to express fully our humanity. When we are in intractable pain and ill, our freedom to do things we are accustomed to do is seriously compromised: we lack knowledge to heal ourselves. Our pain and illness make us dependent. In that state, we must seek out those who profess to be healers. In the wounding of our humanity, pain and illness wreak effect upon the biologic, psychological, and social dimensions of our life. Thus, to be healed as humans, we require not only biological treatments, but care that enhances autonomy and permits the patient to participate in determining what is in his or her best interests according to the patient's own value system. The inequality in the physician–patient relationship places a major burden on the physician for telling the truth, upholding the independent decision making of the patient, being just, and making the clinical transaction as humane as possible.

Pain, and maldynic pain-as-illness somewhat more specifically, dictates that the physician must work with the patient to decide what is in that patient's best interest. If the patient's participation is to be fully human, it must be informed and free—the patient must know his or her situation, what he or she can expect prognostically, what his or her medication can (and cannot) achieve, and particularly, how the patient's functioning as a person will be affected by the capabilities and limitations of both pain and its care.

The decision to undertake a drug regimen must also be free—that is, both without coercion or manipulation by what the physician thinks he or she would do in such circumstances, and without imposition of medicolegal or market influences that are external to the physician–patient relationship. If the patient is to exert autonomy and moral agency, then the patient must be free to reject the drug regimen, and must be equally free to participate in its choice relative to how any such care might impact his or her values and goals. That is not to say that patients can or should freely choose their analgesics. This would be beyond the scope of their expertise and would trump the autonomy of the physician. This is not to imply that physicians should disregard the law. What it does suggest is that the choice of any treatment should be the result of a deliberative engagement between physician and patient—both with respect to their respective roles in the medical relationship (as one who can heal and one who seeks healing), and as human beings.

Obviously, this necessitates that the physician be aware of pain, its effects in the patient, and the effects and side-effects of any and all treatments considered. These requirements may be exceedingly difficult to satisfy. It should be possible, at least, to discuss drug regimens and their effect. But, given the severity, durability, and multifold effects of pain, where does one draw the line between the relief of pain and effects that drugs may induce? Some physicians would hold that the patient's comfort overrides all other considerations, deem all pain worthless and unjust, and thus justify the use of analgesics in an attempt to *rectify* the patient's life. Others might view their role as *gatekeeper*, somewhat more paternalistically steering the patient away from possible burdens and risks through the restricted prescription of an analgesic ladder. The position one takes depends upon what one thinks of the nature and meanings of pain, human existence, and medicine. In a pluralistic society, there is likely to be lack of homogeneity of beliefs on these matters. One may interpret this lack of agreement (erroneously, I believe) to mean that all views are equally tenable, or that the physician always knows what is best, or that social opinion will ultimately dictate what is "right."

There are times that out of humanitarian motives, some may feel impelled to violate autonomy and truth-telling. They fear that the patient's response to knowledge about the intractability of his or her pain might cause greater suffering or loss of hope. But this reduces the patient to a childlike state, just at the time of the most crucial confrontation with vulnerability, suffering, and dependence, and the decisions that pain and illness necessitate. For a long time, physicians held that the knowledge of intractable pain should be withheld from the patient or communicated to the family only. Is this not merely a further degradation of the person who is the patient? Medication certainly should be used to offer optimal relief of pain, but not without first enabling the patient to comprehend his or her situation, and to make his or her own decisions about the extent and amount of pain, and its accompaniments, that are viable for the patient to experience. For each patient, there is a unique element of existential gain and loss that must be considered when addressing what is conferred by pain and its treatment (8,11).

THE PAIN PATIENT

To respect the autonomy of the patient requires telling the truth, keeping promises, and permitting the patient to reject certain treatment regimens, even if the physician considers such rejection ill-advised. It also requires discerning how much of the patient's distress is due to pain mechanisms, and how much is caused by the constellation of psychosocial features that may be both generated by and affecting pain. To adequately deal with these sources of illness demands equal recognition of their causes, effects, and meaning to the patient. To some extent, patient narratives may enable the physician to have access not only to a patient's history but also to the meanings that patients ascribe to certain events, including pain and its effects as illness (12).

Clearly, the treatment for some of these effects is not analgesic drugs. Although such drugs might reduce the patient's experience of illness, their use in this way is not ethically, or even clinically, defensible. Rather, what is required is the time necessary to know the patient, to enable the patient to deal with pain and its illness through compassionate understanding. Psychological and emotional assistance and adjustment in family and social contexts can empower the patient to deal as a human with these effects of pain. It is my belief that the indiscriminate use of analgesics or psychotropics without such discernment violates the canons of not only good ethics but also good medicine.

None of this means that the patient must suffer pain and its effects needlessly, or be denied relief of the degree the patient requires. We are obligated to make the effort to discriminate the varied causes of pain and to offer means of relief addressed specifically to those causes. It is well established that the patient *can* be treated with powerful analgesics to obtain relief; how a specific patient *should* be treated is of more primary ethical concern. What is not ethically defensible are such extremes as allowing the patient to suffer because the physician construes pain to be imaginary or exaggerated, does not want to make the patient *an addict*, or fears to undergo medicolegal sanction or frank litigation.

These are examples of false morality, in that they violate the reasonableness and proportionality that should modulate ethical decisions. When bad things can be overcome by the good of a practice, the obligation to do so becomes clear. Still, even good things can and often *do* conflict with each other; yet, one must be chosen. This type of decision making is not uncommon in medicine and is both therapeutic and moral.

For some, pain may not be purposeless or without good. There are practical and spiritual reasons why some patients may wish to engage pain and suffering, including a deep conviction against medication; fear of drug-induced decrement of vocational or social goals; and an attempt to grow in humility, be witness to courage, or engender fortitude (e.g., see see Chapter 6, in this volume, for a more thorough discussion of these points). To serve these ends, it is believed that pain must be actively and consciously embraced by one's own will. The virtue resides in acceptance, and not the pain and suffering, per se. Bearing pain may even give some more freedom to act.

This is in sharp contrast to others, for whom all pain is a perceived *harm* and must be zealously eliminated. Between these extremes there is a spectrum of responses that each person in pain should be free to express in terms of his or her own value systems. Whether the physician accepts or rejects the patient's values, the physician has a moral obligation to the patient-as-person to enable, as best he or she can, those values to be actualized.

ETHICAL OBLIGATIONS IN CLINICAL CONTEXTS

These ethical obligations must be carefully balanced against the deleterious physiologic and psychological effects of pain. The balance is struck at a different point for each patient, because experience of pain is unique to each person. The ultimate criterion of a humane regimen is the degree to which it allows this uniqueness to be expressed. A humane and ethical regimen of care allows the patient to act as a human—as freely, knowledgeably, and autonomously as the complicated web of his or her clinical illness will permit. The physician can facilitate or negate this by his or her perception of what is scientifically and ethically optimal in the management of pain, and its expressed illness. These perceptions, if erroneous, can constitute serious obstacles to humane and ethical care.

Despite their almost universal use, errors in the clinical management of older and newer analgesics are far too frequently encountered (13). Under- or overdosage results from a variety of causes—the use of standardized regimens of familiar drugs adequate for more acute situations but inadequate for chronic pain; intermittent use without attention to the need to overlap doses; restricting doses to the maximum levels appropriate for nonchronic pain but inappropriate for more durable pain situations; failure to discern how much discomfort is due to pain and how much is attributable to other causes; failure to detect new or unsuspected causes of escalating or persistent pain; inadequate coupling of analgesic agents with psychotropic medication; failure to use the pharmacologically mandatory principles of dose titration, dose variation, and drug rotation; and finally, failure to decide beforehand precisely what degree of pain relief is desirable in this patient—not in some hypothetical patients with these symptoms or this type of pain.

As always, the physician's first ethical responsibility, therefore, is scientific competence in the clinical management of chronic pain. What is often missing, as in so many clinical situations, is an orderly, continuously reexamined plan for action. The physician has the obligation to know as much as he or she can about not only pain, but about his or her patient, especially those social, cultural, and psychological factors that will determine the severity and kinds of response to pain and its attendant illness. This is the first step in gauging intensity and is a necessary guide to the process of care.

The physician must then discern, as far as possible, how much of the patient's overall discomfort is related to those causes that should be treated by means other than analgesia. The physician must not automatically attribute all signs and symptoms to a primary disorder. The clinical constellation may be the result of other events—each of which requires attention, and perhaps different types of care. Physician and patient should decide together what the aim(s) of the treatment regimen should be, how much pain and discomfort the patient can or desires to experience or endure, and what goals he or she may have that are relevant to pain, its effects, and the scope of care provided (for a more complete discussion of the goal-directed pain care, see Chapter 17, this volume).

With these determinants in mind, the physician can design a goal for the treatment regimen; this may include analgesics, psychotropics, and appropriate emotional, spiritual, and family support. The drug regimen must be monitored constantly, and medication doses must be titrated to the desired effect. In the initial assessment and subsequent monitoring, both the patient and his or her family play an important part. The family spends the most time with the patient and is most closely involved in his or her life. Yet, even in those cases where there is complete involvement of the patient and his or her family, there are obstacles of values and attitudes that can impede ethical judgment.

The physician's own attitudes about how pain should be borne are crucial. These include the physician's conception of what is important, his or her tolerance for ambiguity and complaint, his or her enthusiasm or antipathy toward drug use in general and analgesics more specifically, his or her capacity for compassion and empathy (that is, the physician's capacity to apprehend the patient's situation and situatedness in pain)—all can and will influence the physician's sensitivity to patient needs, and his or her acceptance or rejection of the patient's freedom to participate in the design and implementation of the treatment plan. A more in-depth discussion of the relevance of empathy in assessing pain-as-illness is provided in Chapter 5, this volume.

For any regimen to be ethically defensible, the physician's own attitudes and values must be adjusted to the needs of each patient. Yet, when some fundamental moral principle of his or her own is violated, the physician may refuse to adapt to patient need, and can, in fact, validly recuse himself or herself from the clinical relationship. Under these circumstances, the patient should be provided with other clinical resources so as to avoid being clinically abandoned. Thus, while the physician's ethical obligations generally derive from the needs of his or her patient, the physician's promise to be faithful and to help and act in the patient's best interest need not, and should not violate the physician's own moral or professional integrity (14).

CONCLUSION

The physician's ethical obligation to ease pain and the discomfort of its accompanying illness is so fundamental that it is too easily taken for granted. Yet, this is fraught with complexities as soon as we look beyond the simple principle of relief. In sum, I hold that to fulfill the ethical obligations that arise from the fact of pain, the physician's promise to help the patient in pain, and the need to make right and good decisions, the pain physician has certain clear responsibilities. Namely, these are as follows:

1. To be technically proficient in the use of medications that have been demonstrated to be effective in the relief of pain
2. To apply that proficiency in an orderly way with goals agreed upon with the patient, and to ensure that these are constantly reexamined by both the physician and the patient
3. To treat the patient so as to uphold the moral principles of truth-telling, promise-keeping, respect for autonomy and justice, and, in this way, enhance the capacity of the patient to act fully as a human
4. To avoid imposition of the physician's way of living upon the vulnerability and dependency of the person in pain
5. To avoid the patient's imposition on personal and professional integrity of the physician

REFERENCES

1. Leriche, R. 1979. *Surgery of pain (La cirurgie de la doleur)* (A. Young trans.). Baltimore: Williams and Wilkins.
2. Leder, D. 1990. *The absent body*. Chicago: University of Chicago Press.
3. Giordano, J. 2001. The neurobiology of pain. In: *Pain management: A practical guide for clinicians* (6th ed.), ed. R. Weiner, 1089–1097. Boca Raton: CRC Press.

4. Giordano, J., and M. Schatman. 2008. A crisis in pain care—An ethical analysis. Part one: Facts, issues, and problems. *Pain Physician* 11: 6–12.
5. Pellegrino, E.D. 1983. The healing relationship: The architectonics of clinical medicine. In: *The clinical encounter*, ed. E.E. Shelp, 153–172. Dordrecht: D. Reidel.
6. Pellegrino, E.D., and D.C. Thomasma. 1993. *The virtues in medical practice*. Oxford: Oxford University Press.
7. Giordano, J. 2009. *Pain: mind, meaning and medicine*. Glen Falls PA: PPM Press.
8. Giordano, J. 2010. From a neuroscience of pain to a neuroethics of pain care. In: *Scientific and philosophical perspectives in neuroethics*, eds. J. Giordano and B. Gordijn, 172–189. Cambridge: Cambridge University Press.
9. Giordano, J., and M.V. Boswell. 2009. Prolegomenon: Engaging philosophy, ethics and policy in, and for pain medicine. In: *Pain medicine: Philosophy, ethics and policy*, eds. J. Giordano and M.V. Boswell, 13–20. Cambridge, UK: Linton Atlantic.
10. Giordano, J., and M.E. Schatman. 2008. A crisis in chronic pain care—An ethical analysis. Part two: Proposed structure and function of an ethics of pain medicine. *Pain Physician* 11: 589–595.
11. Boeyink, D.E. 1974. Pain and suffering. *J Religious Ethics* 2: 90.
12. Charon, R. 2006. *Narrative medicine: Honoring the stories of illness*. Oxford: Oxford University Press.
13. Ballantyne, J. 2007. Opioid analgesia: Perspectives on right use and utility. *Pain Physician* 10: 479–481.
14. Pellegrino, E.D. 1971. Physicians, patients and society: Some new tensions in medical ethics. In: *Human aspects of biomedical innovation*, eds. E. Mendelsohn, J. Swazey, and I. Taviss, 77–97. Cambridge, MA: Harvard University Press.

16 Children, Maldynic Pain, and the Creation of Suffering

Toward an Ethic of Lamentation

Carlos Gomez

CONTENTS

Editor's Note: This chapter is dedicated to the life, work, and memory of the late Dr. Carlos Gomez; a scholar, teacher, empathic physician, insightful ethicist, and much valued colleague and friend.

FINDING THE WORDS: A PATIENT'S STORY

"I used to cry a lot...started when I was 2. My momma couldn't make me stop, so she used to whack me, and then I'd cry some more." The fellow who told me that was a 35-year-old with sickle cell anemia, referred to my clinic. He was tagged as a *frequent flyer* (i.e., a frequent visitor to the emergency department and the hospital), and that meant that he was a "problem" patient. The house officers in my hospital (including me) would grimace when we saw his name on our admit list, or when we were called to the emergency department (ED) to evaluate him, or when he would intermittently show up in our clinics. As an admitting resident in the ED, we were considered "strong" if we managed to satisfy his need for pain medicine, without having to admit him to the hospital. The unwritten rule, if he managed to penetrate our barriers and get admitted, was "3 days and out"—that is, 3 days of pain medicine, then discharge with an appointment to follow up in a clinic, which he rarely did. When he was admitted, we had a standing set of orders, parameters for the dispensing of pain medication, a standard set of laboratories, and a minimum amount of interaction: his was the one room on the ward we rarely visited, and when we did, we left as quickly as we could.

When his name appeared on my clinic list, I was angry and puzzled. I had worked with terminally ill patients and was used to managing the pain symptoms of those who were dying, but this patient—I will call him Robert—was different. He was not dying, and he had a problem that was not amenable to the sorts of "fixes" that would work quickly. Unlike those with terminal illnesses, moreover, his use of pain medications, especially opiates, was always suspect. For the patient with lung cancer, with metastatic disease to the bone, one could always point to a lesion, visible on radiographs, and justify (even implicitly) what seemed to be enormous dosages of narcotics to control the pain. But when treating a patient with chronic, intractable pain—and in this case, acute on chronic pain—the situation was more nuanced, complicated, and less visible.

Ironically, though, he had a disease with a clear, genetic etiology: a point substitution of valine for glutamic acid on the sixth position of the beta-chain in his hemoglobin—sickle cell disease (1). Moreover, the consequences of this inborn genetic error, vaso-occlusive crisis, or *sickling* of his red blood cells caused by his defective hemoglobin, which, in turn, caused oxygen deprivation, were well known and well documented—multiple infarcts of both bony and soft tissues, necrosis of joints, susceptibility to infections, and myriad other sequelae (2). What the disease means, in reality, for the patient is a stuttering progression of pain, disability, and loss of functional status. It also means confronting a system and a profession that wants "proof" of pain before treating.

In Robert's case, this created a paradox: "I almost liked being real sick, 'cause then they'd do those tests, and when my blood was low…well, I knew I'd get some relief." And when the tests were normal? "Didn't mean I didn't hurt. Just meant I had to fight." And Robert was known as a fighter. He had been "discharged" from our ED multiple times for being "abusive" to the nursing and medical staffs. "Nobody talked to me, you know, I just would go in, they'd see me, get some blood, and make me wait and wait. Sometimes, you know, I'd just leave, 'cause it wasn't worth the wait'" One time, he said, he had been to the ED three times in a 24-hour period, "I knew the sickness was coming on…I just didn't want to wait till it was so bad I was crying. Didn't matter. It's got to be real bad before you get what you need."

Robert eventually left his hometown as a teenager, moved to New York with some relatives, dropped out of school, and would wander the streets in search of his pain relief. "You know, you can get anything you want in the city, if you can pay." He would frequent multiple EDs, with varying luck, would get a temporary job, cobble together enough money to get his "fix" (which was the word he used for street drugs—"medicine" was what he called the same drug when he got it from the hospital), get "stable" for a while, and then start the cycle all over again. He eventually developed bilateral hip necroses, making it impossible for him to work, went back to his hometown to stay with his mother, and became, after a decade away, a patient in my hospital, and subsequently, in my clinic.

I had read Robert's medical history (voluminous at this point) without ever having heard his story. Previously, he had been a patient of mine on a general medicine service when I was a resident, but my job then (or at least, what I saw as my job) was to follow protocol. My "success" was when I would manage the protocol without major incident; get the patient out of the hospital in less than 3 days, and refer him to another clinic. But he always came back, and this time, I was no longer a resident, but an attending, and I could make my own decisions.

So, I decided to listen.

A FRAMEWORK FOR ANALYSIS

Patient to physician, physician to physician, and beyond—medicine is a discipline that flows with human-to-human interaction. Language lies at the heart of its practice—a complaint of calf tenderness, a smile of reassurance, an editorial in a medical journal, a student's hesitant rounds report. In the hospital ward, a physician assesses the symptoms of a patient with congestive heart failure. She may ask the patient about recent changes in the everyday routine. *How are you sleeping? Do you have trouble breathing when you lie down?* She translates her clinical queries to draw out hints of disease progression: *Do you wake up in the middle of the night gasping for air? How many pillows do you sleep on at night?* The patient would need a medical dictionary to understand the shorthand that was jotted down in his hospital chart: *69yo WM with CHF, type II DM, presents with increased orthopnea, PND, DOE.* The care we provide (and are provided by others) is founded in words as much as it is in numerical data.

But *language* is a broad term to use here. As shown above, not all communication utilizes the *same* linguistic tools to reach its end point. An elderly patient in a clinical trial for treatment of refractory rheumatological disease opens a dictionary and finds the word *hospice*: "A house of rest for pilgrims, travelers, or strangers...also, generally a 'home' for the care of the incurably ill or the dying" (3). Her rheumatologist learns that "all patients of #RE58367236-011 will be scheduled for follow-up visits to NIH at 4 week intervals." Each appropriate to its respective situation, these notably different modes of communication are fast at work toward a common goal of alleviating disease, illness, and suffering. Dialogues, stories, regulations, and definitions are fundamental to the complete craft of caring for the sick. If they are not of the same kind, then how can we characterize their uses in medicine? What different *forms* of language take in the undertaking of patient care?

What we need is a model to examine the role of language in medicine. For this, we can turn to James M. Gustafson's depiction of voices of moral discourse (4). In his discussion of medical ethics and literature about medicine, he outlines prophetic, narrative, ethical, and policy voices and their respective vocabularies and linguistic structures. Gustafson uses the term "moral discourse" to encompass a broad range of discourse subjects; in all cases, he explains, one discovers "a concern for human value." Here, we methodically apply these voices of moral discourse to facilitate discussion about the languages of medicine: the patient interview, hospital rounds, the patient chart, legal documents, bioethical arguments, and so on.

According to Gustafson's schema, the *prophetic voice* is the voice of exhortation or command. Not surprisingly, it finds itself throughout ancient texts—Isaiah of the Old Testament (5) and the prophets of the Koran (6). We also hear it in recent history: "A person is a person because he recognizes others as persons" (7). The words are weighted and purposeful. They direct not just the individuals present in context, but entire peoples (often at cultural or ideological odds with the message of the prophetic voice). Bishop Tutu speaks to entire races of Africa, if not to the world. His voice is not argumentative—it does not seek to persuade or convince. Instead, the prophetic voice points toward a greater vision, the embodiment of a truth: "All men are created equal," or "Good health is not a gift to be rationed based on ability to pay" (8). We may or may not agree that all men (and women) are created equal—our opinions are irrelevant. The voice is not inviting us to debate the point or to disagree. It simply resounds with conviction and says, *This is how things are—there are no exceptions.*

There is a different kind of conviction when we have a beginning, a middle, and an end:

> A few light taps upon the pane made him turn to the window. It had begun to snow again. He watched sleepily the flakes, silver and dark, falling obliquely against the lamplight. The time had come for him to set out on his journey westward. (9)

In James Joyce's *The Dead*, Gabriel Conroy is left to ruminate loss while the snow falls over Ireland, including the grave of his wife's dead lover. The *narrative voice* is the voice of myths, legends, and stories.* This is the Greek myth of the harvest goddess Persephone, stolen by Hades of the underworld for 3 months each year, leaving the world with the winter season. It is Milton's *Paradise Lost*, where Satan sits on a hill above Eden and weeps for himself. It is the fairy tale of George Washington chopping down a cherry tree and the true story of Harriet Tubman endangering herself to free American slaves. The power of the narrative lies in the story—its inception, development, and conclusion. Power also comes from the act of *telling* the story. We are taught and retaught with each encounter.

Almost by definition, the prophetic voice booms with the clarity of its message. In contrast, the narrative voice has the potential for much more subtle delivery: "I walk here now...Rather I come and stand...At nightfall...She fancies she is alone...See how still she stands, how stark, with her face to the wall...How outwardly unmoved...She has not been out since girlhood" (10).

Still further removed from these first two voices is the ethical voice. Founded on the use of reason, the *ethical voice* is the voice of philosophy or constitutional law. It may use narratives and even draw from truths delivered by prophetic voices, but its language drives toward argument:

> Socrates is a man. All men are mortal. Socrates is mortal.

We will overlook the reality that Socrates is, by all accounts, dead for thousands of years. Here is perhaps a more pertinent example:

> Those who give considerable weight to patients' rights to determine their own care believe that the patient's informed consent to an action that may cause death is more fundamental than whether the physician intends to hasten death. From this perspective, the crucial moral considerations in evaluating any act that could cause death are

* Note that Gustafson's description of the narrative voice is limited to stories about ethical concerns.

> the patient's right to self-determination and bodily integrity, the provision of informed consent, the absence of less harmful alternatives, and the severity of the patient's suffering. (11,12)

The tone is suasive. The goal of the passage is not to describe a fundamental truth or to teach its readers; it is to convince. To do this, the speaker must construct a rational basis for his or her argument. The speaker must also anticipate the criticism of it and counter it. In the end, we find there is a different kind of truth that comes from this voice, one that is not inherent in its message but has been carefully built, defended from contradiction. It is truth in the form of a philosophical proof, and with it, we can even propose the promise of life after death, as Plato has reminded us at the beginning of this chapter.

The *policy voice* sets into place the conclusions of the ethical voice—this is the voice of codes, protocols, and regulations. If an ethical voice has successfully argued that patients have a right to refuse cardiopulmonary resuscitation, the principles are considered settled by all parties. Next is the task of implementation. A law is written by the policy voice, describing resuscitative treatments, outlining parameters for written refusal, and defining which patients may take such an initiative. For example,

> For the purposes of this part "advance directive" means a written instruction, such as a living will or durable power of attorney for health care, recognized under state law (whether statutory or as recognized by the courts of the State), relating to the provision of health care when the individual is incapacitated. (13)

The policy voice may even surface around intimate and emotionally charged conversations about the end of life. Here is an excerpt from one hospital's guide to managing issues concerning resuscitation:

> **Resuscitation Discussion Guide**
>
> (Based on a study of patients' experiences with resuscitation discussions, funded by the Allina Foundation)
>
> Before you start, note that: Most patients want to talk about resuscitation and may already have clear opinions about their resuscitation choice.
>
> When: Preferably when healthy and/or soon after a serious diagnosis.
>
> Where: Private, quiet, preferably outpatient setting.
>
> Who else is present: Family, friend, spiritual counselor, other professional or no one, as patient requests.
>
> How to start:
>
> "With each of my patients, I always try to discuss their desires about resuscitation..." and/or
>
> "To follow your wishes, I need to know how you want me to take care of you..." and/or
>
> "Let's talk about your goals for treatment, what you want us to do and why..." (14)

There is no question that human communication is as multilayered as our societies. We rely on Greek tragedies and the evening news, pop music and national constitutions, Newtonian proofs, and stop signs. No single form of language will do.

Nevertheless, the components of moral discourse do allow us simple tools to begin to understand how we suffer and how we can measure suffering. These four voices are distinct but by no means exclusive of one another. As we will see, they swim amidst one another, even within the same phrase, translating our inner anxieties into expressions of grief.

Now that we have a framework with which to examine language in the practice of medicine, let us see what voices are at work, where such language has taken us thus far, and what it might suggest about Robert's care.

CHRONIC PAIN AND LANGUAGE

Gustafson's framework is useful for a number of reasons, not the least of which it gives us a tool with which to parse a human experience—pain—and its encounter with a system of care that is meant to attend to it. Robert's case is illustrative. There is a prophetic imperative in American medicine which, at least, allows Robert to enter the "system." He has a disease, identified, described, but not yet curable, that is his "ticket" to some form of care. In other language, the *prophetic* voice of American medicine will say that no patient in acute distress should be turned away from emergent care. In that sense, it serves Robert, but only up to a point. There is also a *policy* voice here, which gives us rules, protocols, criteria for admission, and so forth, which, again, serve some of Robert's needs, but not all of them. There is, similarly, an ethical voice, which posits a sort of rough equality, whether we follow it or not, and insists Robert should be treated no differently than any other patient simply because he is black, for example, or uninsured.

And yet, Gustafson's schema also throws into relief the inadequacy of the system, and how it treats Robert. Even if—and this is a big "if"—we were to grant that the medical system is fulfilling its prophetic, ethical, and policy imperative with Robert as a *patient,* it is failing Robert as a *person.* I say this with no particular sense of smugness or superiority, because I was (and am) as much a part of that system as anyone else. Robert's story—what makes Robert *Robert,* and not anyone else—is intimately tied both to his disease and his illness. In other language, the diagnostic and therapeutic keys, I think, are embedded in the words this man uses to describe his experience as a sufferer of an incurable disease, which has caused—and will continue to cause—him pain.

His voicelessness finds its genesis in childhood—he cannot articulate that he is in pain, for many reasons. One, he lacks the words. But two, who will listen? The child with a debilitating, chronically painful disease is often seen first as a behavioral problem (e.g., with school avoidance). Rather than listening to a narrative that needs explication, parents, teachers, health professionals, and others want to "fix" the problem (14). Thus, in Robert's violent household, the cry that gives voice to his pain is met with more pain, this time inflicted by his mother. Similarly, he finds that to describe his many types of pain, he requires language that is perfectly reasonable to him, but finds no appropriate "objective" correlate. Thus, he uses words and phrases such as "my misery," "my aching," "I hurt all over," which make perfect sense to him, but which fail to move the system around him.

Robert, like enormous numbers of people with chronically painful diseases, is thus isolated in two ways. One, the person in pain is isolated by virtue of the pain, an intensely subjective experience, to which we can react with empathy but not experience ourselves. Two, the experience is further isolating *precisely* because the only way that the person in pain can make their distress manifest is through language (15). But language—words, phrases, grunts, and even well-placed silences—works only when there is an attentive listener. Moreover, when the experience is ongoing, the listener must be willing to hear the narrative repeatedly.

Robert's narrative begins with the early experience of pain, when his sickle cell disease first flared as a 2-year-old. As with any other animal without speech, he wails. But when his expressions of pain are met with violence (from his mother), frustration and anger (from his doctors), or indifference (from the medical system), what begins as pain is manifest as illness and becomes suffering. He is the quintessential bearer of a chronic condition who now *suffers* from a chronic condition. His narrative—the story he tells to himself and about himself—is that his pain signals more isolation, futile attempts at relief, behaviors that are angry and combative, and the ongoing and unmet need to tell *someone* what he is feeling.

Thus, by the time I see Robert again, he has become a *sufferer* of maldynic pain. It dominates his world, influences his interactions with other people, and makes him hostile, afraid, and easily offended. During the short time I worked with Robert after he returned to my clinic, my primary intervention was to listen, and to let him tell the story of his past week. He was always my last patient of the day because his visits took time (sometimes more than an hour), and I could never predict how his visits would affect the rest of my patients in clinic. I also tried, with almost no success, different approaches to his maldynic pain: referrals to a biofeedback center, an appointment with a pain psychologist, and multiple adjuvant medications. The goal was to make him more functional, to avoid repeated visits to the ED, and to integrate him, somehow, into a system that would tend to him, his disease, and the illness he was carrying.

The word *chronic* takes origins from the Greek *chromos,* or "time." The person who bears a chronic illness that causes pain carries the internal narrative of that pain through *time.* It is not episodic. It may fluctuate, modulate, and even abate, but it becomes embedded in the lived experience of the person. Similarly, though, if a chronic condition has this element of "time" embedded in it, then human response—whether from a parent, a teacher, or a health care provider—also requires time. And we work inside a system that demands, for the sake of "efficiency," results with a minimum amount of time. The notion of attentive listening as both a diagnostic and therapeutic approach to chronic pain is, for whatever reason, missing from the training of the vast majority of practitioners in this country, and it has had devastating consequences for these patients we aim to serve. It is, as Giordano notes in this volume, the essence of maldynic pain.

In Robert's case, his lived experience with this disease began with a lamentation, an anguished cry, and then became not a *part* of his life, but almost his entire life. By the time he presents again to a clinic, it is a narrative of illness and suffering, not just pain, that overwhelms Robert and the people around him. Though he was 35, he is still the child that first experienced the fact that he was, literally, born into a condition of pain. Although he might get temporary relief from the pain, it was still at the hardened core of his being.

His story is still mysterious to me. He quit coming to clinic, had a few more episodes in the ED, and then disappeared, or at least could not be contacted. Phone calls, visits from social workers, letters—all went unanswered. One wonders, though, if in a sense, the die had been cast early on: the child who cries in pain and gets silence—or worse—as a response from another human being has already started a story that is difficult, or perhaps, as in Robert's case, impossible to change.

REFERENCES

1. Platt, O.S., D.J. Brambilla, W.F. Rosse et al. 1994. Mortality in sickle cell disease. Life expectancy and risk factors for early death. *New England Journal of Medicine* 330(23): 1639–1644.
2. Pearson, H. 1977. Sickle cell anaemia and severe infections due to encapsulated bacteria. *Journal of Infectious Disease* 136(suppl): S25–S30.
3. *Oxford English Dictionary Vol. 1.* 1971. Oxford: Oxford University Press.
4. Gustafson, J.M. 1996. *Intersections: Science, theology, ethics.* Cleveland, Ohio: Pilgrim Press.
5. *Isaiah* 40:15.
6. *Koran* 47:11.
7. Archbishop Desmond Tutu. From an address given at Bishop Tutu's enthronement as Anglican archbishop of Cape Town, September 7, 1986, Cape Town, South Africa.
8. Kennedy, Sen. E.M. Remarks on health care given at John F. Kennedy Library, April 28, 2002, Boston, MA.
9. Joyce, J. 1991. *Dubliners.* New York: Alfred A. Knopf.
10. Beckett, S. 1984. Footfalls. In: *Collected Shorter Plays.* New York: Grove Press.
11. Quill, T.E., R. Dresser, and D.W. Brock. 1997. The rule of double-effect—A critique of its role in end-of-life decision making. *The New England Journal of Medicine* 337(24): 11.
12. Federal Patient Self-Determination Act Final Regulations, Section 489.100.
13. Cite Resuscitation Guidelines, Copyright 1997.
14. The Mayday Fund. 2009. *A call to revolutionize chronic pain care in America: An opportunity in health care reform.* New York.
15. Scarry, E. 1985. *The body in pain: The making and unmaking of the world.* Oxford: Oxford University Press.

17 Goal-Directed Health Care and the Chronic Pain Patient

*A New Vision of the Healing Encounter**

David B. Waters and Victor S. Sierpina

CONTENTS

INTRODUCTION

The chronic pain patient is one of the hardest challenges facing any physician. Axiomatically, this patient is the focus of the pain physicians' practice; thus, it is vital that the pain physician be able to recognize and respond to the special issues, demands, and difficulties inherent to treating the person in pain. Often distraught and uncomfortable, with lives sometimes reduced to near-inactivity, pain patients frequently approach the physician literally begging for relief from pain. But (not unlike most people who have been reduced to begging) they are also frequently overtly or covertly angry, and disgusted with the limitations of medicine (as either science or art!). They desperately need help but may distrust or suspect the physician of not

* This chapter has been adapted and reprinted, with permission from Waters, D.B., and V.S. Sierpina, 2006, Goal-Directed Health Care and the Chronic Pain Patient: A New Vision of the Healing Encounter, *Pain Physician* 9: 353–360.

actually knowing and/or caring about them. They frequently seek narcotics, because that is the only relief they have known, and often in much higher doses than physicians are comfortable prescribing. They commonly say that they will *do anything* to get past the pain, but the only thing that seems likely to happen is that they will take the medications offered. Other approaches (e.g., physical therapy, mind–body techniques, acupuncture, massage, and lifestyle changes) may be deemed impossible because of the pain, or simply not happen for a variety of economic or circumstantial reasons. Physicians often respond with a variable mix of concern, disbelief, or negative emotions, and often the patients' cycle of anxiety and distrust begins anew.

Both the pain physician and chronic pain patient are in a dilemma: both want the pain to be managed as effectively as possible, with improved quality of life—and yet, physician and patient often have vastly different ideas of how to achieve these ends. The patient's wish is a medical intervention that will make him or her feel better (i.e., completely alleviating the pain, and often the resolution of the life-effects that pain has incurred) quickly and hopefully permanently. The physician knows that such a wish can be both unlikely and potentially dangerous. If we can *take the pain away*, it is often at the cost of incurring considerable burdens and risks: danger of addiction, loss of clarity and effectiveness, potential for ever-rising needs for medications, and so forth. Thus, while it appears to the patient that he or she is asking for what seems to be a simple intervention; the physician recognizes its complexity and problematic potential.

How might we change the discourse from that of a supplicant asking an expert to make his or her life better to two people with different stakes facing a dilemma together given that (1) patients often feel as if physicians do not understand or care about the depth and extent of their pain, and (2) physicians often feel that patients have no awareness of the complexities represented by the escalating use of narcotics for pain relief? We believe that a new paradigm is needed that facilitates active physician–patient relating, that establishes the physician–patient dyad as a team, in which each individual brings relative expertise and different insights to the same problem.

The current paradigm of acute health care situates patients and physicians within a superficial scheme of seeking symptom reduction or relief. This paradigm establishes an isolated, "do-something-now approach" that decontextualizes the pain and puts the physician in the position of either doing something that is immediately helpful or not. "I am coming to you with a serious issue—it's your job to make me feel better" is the unwritten implication within this encounter, and this explicitly assumes the tenor of an overt contractual relationship that is built upon a unrealistic premise, and foils the intentions and abilities of both physician and patient. For even when pain is appropriately redefined and treated as a chronic illness, it is common for the pain patient to regard it as an acute issue requiring an immediate solution, and reinstigate the unrealistic demands upon the physician, and the clinical relationship.

GOAL-DIRECTED HEALTH CARE (G-DHC)

We propose a new approach, Goal-Directed Health Care (G-DHC), for the care of the chronic pain patient (as well as the care of other chronic illness). G-DHC alters the physician–patient relationship in several dimensions: (1) greater equality between physician and patient through (2) more shared responsibility; (3) development and maintenance of a larger, longer-term context to care rather than the "make me feel better now" and "turnstile" mentalities of patients and physicians, respectively; and (4) a shared understanding of the patient's fundamental life goals as a major basis for medical decision making. The incentive for G-DHC is relatively straightforward: help the patient articulate his or her primary life goals (inclusive of, and beyond the pain/illness experience) as a basis for evolving pain control strategies that serve those goals.

G-DHC moves away from the idea of the physician simply "taking care of" the patient, and moves toward a shared decision-making process based on the physician's teaching about what will (or might) make the most difference to the patient. It replaces symptom reduction—the unstated but actual goal of most medical encounters—with progress toward larger goals that embody patient values. It implies that the physician expects less overt control, assuming instead the role of knowledgeable consultant and supporter while the patient is more empowered to make informed decisions. Those decisions may or may not appear to be sensible—at least at face value—to the physician, or be the ones that the physician wishes the patient would make. But ultimately, such decisions reflect patients' wishes and choices, and also rest upon the patient assuming some active level of responsibility for the consequences of (informed) choices that are made.

G-DHC: A BACKGROUND

A search for previously published literature on G-DHC using Ovid, Med-Line, CINAHL, and PsychINFO databases (with query terms of *goal-directed medicine, goal-oriented medicine, patient-oriented medicine, chronic pain, quality of life, physician–patient relations*) yielded no specific references related to G-DHC as we present here. Even though much has been written about specific disease-oriented goals for pain management, the risks of analgesic abuse as well as of undertreatment and other patient-centered care issues (1–3), there is not a substantial medical literature that addresses the higher-order life goals that comprise G-DHC. Instead, the stage-setting for such an approach is found in the human performance and psychological literature. The flavor of this has been established by Earl Nightingale's definition of success as "… the progressive realization of goals that are worthwhile to an individual" (4). For the patient in pain, it would seem that getting rid of, or controlling pain is one of the highest priority goals, and that any other goals are subsumed by this. Yet, this leaves the patient and physician in a quandary: if we adopt Maslow's hierarchy (5), for example, the pain patient is "stuck" at one level until that need (i.e., freedom from pain) is satisfied. Approaching the problem nonhierarchically, but rather heuristically, allows the vector of change to move to transcendence. This is necessary for the patient to move beyond a sole focus on the biological, social, psychological, and even spiritual dimensions of the seemingly overwhelming pain experience. This kind of change in referent is what psychiatrist Viktor Frankl observed and detailed in his book, *Man's Search for Meaning*. While imprisoned in a Nazi concentration camp, Frankl noted that those fellow concentration camp inmates who survived did so because they seemed to have had a purpose and meaning beyond the focus on the horrific, day-to-day experiences of the camp (6).

Many pain patients surely see themselves, at times, as likewise imprisoned by the constraints imposed by pain upon their body and activities. To survive and thrive in spite of this requires a transcendent or transformational process. Such a dynamic of change and change agency is well described in the works of Wilber (7,8), Quinn (9), and Beck and Cowan (10) which purport that value derives from increased self-awareness, enlargement of consciousness, and the choice to grow and transform not only in the face of pain and suffering but because of it. However, depending on the psychological readiness of the pain patient, the kinds of transcendent goals and the steps to reach them might be very different. A person afflicted with pain who sees it as a "punishment" may be willing to endure it until a spiritual solution is found, but one who is frustrated by his or her pain from achieving career goals might be best motivated by an alternative career path that can be managed despite the painful state. Recognition of these goal domains, as well as the patient's precontemplative-, contemplative-, and action-readiness to change is essential (11). This approach requires "motivational interviewing" techniques as well. A key to motivational interviewing is "... allowing patients to explore their own ambivalence for change" (11).

Patients' attitudes are shaped by their own words, not by those of the practitioner. This allows for internal examination that can be a powerful trigger for change (12); providing a familiar, though incompletely expressed context for the G-DHC interview and subsequent process. Discussions of cultural psychiatry, anthropology, ethnic, and cross-cultural issues in the medical context address the challenges to the provision of care and overall goal-setting when language, traumatic experiences, and values of the care provider and the patient or family differ substantially (13–15).

Confounding the problem of communication about patients' life goals is the pluralism of contemporary society that has altered (to some degree) the fundamental, moral concordances between physicians and patients. The medical ethics community has struggled with this issue, as it directly impacts the question of moral agency (16) as germane to the proper role of the physician in the therapeutic relationship. Reconstruction of the relationship between physician and patient is currently called for as possible and necessary (17–19), and this call may be addressed to some extent by the model of G-DHC that we are proposing.

Finally, it is crucial that clinicians provide continuous, positive support for even the most unexplainable symptoms that a patient may be experiencing, for failure to do so could incur far-reaching negative effects upon both the clinical encounter and the subsequent physician–patient relationship, as a whole. It was recently shown that patients' coping skills and abilities are enhanced when physicians overtly express direct interest in providing ongoing health-related support (20). Interestingly, even though 62% of patients saw clinician support as vital, only 2% of clinicians saw it as instrumental to patient-centered clinical interaction. In fact, substantial clinician dissatisfaction is centered upon the failure of the *expert knowledge* model of care when there is persistent inability to resolve unexplained symptoms, including pain. G-DHC can foster better communication about expectations for patient coping, clinician support, and functional improvement, by helping align therapeutic goals and anticipations with the broad and powerful traction of life goals. Indeed, these may often transcend purely biological goals (21) and enter into realms of patient values that are experiential (and may have significance for patients on a more spiritual level) (22). Thus, the clinician must have or develop resources and/or expertise to refer to clergy, chaplains, or others trained and skilled to address patient life goals and needs at this level.

HEALTH GOALS AND LIFE GOALS

As described, G-DHC reframes the sickness encounter into one of hope and possibility. The patient is led through a brief process of reviewing and exploring his or her most important life goals, beyond the sick role, and for which he or she wants to regain health and wellness. Health goals are always separated from life goals, and the rule of thumb is that all health goals must serve one or more life goals, or there will be little or no change-energy behind them. Further, health goals can, and often will change from encounter to encounter, while the life goals of the patient may not change to any significant extent.

For the pain patient, this focus changes the conversation from the goal of *being free of pain*, or *getting a refill on my narcotics*, to those that are more existential and essential regarding what matters in his or her life. What the patient would actually do with life if medical needs were met becomes the central issue. For what end does he or she want to be well? What really matters to and in his or her life?

Questions that attempt to identify life goals are important, are not trivial, and include the following: *What do you care most about in your life, and why? What is the most valuable and enjoyable part of your life? What is most worth living for? What, if you could not have it, would make life less (or not) worth living? What do you want to be able to do that you can't do now? What do you worry about not being able to do in 2 to 5 years that is important to you? How long do you imagine you will live?* and *How important is it to you to live long?*

These kinds of questions reframe and recontextualize the experience of pain and help to reveal and/or illustrate different horizons of possibility. The goal is to get the patient beyond immediate symptom relief so that he or she can make intelligent, informed choices based on the most fundamental individual desires. (See Table 17.1.)

The treatment of patients' pain represents one of the most effective uses of G-DHC. Pain patients become focused upon symptom (pain) removal at almost any cost. Having lost their enjoyment of life to the pain, they become narrowly perseverant upon the outcome of "not hurting," without any particular awareness of what such freedom would allow. As a result, the freedom from pain becomes an end in itself, and any method toward this end will suffice. The goal of *no pain* or *less pain* is an abstract, impersonal goal that may never be fully achieved. As well, it is a poor goal, because even if it is reached, it is continuously self-monitored for how long it will last, and invariably it will not last long enough. In contrast, functional goals (e.g., "I want to work in my garden," "we want to be able to visit the kids") are healthier, more positive, and more attainable.

In addition to an intense desire for the physician to passively remove or alleviate pain, patients often develop the fantasy that the physician can take the pain away, if

TABLE 17.1
Sample Questions for Goal-Directed Health Care

What do you care most about in your life, and why?
What is the most valuable and enjoyable part of your life?
What is most worth living for?
What, if you could not have it, would make life less (not) worth living?
What do you want to be able to do that you cannot do now?
What do you worry about not being able to do in 5 years that is important to you?
How long do you imagine you will live?
How important is it to you to live long?
What is your greatest dream, what do you still wish to accomplish in your life?
If you had 6 months to live, how would that change what you are doing now?
In response to any answer to the above questions: Why? Repeat as needed to make clear the difference between health goals and life goals (e.g., why would you want to have no pain—what would you do once pain-free?).

only he or she *would*. The patients' imagined role in this scenario is that they need do nothing except be available for treatment. The result is that patients have neither a healthy goal in mind, nor a healthy process to achieve more realistic goals (i.e., what choices must be enacted to achieve goals by working together with the pain physician). The blind pursuit of an effective (viz. magic) cure becomes an entire undertaking. However, with G-DHC, the process has a much broader focus, recognizes and acknowledges a healing versus solely curative model of health care, and changes the question from *Will the physician cure/save me?* to *Can I/Will I do the things that will make a difference in my life?*

Pain patients often epitomize the person who turns himself or herself over to the prospect of passive, curative intervention that is often impossible to provide. The underlying message is, *make me feel better*, yet, the underlying reality is that the means to fully achieve such an end are sometimes, if not frequently, not within the control of the physician. By sheer circumstances of time, it becomes obvious that the patient is 95% responsible for what happens beyond the boundaries of the clinical encounter. Yet the physician is often the one motivated to try another test, a louder urging for cooperation or admonition against certain behaviors, and a search for new or better treatments—all with the concomitant feeling of failure when these results cannot be realized.

This paradigm reflects a fundamental misallocation of responsibility. The idea of the patient doing something different or better often becomes a vain wish for the physician, and therefore becomes something of an idealized expectation that can generate frustration for both patient and physician.

CHANGING THE DIALOGUE

However, such expectations persist; not least because far too often the physician is rendered professionally impotent without the explicit cooperation of the patient. Yet,

TABLE 17.2
Differences in the Narrative between Problem-Oriented Health Care and Goal-Directed Health Care

Narrative in Problem-Oriented Health Care	Narrative in Goal-Directed Health Care
Eliminate pain	Find what important thing(s) pain prevents
Pathography (story of illness)	Future biography
Problem focused and immediate symptom focused	Goal focused
Emphasizes problem solving and symptom reduction	Emphasizes creativity and establishment of important, life-giving realities
History; repetitive patterns of dysfunction	Future; possible scenarios of vital activity
Regrets, fears, doubts	Hope, aspirations, possibilities
Push toward health by physician often responded to with patient resistance	Pull to health by patient as part of a team

physicians frequently do not know how—or often even if it is acceptable—to ask the patient for improved cooperation in order to become a reciprocal partner in the process of long-term care. When and where it feels acceptable, the physician is often reduced to challenging and/or pleading the case—things that simply do not work well (11).

How then can we evoke change so as to allow for a better cooperation, shared responsibility, and more effective patient involvement in pain care? It is in this light that G-DHC was specifically developed as a way to change the discourse with chronically ill patients. The new discourse basically begins by determining what the patient values (in life), and what specific, objective life goals arise from these values and are explicitly desired. This moves the physician to a role of understanding the patient's values and goals, and informing and recommending those ways that these could be best achieved, and helping the patient to sift through various choices and options. But ultimately, the patient must be responsible for his or her choices, and the execution of the acts toward such choices. In other words, it communicates to the patient that *You must recognize and choose what your goals are, what to do, and whether or not you will do it. I cannot be responsible for that, and I cannot do things for you.* (See Table 17.2.)

In teaching patients about this process, the metaphor of a mountain-climbing guide is often used. The guide (i.e., the physician) is responsible for knowing the routes and equipment and schedules for getting to the top, but he or she cannot carry anyone up. Instead, he or she can help the climber (i.e., the patient) judge how high to scale, what is needed to make the climb, and what will favor a good probability of (relative and self-relevant) success. As well, the guide can relate what has happened to others who have succeeded or failed at the ascent, but in the end, each climber must do the work to make it to the goal (which, incidentally, need not be the summit). This represents a fundamental shift of responsibility in choice that we feel is both long overdue, and which the contemporary medical system has not taught physicians how to negotiate.

A NEW VISION OF THE HEALING ENCOUNTER

Reformulating the physician–patient relationship along these lines is difficult. It challenges expectations of both the patient and the physician (or other health care provider). While this approach regards the *readiness to change model* (1), it also critically relies upon the willingness to redefine the topography of the healing relationship. Although patient centered, it moves the expectations from the "fix me" perspective, to a more holistic, shared, and integrated approach to patient-focal goals, over and above simple passive, symptomatic control. For the physician (or other healer), it reverses the polarity from push or pull, in which the physician and patient are in tension about what is important, to a new, shared, intersubjective dynamic. The new dynamic states: *Tell me what really matters to you, and I will give you the information and viable tools that help you to decide what is worth doing and commit to a path to get it done.* This means that the patient is clearly a principal agent in the change process. The exhausting urging, prompting, and lecturing by the physician are no longer expected to be the catalysts for change. The physician's role becomes one of clarifying choices and how they relate to the patient's avowed life goals, and providing the medical (i.e., intellectual and technical) and relational support required for the patient in his or her difficult task.

The patient's new role is to redefine the question of what he or she cares about, wants, and will work for. Because the focus is on desired positive outcomes, not merely on symptoms and problems, the patient's autonomy and self-responsibility are enhanced rather than undermined. We are working in the patient's area of expertise (e.g., what he or she cares about, and what he or she will do to achieve such goals) and utilizing our own expertise as well (e.g., what is wrong with the patient and what can we do about it). Instead of being in constant tension with the physician, patients are reminded again of the goals that they have defined as important, and are guided so as to remain focused on what they most wish to attain, in specifically functional terms. Esoteric or obtuse goals, such as "I want to have a happy life" would not be helpful; instead, it is more important to help the patient understand what makes for a happy life (e.g., family, work, hobbies, achievements) and how objective goals within these domains might best be realistically achieved, via a process of patient and physician reciprocally sharing tasks and responsibilities within the respective domains of expertise and competence. The following cases provide examples of how G-DHC can be utilized to foster such a process.

Case 1

The patient was a 62-year-old, female, chronic pain patient who was in a protracted battle with her physician for more narcotics to manage her "unbearable pain." When the referring physician, Dr. X, asked for a G-DHC consult, after explaining to her the concept of G-DHC, the conversation proceeded as follows:

PHYS: So, what do you really care about? What would you do if you didn't have the pain?

PT: I just want to get rid of the pain. I can't stand it anymore.

PHYS: Of course you do. What I am interested in is what you would do if you didn't have the pain. What matters to you so much that you would be willing to work toward it?

PT: I am not sure what you mean, Doctor. I really care about getting rid of this pain and getting on with my life.

PHYS: And what would be the most important things to you if you could get on with your life?

PT: My kids and grandkids are the most important things in the world to me. I would like to care for my grandkids so I have more time with them and so their parents can work and make some money instead of paying for day care.

PHYS: So, you'd really want to step in and take an active role with the younger generations? Really make a difference in their lives?

PT: Exactly. And it would be better for me. But I can't do it with this pain.

PHYS: But you'd have to be pretty clear-headed and on top of things, yes?

PT: You bet. Four kids under 10, and two of 'em not in school yet. It's a lot.

PHYS: So, if the medication made you foggy, you couldn't even do it! That would be a waste.

PT: That's true. But I can't do it with all this pain either.

PHYS: Right. I get that. So we need to find a way to ease your pain that lets you do what you care about doing—that lets you feel good enough to do it, but doesn't make you so foggy you're not responsible.

PT: Well, couldn't the Oxycontin do that?

PHYS: Not from what I hear from your doctor about the kind of dose you're on. No one—not you, your kids, or Dr. X—would feel good about you tending the grandkids on that much medicine.

PT: Whew! They didn't tell me that. Dr. X. never mentioned that. That's not good.

PHYS: That's because you and Dr. X never talked about why you wanted to get over the pain, you just worried about the meds and the dose and all. We are trying to get better about putting that together with the why, so we're aiming people toward what matters to them.

PT: So, what do I do? I'm in a box here. I didn't realize…

PHYS: It's not for me to say. I do know this: there are a lot of things you can do for pain that have no bad side effects of that kind. What Dr. X and you and I need to do together is to figure out what you can do for the pain that makes your goal more possible.

The case progressed with Dr. X presenting the patient with a variety of possible approaches that could help with her pain and her mobility. These included joining a mind–body group working on symptom reduction through increased mindfulness, deep breathing, and imagery. She also enrolled in a water aerobics physical therapy program for 12 weeks. Throughout the process, there was dialogue about her choices and her desired outcomes, about ways her children and grandchildren could be part of her recovery, and her larger pain treatment (taking walks together, throwing a medicine ball with her, etc.). The biggest change was in the process between the patient and Dr. X, where the constant struggle for more and more narcotics changed substantially, and the patient started to take more interest in managing the pain by using the lowest effective dose, as recommended and needed.

Case 2

The patient was a 50-year-old female who came to the physician taking 600 to 700 mg of long-acting morphine daily for severe arm pain. The pain resulted from a brachial plexus injury during lymph node dissection for breast cancer. She had been under the care of the pain clinic that kept her on maintenance doses such as this, but she wished to find alternatives to being on such high doses. Specifically, she was interested in acupuncture, or other less pharmacologically based means of pain control. In reviewing her case, the physician recognized that the pain had a well-documented anatomical cause, and also came to believe that the patient was sincerely interested in reducing her pain medication or even eliminating it. She was unemployed because of the pain and felt that there was little she could do because the pain medication made her unable to concentrate.

PHYS: So, what is it that you really see as the best outcome in this situation?

PT: I just want someone to believe all the pain I am in. The folks in the pain clinic look at me as if I am a junkie, and I don't like that.

PHYS: Well you are on rather high doses of medication.

PT: I know and I'd really like to cut back. That's why I have an interest in acupuncture.

PHYS: So, what is your goal if we reduce your pain and your medications? How would you like your life to be?

PT: I want to start my own tax consulting business. I was working as an accountant before my surgery, you know. Also, I'd like to be able to take better care of my mother, as she is well over 80 now, and I feel like I can't help her much because of my problems.

PHYS: All right, so you want to be free enough from pain and the effects of medications that you can be clear enough to return to your profession and to help your mom remain independent.

PT: Yes, that's it.

PHYS: I respect and honor that and believe we can work together. Let's schedule an appointment for acupuncture and talk about a gradual decrease in your morphine dosage.

PT: You will give me my medications won't you?

PHYS: Yes, but only as much as you need to control your symptoms and at the same time be able to function toward your life goals. Is that fair?

PT: Yes, thank you. You don't make me feel ashamed of myself for being on medication.

PHYS: But that's because *you told me* what you really want, and it's my job to help you achieve that. Let's schedule your next appointment and let me fill out my triplicate for a dose that is 10% to 25% less than what you are on now. You must stop getting prescriptions from the pain clinic or elsewhere while we work on adjusting your dosage.

PT: Will the acupuncture help me to do that?

PHYS: We will give it a try. The main thing is that we are going to try some new things to help you move closer to what really matters to you.

Within 3 months, the patient had reduced and maintained her daily usage of MS Contin® from over 600 mg to 180 mg. She was spending more quality time with her mother and had enrolled in evening classes at the community college to expand her knowledge of tax law. While she could not afford regular acupuncture after the first few helpful sessions, she maintained a more consistent, active way of managing her pain and her life.

CONCLUSIONS

As these cases illustrate, G-DHC can involve aspects of both patient-centered care and relationship-centered care. Even a physician who has known a patient for many years can learn entirely new dimensions and aspects of a patient's reality and worldview by asking questions about, and then sincerely listening to the answers regarding patients' life goals. This type of interaction and the dialogue it produces vastly enrich the patient's opportunity to view himself or herself as being made "whole" as consequential to the healing process, and as having played an important role in that reintegration. It provides an opportunity for the physician to enter enhanced levels of empathy, compassion, and understanding. It recenters the relationship in the precise question of what values, goals, and choices are important to the patient, and what the patient must do—by working together with the physician—to achieve these goals.

We hope to replace the scenario of frustrated physicians and patients involved in an unproductive and nonhealing relationship. In G-DHC, physicians act as change agents to allow the patient to commit himself or herself to healthier behaviors by choice. It is clearly the patient's goals that matter. However, in this model, physicians are no longer merely "taking care of" patients, but rather are helping patients to take responsibility for being a partner in their care, as they move toward realizing important life goals. Or not, for if patients choose not to acknowledge this role and accept responsibilities for cooperation and change, then physicians must acknowledge that this is both the patients' choice and their responsibility. But will the dynamic of *fix me*, *refill my medications*, and *take away my pain* start anew at this point? Perhaps not; if the topography of the therapeutic relationship has sufficiently changed, the patient now better understands that the physician is acting from a beneficent imperative, and not a position of antagonism, hostility, or mistrust. It becomes clear(er) that the physician really desires that the patient move to his or her realistically addressed and recognized personal goals, beyond the imposition of pain and the issues it manifests. This is a new level of relationship-centered care for both patient and physician.

What is different about G-DHC is that if the patient chooses not to pursue healthy choices, it is seen as a distinct choice, with consequences, and the physician is not expected to assume sole responsibility for "making the patient better." By helping the patient to identify his or her most important life goals, physicians can work to relate and realize the connection(s) between patients' life goals and health goals. This allows physicians to enhance the quality of their patients' lives, and may lead to enhancing the quality of the healing encounter, and ultimately to the satisfaction of both the patient and the physician. Wellness—in the most literal sense—becomes the goal instead of avoidance of pain or the problems of illness-associated morbidity, and pain medicine becomes better able to embrace a healing role for those patients in whom pain may not be able to be cured.

ACKNOWLEDGMENTS

This chapter was written with support of the NIH/NIA Mind–Body Exploratory and Development Grant #1R21 AG023951-01. The opinions expressed are those of the authors and do not necessarily reflect the view of the National Institutes of Health or the National Institute on Aging. We wish to thank Julie Trumble, Head of Reference at the Moody Medical Library at University of Texas Medical Branch (UTMB) for her tireless assistance in researching the background for this chapter.

REFERENCES

1. Verbeek, J., Sengers, M., Rikemens, L., and J. Haafkens. 2004. Patient expectations of treatment for back pain: A systematic review of qualitative and quantitative studies. *Spine* 29: 2309–2318.
2. Wagner, E.H., Bennett, S., Austin, B.T.G., Breene, S.M., Schaefer, J.K., and M. Vonkorff. 2005. Finding common ground: Patient-centeredness and evidence-based chronic illness care. Third American Samueli symposium, developing healing relationships, April 21–22. *J Alternative and Complementary Med* 11(suppl. 1): S7–S15.
3. Tresolini, C.P., and the Pew-Fetzer Task Force. 1994. *Health professions education and relationship-centered care.* San Francisco: Pew Health Professions, Commission.
4. Nightingale, E. 1973. *Lead the field* (audiotape series). Chicago: Nightingale-Conant.
5. Maslow, A. 1970. *Motivation and personality* (2nd ed.). New York: Harper & Row.
6. Frankl, V.E. 1984. *Man's search for meaning.* New York: Simon and Schuster.
7. Wilber, K. 2000. *Integral psychology.* Boston: Shambhala.
8. Wilber, K. 2005. *The integral operating system, Version 1.0.* Boulder: Sounds True.
9. Quinn, R.E. 2000. *Change the world: How ordinary people can accomplish extraordinary results.* San Francisco: Jossey-Bass.
10. Beck, D.E., and C.C. Cowan. 1996. *Spiral dynamics: Mastering values, leadership, and change.* Malden, MA: Blackwell.
11. Rollins, S., Mason, P., and C. Butler. 1999. *Health behavior change. A guide for practitioners.* Edinburgh: Churchill Livingstone.
12. Rakel, D. 2003, Motivational interviewing techniques. In: *Integrative Medicine,* ed. D. Rakel, 732. Philadelphia: Saunders.
13. Lewis-Fernandez, R., and A. Kleinman. 1995. Cultural psychiatry. Theoretical, clinical, and research issues. *Psychiatr Clin North Am* 18: 433–448.
14. Fadiman, A. 1997. *The spirit catches you and you fall down: A Hmong child, her American doctors, and the collision of two cultures.* New York: Farrar, Straus, and Giroux.
15. Buck, T., Baldwin, C.M., and G.E. Schwartz. 2005. Influence of worldview on health care choices among persons with chronic pain. *J Alt Compl Med* 11: 561–568.
16. Giordano, J. 2006. Moral agency in pain medicine: Philosophy, practice and virtue. *Pain Physician* 9: 41–46.

17. Pellegrino, E.D. 2006. Toward a reconstruction of medical morality. *Am J Bioeth* 6: 72–75.
18. Veatch, R.M. 2000. Doctor does not know best: Why in the new century physicians must stop trying to benefit patients. *J Med Philos* 25: 701–721.
19. Veatch, R.M. 1998. The place of care in ethical theory. *J Med Philos* 23: 210–224.
20. Nordin, T.A., Hartz, A.J., Noyes, R. et al. 2006. Empirically identified goals for the management of unexplained symptoms. *Fam Med* 38(7): 476–482.
21. Gallagher, R.M. 2004. Biopsychosocial pain medicine and mind–brain–body science. *Phys Med Rehabil Clin North Am* 15: 855–882.
22. Giordano, J., and J. Engebretson. 2006. Neural and cognitive basis of spiritual experience: Biopsychosocial and ethical implications for clinical medicine. *Explore* 2: 216–225.

18 The Problem of Pain and the Moral Formation of Physicians

F. Daniel Davis

CONTENTS

INTRODUCTION

Pain, the most common reason that patients seek the care of physicians, is ubiquitous. Recent years have witnessed significant strides in our knowledge of pain's basic mechanisms as well as progress in both conventional and alternative therapies (1). Curricula in pain medicine have been developed and disseminated (2), and new standards in pain management have been promulgated, for example, by the Joint Commission on the Accreditation of Healthcare Organizations (JCAHO). And yet, pain-*un*treated, *under*treated, and *mis*treated remains a problem of significant scope and complexity. It is a problem, of course, for patients who suffer pain. It is a problem for society as a whole, with annual economic costs estimated at $100 billion by the American Academy for Pain Medicine. And pain is a problem for the profession of medicine, whose very status *as a profession* is undermined by well-substantiated evidence of physician neglect, indifference, ignorance, and bias toward the problem of pain.

As numerous critics have observed, there are multiple barriers to effective treatment of pain (3). Pain relief is often of low priority in the clinical management of "other" patient problems. Clinicians often lack the requisite knowledge and skills for assessing and treating pain, and health systems rarely hold them accountable for these deficiencies. And there are well-entrenched prejudices about the risks and the harms of prescribing and using opioid analgesics and, yes, about the reality of pain. All of these barriers are, no doubt, "real," but I wish to offer, here, a different, somewhat speculative perspective on *why* pain continues to be *un-*, *mis-*, and *under*treated on such a wide scale. I wish to argue that the problem of pain, so defined, is a problem in the moral formation of physicians. As such, the problem of pain is a symptom of a deeper malaise in the culture of medicine, the immediate setting in which this ongoing process of formation occurs. If this diagnosis is correct, it suggests—and I will argue—that such educational initiatives as new or revised curricula in pain medicine may be useful as forms of therapy *at the symptomatic level,* but they are unlikely to be radical enough to effect a cure of the problem.

The perspective from which I write is one that can be mapped, so to speak, at the intersection of several commitments and experience-based observations and convictions. One such commitment is to an essentialist definition of the ultimate ends of medicine, rooted in a phenomenology of the physician–patient relationship. Another related commitment is to a vision of medicine as an inherently moral endeavor and, thus, of medical education as a process of moral formation. The observations and convictions that are integral to my perspective have been gathered and formed over nearly 25 years of experience in academic medicine, many of which have been spent thinking about and pursuing curricular reform in undergraduate medical education. Like many others who have done likewise, I have long been impressed and fascinated by the obdurate resistance of medical educators to change, which, if successful, is usually incremental in scope and relegated to the margins of a curriculum that all too often resembles an overstuffed closet, its shelves bulging with accretions of this or that pedagogical innovation. In time, I came to believe, and still believe, that one source, if not *the* source, of this resistance is the link between the process of professional identity formation that every physician undergoes in becoming a physician, on the one hand, and the development and renewal of a distinctive culture of medicine, on the other. The culture of medicine is not something abstract and monolithic, standing over and against physicians, other health professionals, and, yes, patients: although its origins are complex and multiple, it is sustained through the attitudes and behaviors reflective of the distinctive identity that physicians form and adopt as a tacit condition of their membership in the profession. Fundamental change in the culture of medicine turns on fundamental change in the individuals in and through whom this culture lives—and as any psychotherapist can testify, change of this nature is a daunting task for any individual. It is in this light that I see the problem of pain as a problem in the moral formation of physicians. My task now is to describe what I see and how and why I see it the way I do.

PAIN, ILLNESS, AND THE ENDS OF MEDICINE

Our shared humanity is evident in countless ways, but among the more salient is the experience of illness. You in your way, and I in my own, have intimate familiarity with that state that each of us calls *health*, that state in which self and body exist in harmony, oblivious to the ever-present risk of rupture and discord, and relatively unimpeded in the pursuit of daily life. And I in my way and you in yours have knowledge of how it feels to leave that state, if only for a moment or a day or a week, and to enter an altogether different state, the state of being ill. To one degree or another, illness disrupts my usual, habitual way of being and acting in the world, a way that is both like and different from your way of being and acting in the world. Thus, the experience of illness is utterly unique for me, just as it is for you. And when each of us, and his or her way, becomes ill, each of us becomes a patient—one who suffers disintegration of the experienced unity of self and body and destruction of the dynamic process of being in the world and acting in and on it such that the aspirations and intentions of the self are realized through, rather than frustrated by, the body.

Pain is often a harbinger of the transition from being healthy to being ill for many if not most of us. And like illness, the existential state that it portends and to which we give witness, the experience of pain is unique: pain is *subjective*, to use a word that is often invoked to dismiss or devalue an experience or a form of evidence, especially in the context of medicine. To be sure, pain involves structures and processes (i.e., nociceptors and nociception) that are common to all sentient beings and are susceptible to objectification and measurement; but the structures and processes do not constitute the totality of the experience of pain. There is "more" to pain, as Cassell explains:

> The actions of humans in response to pain generally take into account the location, severity, cause and anticipated course of the pain. Knowledge and judgment are required. Further, reactions to pain range from the momentary to well laid future plans. While the former may depend on reflexes, the latter do not. Pain is this entire process of sensing, interpreting, and modulating the nociceptive process, assigning cause, anticipating course, and determining response. As a consequence, it is obvious why it is a source of confusion that human pain does not exist without sentience. Unconscious or comatose persons may demonstrate nociceptive reactions such as reflex withdrawal from noxious stimuli or rises in pulse and blood pressure. Consciousness as awareness of discriminate stimuli, however, is required for the full experience of pain. Thus, a useful working definition of pain is the experience reported in the statement, "it hurts." (4)

The uniqueness, the subjectivity, of illness and pain have been a source of confusion, as Cassell notes, *and of controversy* in Western medicine for more than two millennia. The confusion has been spawned by a metaphysics that begins with the divorce of body and self *as established fact* and then goes on to struggle with the conundra of how causes in one have effects in the other, of how the one stands vis-à-vis the other in the hierarchy of being, of how thoughts and feelings, as expressions of the self, are engendered in and affect the body. And the controversy has been sparked, in part, by the question: "Which is the proper 'object' of the physician's knowing and doing?" The illness and all of its manifestations, including pain, in this patient? Or, the "underlying" pathology, the disease, that gives rise to these feelings of illness and pain? This question bedeviled the authors of key treatises in the *Hippocratic Corpus,* written some 2,000 years ago, and it continues to bedevil physicians and patients today.

It is a question of ontological, epistemological, and teleological significance. For in clarifying the physician's focus, we not only identify the proper objects of the physician's knowing and doing, we also indicate *how* the physician will know and do what he or she knows and does, and with what aims or ends in mind. From the perspective from which I write this essay and engage this question, it is to the world of everyday experience that we must turn if we are to obtain an edifying grasp on the primordial soil from which medicine arises—on the fundamental human need to which it is responsive and to which it owes its existence as the sort of special human activity that it is. It is perhaps an abiding trait of our humanity that we only come to consciousness of something essential to our humanity when we sense its absence or loss, and so it is with illness and health. When we are ill, we become more acutely aware of our *need* for the state of wholeness, for the lived integrity of self and body, that we call *health.* This fundamental human need is an *ontological* need, as are our needs for food, water, shelter, and affiliation, be it with other human or sentient beings or with God: our very being is dependent upon the fulfillment and satisfaction of our ontological needs.

And as long as humans have had the capacity to communicate to others the ontological need that makes itself known in illness, there have been, among us, individuals who have claimed the right, the privilege, the expertise, or perhaps the mere inclination to be responsive to this need. That is to say, as long as human beings have fallen ill, there have been healers, who seek to lift us up from that state and restore us to health, who profess to have the ability to heal and promise to use this ability for this end. Thus, from a phenomenological perspective, medicine originates—again and again, as it were—in the patient's experience of illness, in the healer's promise to heal, and in their relationship, oriented as it is to the shared aim or end of healing. This perspective has been given its most eloquent and insistent voice in the work and writing of Pellegrino:

> Medicine is a special moral enterprise because it is grounded in a special personal relationship—between one who is ill and another who professes to heal. Illness is an altered state of existence arising out of an ontological assault on the humanity of the person who is ill. Healing is a mutual act that aims to repair the defects created by the experience of illness. The moral authenticity of the healing act is thus measured by the fullness with which it remediates the afflicted state that illness represents. (5)

The achievement of this end, in any given relationship between a physician and a patient, is dependent upon the physician's possession and mastery of forms of knowledge and skill that have, since the dawn of Western rational medicine, been distinguished as specialized—that is, as instrumental to medicine's definitive aims of working with and through that unity of self and body that illness disrupts and healing seeks to restore. Pathophysiology, the knowledge of the structure and functions of the human body in health and disease, is central among these forms of knowledge, as is the skillful deployment of this knowledge in diagnosis and therapy.

But the medium of this specialized knowledge and skill is the relational engagement of the physician with the patient in all the relevant aspects of his or her uniqueness—that is, with the utter singularity of this patient's experience of illness, which cannot be assimilated to this or that nosology by means of either inductive or deductive reasoning. No, the only means to the end of apprehending the particularity of the patient and his or her experience of illness is the physician's capacity for *compassion.* Compassion is the inclination and the ability *to suffer* with the patient—to be open to and concerned about this patient's lived experience of illness. Compassion mitigates the inherent inequality of knowledge, skill, and thus power in the relationship between the physician and the patient, for in suffering with the patient, the physician gives witness to the humanity that both share, ultimately acknowledging that morbidity and mortality are our common, equalizing fate. Compassion has both epistemic and affective dimensions. It entails a kind of knowing, secured through experience of self and others—a knowing of how to discern, weigh, and take the measure of a patient's experience of illness. And it entails feeling, communicated through word and gesture. Because compassion, in its epistemic and affective dimensions, is formed through habit and manifested through disposition, it is one of the cardinal clinical virtues of medicine.

That medicine originates in the patient's experience of illness and in the physician's promise to heal, and that realizing the end of healing demands knowledge, skill, and such virtues as compassion are convictions that have long been integral to the ideals of the profession. As I will later suggest and argue, many of the "movements" for reform—of medical education and medical practice—that have emerged in the last several decades are animated by these ideals, which they seek to realize, more fully and, as it were, against the grain of what all too often really happens in the formation of physicians and day-to-day care of patients. For all too often, what really happens is that physicians are formed—epistemically, morally, and affectively—to be masters of applied human biology. As such, their specialized knowledge and skill are deployed in the service of a narrower aim than the end of healing (i.e., the aim of treating and curing disease, understood as a biological entity that can and must be abstracted from the individualizing features of any given patient). The physician is trained *to see through,* rather than fully apprehend, the patient's experience of illness, in order to grasp and treat the "really real," the disease. That which is subjective, including the phenomena of pain, has value only insofar as it points to and illuminates the underlying pathology.

THE CULTURE OF MEDICINE AND THE MORAL DEFORMATION OF PHYSICIANS

The ideals emerging from an essentialist understanding of medicine's ends and from a phenomenology of the patient–physician relationship have long coexisted, more or less in tension, with the attitudes and behaviors that prevail in the culture of medicine today. In the *Hippocratic Corpus,* for example, one finds a rationalist dogmatic philosophy of medicine, which privileges an almost exclusive focus on pathology, alongside a more authentic Hippocratic empiricism, which, by contrast, enjoins the physician to anchor his or her diagnosis and therapy in the particularizing features of the patient. Throughout the subsequent history of Western rational medicine, this tension persists and evolves, shaped by broader social and cultural forces, but particularly by advances in the sciences and technologies of medicine.

Beginning in the seventeenth century and culminating in the mid to late nineteenth century, however, a series of developments began to accentuate the tension in decisive ways. The work of Thomas Sydenham in seventeenth-century England marked a crucial stage in this process: inspired by innovations in the classification of plants, he turned medicine's attention to discovering the patterns of symptoms occurring in different patients and sought to create a taxonomy of disease based on these patterns. Symptoms of sickness in human beings are of two types, he argued: pathognomonic and idiosyncratic. The former are symptoms shared by individual patients and constitute the definitive patterns of separable disease entities, which Sydenham conceived as real beings or entities in the order of nature. The latter are symptoms that are unique to particular patients; they cannot be assimilated to the symptom patterns that reveal disease entities, the "really real," and are not, therefore, the legitimate focus of the physician. Sydenham's advances in the classification of human disease were an integral force in an evolutionary process that accelerated, many years later, in the nineteenth century, with the germ theory of disease, the spectacular growth and increasing specialization of the basic biomedical sciences, and the formation of specialized clinics for the treatment of disease. Although the image of the physician as a scientist, whose specialized domain within the order of nature is human biology (and particularly, pathophysiology), has an ancient provenance, it drew unprecedented sustenance from Sydenham and these subsequent developments. New knowledge and new technologies began to equip the physician with a degree of practical efficacy that earlier eras had only dreamed of: cure of human disease (e.g., of infectious disease) became possible and determinative of medicine's and the physician's self-understanding.

In this brief historical sketch, it is important to pause and take note of the fact that these developments in the science and technology of medicine unfolded over a period of years, in relative independence from the day-to-day practice of medicine. Before the fruits of these developments were widely disseminated in and became critical standards for medical practice, physicians had little beyond individualized care and comfort to offer their patients, usually in the intimate setting of their homes. Although most appealed to their knowledge of the science of medicine, they sought to excel in what was and always is, at least potentially, attainable in relationships with patients—that is, in what we today continue to call *the art of medicine*. Remnants of this state of affairs survived at least until the mid-twentieth century and, in isolated locations in the United States, may still exist today. In contemporary discourse about medicine, the concept of *bedside manner* bears the somewhat nostalgic imprint of this era, memorialized in Sir Samuel Luke Fildes' painting, *The Doctor*.

In the United States, the achievements of the hard-won, centuries-long progress of medical science and technology were consolidated, and further advanced, with the post–World War II institutionalization of research, education, and practice. The Flexnerian Revolution in medical education, inaugurated in 1910, was no longer revolutionary and had become a nascent target of critique, precisely because of its grounding in what had come to be known as the *medical model*. And the preferred locus of medical education had shifted to those conglomerates known as academic medical centers, which embraced and sought to integrate within their institutional structures expansive programs of basic and clinical research and highly sophisticated, increasingly technology-driven programs of tertiary patient care. The resulting gains were and are undeniable, but so, too, is the fact that they have been accompanied by a profound loss—a loss that Reiser aptly describes as "the eclipse of the patient." Like its heavenly parallels, this eclipse has been gradual and, if Kleinman is correct, emanates from a structural fault inherent in Western medicine. His argument is worth quoting at some length. According to Kleinman, Western *biomedicine* is different from Chinese, Hindu, or Islamic medicine:

> [It] differs from these and most other forms of medicine by its extreme insistence on materialism as the grounds of knowledge, and by its discomfort with dialectical modes of thought. Biomedicine also is unique because of its corresponding requirement that single causal chains must be used to specify pathogenesis in a language of hard structural flaws and mechanical mechanisms as the rationale for therapeutic efficacy. And particularly because of its peculiarly powerful commitment to an idea of nature that excludes the teleological, biomedicine stands alone. ...In the biomedical definition, nature is physical. It is knowable independent of perspective or representation as an "entity" that can be "seen," a structure that can be laid bare in morbid pathology as a pathognomonic "thing." Thus, a special place is given to the role of seeing in biomedicine, which continues a powerful influence of ancient Greek culture. Biology is made visible as the ultimate basis of reality which can be viewed, under the microscope if need be, as a more basic substance than complaints or narratives of sickness with their psychological and social entitlements. The psychological, social, and moral are only so many superficial layers of epiphenomenal cover that disguise the bedrock of truth, the ultimately natural substance in pathology and therapy: biology as an architectural structure and its chemical associates. The other orders of reality are by definition questionable. (6)

Western biomedicine, Kleinman goes on to contend, values disease without suffering and offers treatment without healing:

> This radically reductionistic and positivistic value orientation is ultimately dehumanizing...through its insistence on the primacy of definitive materialistic dichotomies (for example, between body/mind [or spirit], functional/real diseases, and highly valued specific therapeutic effects/discredited non-specific placebo effects) biomedicine presses the practitioner to construct disease (disordered biological processes) as its object of study and treatment. There is hardly any place in this narrowly focused therapeutic vision for the patient's experience of suffering. The patient's and family's complaints are regarded as subjective self-reports. The physician's task, whenever possible, is to replace these biased observations with objective data: the only valid sign of pathological processes because they are based on verified or verifiable measurements. Thus, doctors are expected to decode the untrustworthy story of illness as experience for the evidence of that which is considered authentic, disease as biological pathology. In the process, they are taught to record experience, at least the experience of the sick person, as fugitive, fungible, and therefore discreditable and invalid. Yet by denying the patient's and the family's experience, the practitioner is also led to discount the moral reality of suffering—the experience of bearing or enduring pain and distress as coming to terms with that which is most at stake, that which is of ultimate meaning, in living. (6)

The focus on disease, exquisitely refined and sharpened through ever-increasing specialization; the reliance on sophisticated technologies for revealing the often hidden or obscured locus of disease within the body; and the consequent displacement or eclipse of the patient: these are among the dominant values of the culture of Western medicine. And as I noted at the outset, it is within this culture that the moral formation of physicians occurs. In undergraduate medical education, that process of formation occurs, in part, in the succession of educational experiences that make up the formal curriculum of courses and clerkships, from year one to year four. But, as an ever-growing body of literature has sought to demonstrate, the experiences through which these dominant values are communicated and inculcated occur within what has been dubbed *the hidden curriculum* (7). It is in the interstices, as it were, of the formal curriculum—in the unscripted interactions between medical students, residents, and attending physicians and in the explicit as well as implicit cues communicated via these interactions—that the culture of medicine is effectively transmitted and sustained. This is not to say that this culture is utterly devoid of the ideals of healing, of aspirations to relational engagement with, and compassion for patients. It is something of a paradox and certainly cause for hope that these ideals and aspirations survive, not only in individual physicians but also as the stimulus to a whole constellation of movements to reform medical education and practice. But the dominant values of the culture of medicine remain just that—dominant.

As I stated in the introduction, my perspective in this essay has been shaped by my own experiences as a medical educator—experiences in which I have witnessed the persistence and vitality of the ideals and aspirations, as well as the awesomely powerful influence of the culture's dominant values on the moral formation of physicians. Over the course of 8 years, I interviewed more than 450 applicants to medical school. When asked the standard question, "Why do you want to be a physician?", very few failed to cite their strong desire to help other human beings in need and their commitment to the care and healing of their future patients. Many explained, as well, that they had an intense interest in science (usually biology or chemistry), and, often, that they were fascinated with the workings of the human body. Only midway through the course of those 8 years did I realize how successful the culture of medicine, working through the hidden curriculum, could be in modulating these initial motivations—in amplifying, promoting, and satisfying the intellectual motives, while devaluing, dulling, or diminishing the affective ones.

By that time, I had several seminal experiences with students a little farther down their paths to becoming physicians—that is, at various points in the third year, after their immersion in clinical medicine. The most memorable was also the very first of these experiences, and, as will soon become clear, it was an experience of direct relevance to the subject of this essay. Katy was a student I had known since her first day of medical school; she was intelligent and thoughtful, highly articulate and warm, and passionate about her commitment to *cura personalis*, the care of the whole person, which is central to the Jesuit philosophy of education that my institution embraced and trumpeted as the philosophy of medical education and medical practice. She was a regular visitor in my office during her first and second years, when she was preoccupied with her studies of the basic biomedical sciences, but after she started her third year, our meetings were infrequent and by chance. One day, Katy appeared at my door, looking worn and distraught. I invited her to take a seat and tell me about what was of such obvious concern to her. "They want me to become someone I swore I'd never become." She uttered the statement very deliberately and firmly, her eyes welling with tears. I asked her what she meant, and she went on to tell me the story of the patient she had been caring for as a member of the team in surgical oncology, the clinical rotation she was on at the time. The patient was a middle-aged woman, as I recall, hospitalized for surgery and in the midst of the slow, difficult, and painful recovery. Katy visited the woman daily, meticulously writing up her notes, and conveying to the resident the woman's persistent complaints about unrelieved pain. At first, according to Katy, the resident simply nodded and then moved on to question her about other aspects of this, and then the next, and then the next case. Because she reviewed the patient's chart every day, Katy knew that no changes were made in the patient's medication and that no efforts were made to treat her pain more effectively. One day, Katy became more insistent with the resident about the woman's pain, stating her conviction that this apparent neglect seemed cruel. The resident response was cold and unequivocal: he told Katy that the patient's pain was "of no use" in making the "important decisions" about her "clinical management," that patients often exaggerate the symptoms, and that "doing what they want us to do" could be "worse" than the "pain that they think they're having," especially if they "get hooked."

As I went on to learn, this experience with the surgical resident was not the first time Katy had been counseled to focus, in her patient workups, on information and data that "really matter" in clinical judgment and leave the psychosocial dimensions of human illness to the chaplains and social workers. This particular experience, however, brought Katy's growing discomfort to a point of crisis: she felt she was being forced to set aside the ideals and beliefs that had inspired her to become a physician, *in order to become* a physician. Although unique in its details and in its impact on her, Katy's story is an all too familiar one. Over the last few decades, a distinct subgenre of memoir has emerged, created by physicians recounting their own crucibles of *professionalization*, the morally formative process by which their own selves—better yet, their own *souls*—were transformed. Abraham Verghese, Peri Klass, Samuel Shem, and Raphael Campo are just four among the many physician-writers who come to mind. In reading their accounts of this process, however, it is difficult to avoid the conclusion that the process is not so much one of moral formation as it is one of demoralization. The process is narrowly focused on the acquisition of scientific and technical competence oriented to the cure of human disease. This aim, rather than the end of healing for patients, is ethically normative for the practice of medicine and the patient–physician relationship.

PATIENTS, PAIN, AND PROSPECTS FOR CHANGE IN THE CULTURE OF MEDICINE

I am not a historian of medicine; thus, I am ill-equipped to elucidate the historical significance of something that, nonetheless, strikes me as remarkable about contemporary medical education and medical practice. I am referring, here, to the confluence, over the last several decades, of several "movements" for reform of medical education and medical practice. Although born of different circumstances and addressed to different issues, questions, or problems, these movements ultimately converge, for each, in its own way, seeks to realize, more fully, the ideals to which the profession has always claimed to aspire. A detailed account of each is beyond our present scope, but in identifying each, I will try to illustrate its convergence with the others:

Patient-centered care: More philosophy than movement, patient-centered care is neither novel in practice nor complicated in theory. The simplicity of the core idea is elegantly expressed in Francis Peabody's famous injunction: "the secret of the care of the patient is in caring for the patient" (8). Peabody spoke and wrote these words in 1927, at a time when the forces of science and technology were beginning to reshape medical education and medical practice—that is, in the early stages of the "eclipse" of the patient. Today, patient-centered care embraces initiatives at all levels of systems of care, including that of the patient–physician relationship (which, properly speaking, may not actually constitute a separable level of care but should be understood as its hub, as its very reason for being). At this level, the philosophy centers on the physician's commitment to a clinical method for not only diagnosing and treating disease, but also exploring and determining the significance of the experience of illness for *this* patient. Although the process of diagnosis is usually one of classifying the patient's problem (i.e., determining where, in the nosological schema, the problem belongs), with the patient-centered clinical method, diagnosis is also the means by which the physician comes to understand the significance of this disease, this experience of illness, for this patient, in the context of his or her life history and life plans. Subsequent steps in the method entail clarifying the goals of treatment, negotiating in developing the plan of care, incorporating measures for prevention and health promotion, and, in short, sustaining a relationship rooted in a shared end of healing (9).

Cultural competence: The cultural competence "movement" has marked affinities with the philosophy of patient-centered care in that both seek to redeem the experience of illness and to place the whole patient—self and body—at the focal point of the clinical method, of the physician's knowing and doing, and of organizational contexts in which patient–physician relationships are formed and sustained. Cultural competence, however, seeks to achieve these aims by throwing into relief and eliminating the significance of racial, ethnic, and religious differences between the patient and the physician. The growth of this particular movement, which seeks reform in education as well as in practice, has been fueled by mounting evidence of neglect, bias, and distortion in the clinical judgment of individual physicians and entire "systems" of care. Some critics have learned that cultural competence may play on ever-present tendencies to assimilate particular patients—concrete, unique individuals—to "the general," to niches in some taxonomy, based on differences in race, socioeconomic strata, ethnicity, gender, sexual orientation, age, or functionality. But a more deft understanding of the place of cultural competence in patient-centered care, along with modes of practice informed by this understanding, should help to keep this tendency at bay.

Professionalism: One could argue that the questions of professionalism have been thematic for medicine for two millennia, at least that is arguably the case for Western rational medicine. By *questions of professionalism*, I mean, for example, *What is a profession? Is medicine a profession? What significance does medicine's status* as a profession *have for education and practice? What significance does it have for relationships between physicians and patients?* In the last decade or so, however, engagement with these questions has intensified for multiple reasons. One has to do with the impact of managed care on the physician–patient relationship, specifically, the intrusion of financial factors as *perverse incentives* in the physician's clinical judgment. "The Patient–Physician Covenant: An Affirmation of Asklepios," published in *The Journal of the American Medical Association* in May 1995 and written by an exceptionally distinguished cadre of physicians (10), is a manifesto of the movement and reflects an explicit concern with the corrosive effects of a health care system increasingly driven by considerations of cost, especially the effects on the physician's character and on the patient's trust. With a similar impetus, the Tavistock Group followed in 1997 with a "Shared Statement of Ethical Principles" (11) and 5 years later, in 2002, a new Charter of Medical Professionalism was unveiled by the American Board of Internal Medicine, the American College of Physicians–American Society for Internal Medicine Foundation, and the European Federation of Internal Medicine (12). The covenant, the principles, and the charter all seek a realignment of the profession and of medical practice with the ideals that have been central to its ethical self-consciousness since, some would claim, the era of Hippocrates, when the patient's good was recognized for what it is: the moral center of the activity of healing.

These initiatives have direct and interrelated parallels with developments in medical education, where the questions of professionalism include "How and in what settings and formats should professionalism be taught?" "How should professionalism be assessed?" And, in assessing professionalism, what is being assessed: *knowledge, skills, attitudes,* or *behavior*? In recent years, there has been an exponential increase in the scholarly literature devoted to asking, exploring, and answering these questions. As the literature shows, professionalism is understood in a variety of ways: in some, the physician is envisioned as practicing his or her profession in accord with either some set of principles or some theory of obligation: the Tavistock statement and the Charter of Medical Professionalism are examples of this particular cast to the definition of professionalism. But many of these understandings of professionalism, as well as others, also see certain clinical virtues, elements of the physician's character, as elements of professionalism. The authors of the covenant are anxious not only about the loss of the moral center of medicine but also about the loss of the moral center of the individual physician—that is, the physician's character, as the basis on which trust between patient and physician is built. And in another leading expression of this movement, the American Board of Internal Medicine's Project Professionalism, professionalism is composed of altruism, accountability, excellence, duty, honesty and integrity, and respect for others—all virtues, all tightly tethered, in theory and practice, to the ideals and ends of the profession (13). But can one be taught and learn the elements of character; in the words of Plato in the *Meno*, "can virtue be taught?" And, how do we know that virtue has been learned by the student, the resident, and, yes, the attending physician? Perhaps teaching and learning are, in the end, misnomers for the process of moral formation that spans any given physician's lifetime and that finds its dynamic expression in the physician's character, the force animating the physician's words and deeds, be they virtue or not. But that does not, in any way, mitigate the demand for the character as the essence of professionalism to be assessed in physicians at all stages of their formation as healers. We tend to assess what we value: if we value professionalism in medicine, we will not find the prospect of taking the measure of an individual's character either so offensive or so resistant to realization, in practice, that we content ourselves with measures of knowledge or skill and abandon character.

Competency-oriented medical education: Questions about the pedagogy of professionalism in medicine are also prominent in our penultimate reform movement—that is, in competency-oriented medical education. In the United States, in 1998, two organizations—The Accreditation Council on Graduate Medical Education (ACGME) and the Association of American Medical Colleges (AAMC)—launched initiatives focusing on the outcomes and objectives of medical education. The ACGME's outcomes project is a blueprint for fundamentally altering the process of accrediting programs of graduate medical education by defining and assessing competence, broadly defined, in terms of certain elemental outcomes, including, for example, a demonstrated "commitment to ethical principles," a "sensitivity and responsiveness to patients' culture, age, gender, and disabilities," and "respect, compassion, and integrity; a responsiveness to the needs of patients and society that supercedes self-interest; accountability to patients, society, and the profession; and a commitment to excellence and on-going professional development" (14). And there are clear resonances between the outcomes project and the AAMC's Medical Schools Objectives Project, which gives pride of place to altruism—along with knowledge, skill, and duty—as an overarching objective of teaching and learning.

These two initiatives are reflective of what some have called a paradigm shift in medical education. The reigning paradigm—the structure/process paradigm—traces its origins to Flexner, and in the design of the curriculum, emphasizes an evolving division of the basic and clinical sciences of medicine and a sequence in which they are ordered and studied by those who aspire to membership in the profession. The recently ascendant paradigm—the competence-oriented paradigm—emphasizes outcomes of the educational and morally formative process, looking to both traditional and innovative approaches to teaching and learning but also for evidence, including evidence of character. And one of the first signs of its ascendancy was the change, instituted in 2005, in the U.S. Medical Licensing Examination (USMLE). In that year, for the first time, students sitting for Step 2 of the USMLE had to undergo a 12-station, standardized patient-based assessment of their history taking, physical examination, and interpersonal skills. How to broaden the assessment of competence to include professionalism—that is, character—is now a question under some debate.

Palliative and end-of-life care and pain medicine: At this juncture, the correlates that I have attempted to uncover among these reform movements should be clear. Among others is what I believe to be a kind of shared longing for that which has been lost with the eclipse of the patient: the possibility for healing and the attainment of a form of excellence, moral excellence, or character, evident only in relationships with patients and in the physician's words and deeds—in the physician's behavior. That longing, I would argue, is present in the works of many of the leading exponents of palliative care and a movement to a form of care at the end of life, including, for example, Cicely Saunders, Ira Byock, Timothy Quill, and Joanne Lynn. It is useful to recall, here, that end-of-life care is situated within a particular phase of life, the process of dying and death, and that palliative care is the philosophy that informs the physician's giving of care at the end of life. At the same time, that embodies knowledge, skills, and attitudes that are instrumental to healing at early as well as later points on the trajectory of the patient's illness, beyond the context of the end of life.

In the history of end-of-life care reform and of palliative care, a critical milestone is the 1995 publication of the findings from the Study to Understand the Prognosis and Preferences of Research and Treatment (SUPPORT). This study of 9,000 seriously or terminally ill patients in several major academic health centers found sobering, disturbing evidence of mis-, un-, or undertreated pain in significant numbers of patients; widespread disrespect for patient values and preferences; fragmented, ineffective communications with patients; and systems of processes of care marred by depersonalizing technologies. Since then, other studies have gone on to explore and chart the evidence for one or more of these findings. Although in subsequent years, research has spawned model curricula and more effective approaches to skills acquisition and assessment, evidence of deficiencies in the effective treatment and care of the patient with pain persists.

Pain medicine, as a specialty and clinical medicine, is an outgrowth of this gradual but transformative recognition of the problem of pain—for the patients who suffer it, the physicians who treat it, and the broader institutional and social context in which patients and physicians find themselves in relation. The emergence of pain medicine as a specialty has been accompanied, in part, by a related process—that is, the formulation of specialized, model curricula and programs for the acquisition of skills for interpersonal communication, history taking, and advanced care planning, for example. New tools for communicating about pain have been developed and, in light of our earlier account of the challenges presented by the subjectivity of the experience of pain, their aims are worthy of passing note. Wong-Baker faces, quality-of-life scales, and the CRIES pain intensity scale are aimed at bringing the subjective experience of pain to intersubjective understanding: none quantifies pain, per se, and none purports to measure and thus objectify pain in gradations along a single or double axis. That is to say, they are tools for elucidating and validating *this* patient's experienced pain.

What are the prospects for ensuring that more patients benefit, concretely, from such elucidation and validation of their experiences of the problem of pain? There are signs and causes for hope. New curricula, empirical evidence, and paradigmatic shifts in the definition of competency are all important constituents of a program for fundamental change. But such programs have to penetrate, as it were, the hard shell of the established culture of medicine, whose dominant beliefs and values are transmitted via the identities—and thus, through the words and deeds—of physicians in an ongoing process of moral formation. That process unfolds not so much through the aspiring physician's passage through the formal curriculum but in his or her interactions with others whose words and deeds are both decisive and exemplary. The only question is decisive and exemplary of what—the ideals that have always animated the nobler impulses of medicine, or the dominant values of the culture of medicine, which relegate the patient, and along with the patient, the experience of pain, to the periphery of the physician's concern?

REFERENCES

1. Meldrum, M.L. 2003. A capsule history of pain management. *Journal of the American Medical Association.* 290: 2470–2475.
2. Turner, C.H., and D.K. Weiner. 2002. Essential components of a medical student curriculum on chronic pain management in older adults: Results of a modified Delphi process. *Pain Medicine* 3: 240–252.
3. Rich, B.A. 2000. An ethical analysis of the barriers to effective pain management. *Cambridge Quarterly of Healthcare Ethics* 9: 54–70.
4. Cassell, E.J. 1999. Pain, suffering, and the goals of medicine. In: *The goals of medicine: The forgotten issues in health care reform,* eds. M.J. Hanson and D. Callahan. Washington, DC: Georgetown University Press.
5. Pellegrino, E.D. 2001. Being ill and being healed: Some reflections on the grounding of medical morality. In: *Physician and philosopher. The philosophical foundation of medicine: Essays by Dr. Edmund Pellegrino.* eds. R.J. Bulger and J.P. McGovern, 32–45. Charlottesville, VA: Carden Jennings.
6. Kleinman, A. 2000. What is specific to Western medicine? In: *Companion encyclopedia of the history of medicine, vol. 1.* eds. W.F. Bynum and R. Porter. London: Routledge.
7. Hafferty, F.W., and R. Franks. 1994. The hidden curriculum, ethics teaching, and the structure of medical education. *Academic Medicine* 89: 862–872.
8. Peabody, F.W. 1927. The care of the patient. *Journal of the American Medical Association* 88: 877–882.
9. Stewart, M. et al. 1995. *Patient centered medicine: Transforming the clinical method.* Thousand Oaks, CA: Sage.
10. Crawshaw, R., Rogers, D.E., Pellegrino, E.D. et al. 1995. Patient physician covenant. *Journal of the American Medical Association* 273: 1553.
11. Berwick, D., Hiatt, H., Janeway, P., Smith, R. 1997. An ethical code for everybody in health care. *British Medical Journal* 315: 1633–1634.
12. ABIM Foundation, ACP-ASIM Foundation, European Federal of Internal Medicine. 2002. Medical professionalism in the new millennium: A physician charter. *Annals of Internal Medicine* 136: 243–246.
13. ABIM Foundation, Project Professionalism. Accessible at: http://www.abim.org/pdf/profess.pdf.
14. Accreditation Council on Graduate Medical Education. ACGME General Competencies. Accessible at http://www.acgme.org/outcome/comp/compFull.asp#5.

Index

H

I

J

K

L

M

Q

R

S

T

Z

FIGURE 7.3 School, Umbrian. *The Man of Sorrows,* c. 1260. Part of a diptych. Egg tempera on wood, 32.4 × 22.8 cm. Bought, 1999 (NG6573). National Gallery, London, Great Britain. (Image Copyright © National Gallery, London/Art Resource, New York.)

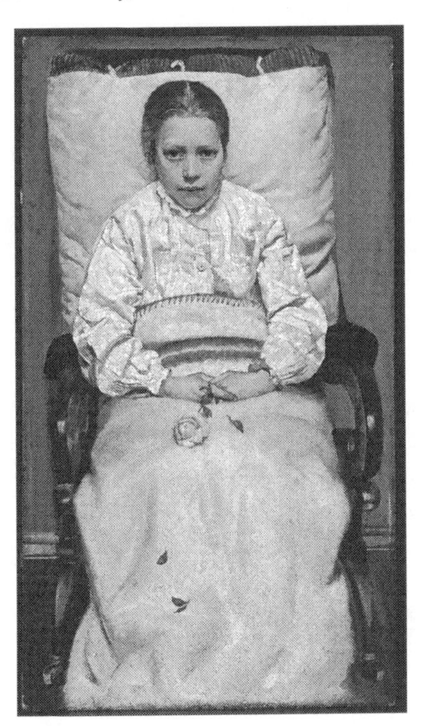

FIGURE 7.6 Christian Krohg, *Sick Girl,* 1880–1881. Oil on canvas, 103.5 × 51.4 cm. NG.M.00805. Nasjonalmuseet for kunst, arkitektur og design/The National Museum of Art, Architecture and Design, Oslo. (Image Copyright © The Munch Museum/The Munch-Ellingsen Group/ARS 2010.)

FIGURE 7.7 Edvard Munch, *Sick Child,* 1885–1886. Oil on canvas, 119.4 × 119 cm. NG.M.00839. Nasjonalmuseet for kunst, arkitektur og design/The National Museum of Art, Architecture and Design, Oslo. (Image Copyright © The Munch Museum/The Munch-Ellingsen Group/ARS 2010.)

FIGURE 7.8 Edvard Munch, *The Scream,* 1893. Oil, tempera, and pastel on cardboard, 91 × 73.5 cm. NG.M. 00939. Nasjonalmuseet for kunst, arkitektur og design/The National Museum of Art, Architecture and Design, Oslo. (Image Copyright © The Munch Museum/The Munch-Ellingsen Group/ARS 2010.)

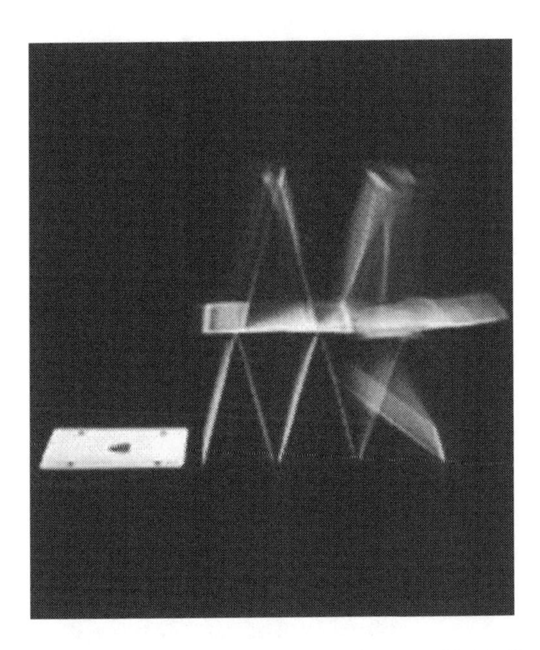

FIGURE 8.1 *Pack of Cards*. (From *Perceptions of Pain*, pp. 78–79. Reproduced with the permission of Deborah Padfield © 2003 and her colleagues [Deborah Padfield with Robert Ziman-Bright].)

FIGURE 8.2 *Red Hot Wire*. (From *Perceptions of Pain*, p. 31. Reproduced with the permission of Deborah Padfield © 2003 and her colleagues [Deborah Padfield with Linda Sinfield].)

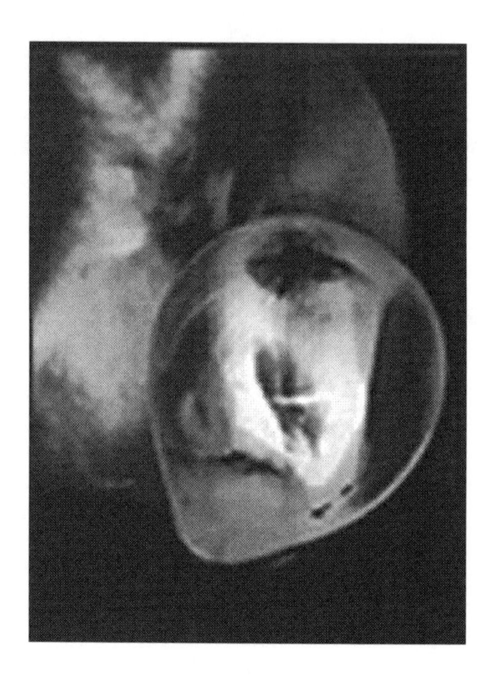

FIGURE 8.4 *Face and Balloon Off Side.* (From *Perceptions of Pain*, p. 58. Reproduced with the permission of Deborah Padfield © 2003 and her colleagues [Deborah Padfield with Rob Lomax].)

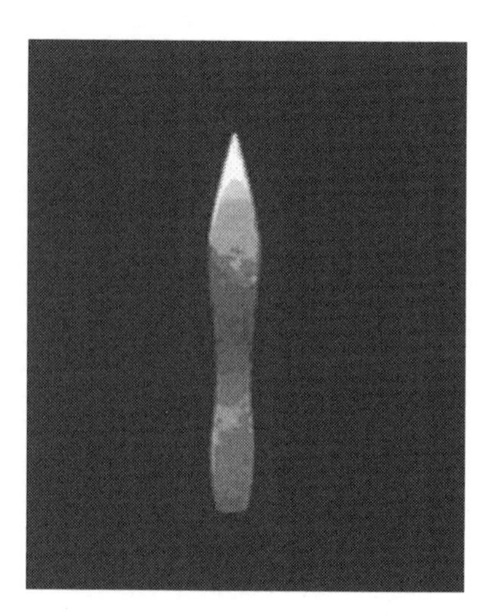

FIGURE 8.6 *Red Dagger.* (From *Perceptions of Pain*, p. 61. Reproduced with the permission of Deborah Padfield © 2003 and her colleagues [Deborah Padfield with Frances Tenbeth].)